Henry Sinclair Hall, F. Haller Stevens

An Elementary Course of Mathematics

Comprising Arithmetic, Algebra and Euclid

Henry Sinclair Hall, F. Haller Stevens

An Elementary Course of Mathematics
Comprising Arithmetic, Algebra and Euclid

ISBN/EAN: 9783744750592

Printed in Europe, USA, Canada, Australia, Japan

Cover: Foto ©berggeist007 / pixelio.de

More available books at **www.hansebooks.com**

AN ELEMENTARY
COURSE OF MATHEMATICS

ARITHMETIC, ALGEBRA AND EUCLID

BY

H. S. HALL, M.A., AND F. H. STEVENS, M.A.

MASTERS OF THE MILITARY SIDE, CLIFTON COLLEGE

MACMILLAN AND CO., Limited

NEW YORK : THE MACMILLAN COMPANY

1899

PREFACE.

THE main purpose of this text-book is to provide in a single
and inexpensive volume a short course of Arithmetic, Algebra,
and Euclid specially adapted to the needs of a large and
increasing class of students : namely, those who after leaving
school desire to continue their study of Elementary Mathematics, partly with help derived from evening classes, and
partly by means of private work at home.

The majority of such students work under somewhat difficult
conditions. They have already received a certain training in
Arithmetic, but their mathematical education has gone no
further ; and the problem which meets them is how—with
very limited instruction and leisure—to maintain what they
have already learned, and at the same time to acquire from
the beginning a sound knowledge of Elementary Algebra
and Geometry. What is needed by such learners is first of
all a course of varied and graduated Exercises in Arithmetic ;
then a careful exposition of the two new subjects, Algebra
and Euclid, arranged and treated in such a way as to smooth
the path of a beginner, and to encourage him to overcome
difficulties by relying on his own industry and intelligence.

These considerations have been kept in view in compiling the
following pages.

Accordingly the section on Arithmetic is not intended to
take the place of an ordinary text-book, but to supplement
lectures and aid in the work of revision. It provides a series
of progressive examples so arranged as to distribute the
necessary matter over a winter session of some thirty weeks,

the examples in each Exercise being graduated with a view to facilitating home work as much as possible. The later Exercises have been prefaced by a few notes and hints together with solutions of some typical questions. These have been selected with great care, and it is hoped that they will sufficiently recall and illustrate the more important methods required in the Exercises which follow.

The section on Algebra assumes no previous knowledge; and those parts of the subject with which it deals are treated from the first as fully as is consistent with its elementary purpose. It covers all the ground usually taken before Quadratic Equations, and will be found complete as a first text-book for beginners within the limits indicated.

The geometrical section contains Euclid's First Book with Exercises and Additional Theorems. This is based upon the Authors' edition of Euclid's *Elements*; but the text has been thoroughly revised, and some additional notes and test questions have been introduced. Deductions and Exercises on which it is desired to lay special stress, as being in themselves important geometrical results, have been italicised.

H. S. HALL.
F. H. STEVENS.

May, 1899.

** *In the Arithmetic Section examples taken from the Examination Papers of the Science and Art Department will be found distinguished by an asterisk at the end of the different Exercises.*

CONTENTS.

ARITHMETIC.

ALGEBRA.

EUCLID.

BOOK I.

ARITHMETIC.

First Week. *Compound Quantities. Primary Rules.*

Examples I.

1. Having a balance of £269 at my bank, I draw cheques for £75. 15s. 5d., £37. 7s. 1d., and £42. 4s. 2d. How much have I left?

2. In four successive years a business cleared £401. 11s. 10d., £491. 8s. 8d., £528. 6s. 11d., and £719. 4s. 3d. respectively. What were its average annual profits?

3. Multiply £107. 8s. 7d. by 59.

4. Divide £1058. 15s. 9d. by 142.

5. Express in feet 1 furlong 13 poles 3 yards.

6. Reduce 432,109 ounces to tons, cwts., etc.

7. To what sum will a daily payment of 1s. 6½d. amount in a year of 365 days?

8. How many grains are there in 52 lbs. 1 oz. Troy?

9. How many payments of £6. 5s. 10d. could be made from a sum of £2497. 15s. 10d.?

10. Reduce 74,327 sq. yds. to acres, roods, poles, etc.

11. Find the value of 1592 articles at £1. 17s. 6d. per dozen.

12. How many pieces of calico, each 4 yds. 2 ft. 6 in. long, can be cut from a roll of 50 yards, and how much will be left?

13. A man wishing to dispose of 300 articles, sells 122 for £94. 11s., and the rest at the rate of ¼d. less each. How much does he get altogether?

14. A grocer mixes 12 lbs. of tea which cost him 2s. 4d. a lb. with 10 lbs. at 2s. 6d., and 8 lbs. at 2s. 8d.: if he sells all the mixture at 2s. 9d. a lb., what is his gain?

15. If 3 tons 6 cwt. of tea are bought at £10. 14s. 8d. per cwt., and sold at 2s. 3d. per lb., what is the total gain?

*16. A carriage wheel is 8 ft. 3 in. in circumference; how many revolutions does it make in driving 7 miles and 1331 yards?

*17. Copper weighs 550 lbs. and tin 462 lbs. to the cubic foot. What will be the weight of a cubic foot of a mixture of 6 parts copper to 5 parts tin?

*18. Four hundred calendar years contain 146,097 days. What would be the length of a calendar year in seconds, supposing all such years are alike?

E. C.   A   Œ

Second Week. *Further Questions on Compound Quantities.*

Examples II.

1. In the first four months the takings of a business are respectively £335. 2*s.* 1½*d.*, £371. 15*s.* 11*d.*, £401. 11*s.* 5½*d.*, and £446. 11*s.* 6*d.* What must be the average takings for the remaining months in order that the total turn-over for the year may be £5000 ?

2. In 2483267 sq. inches, how many roods, poles, square yards, etc. are there ?

3. The cost of 35688 articles is £4730. 10*s.* 4½*d.*; at what rate is this per dozen ?

4. How many times is 1 cwt. 2 qrs. 22 lbs. 4 oz. contained in 8 tons 13 cwt. 1 qr. 1 lb. 8 oz.?

5. The net earnings of a company for the past year are £28,651. 4*s.* 6*d.* After carrying forward £2463. 14*s.* 6*d.* to next year's account, a dividend of 17*s.* 5½*d.* per share is paid. How many shares are there ?

6. How many inches are there in 2 mi. 1 f. 37 p. 3 yds ?

7. A tradesman buying articles at the rate of £3. 7*s.* 6*d.* per score, sells at £2. 2*s.* 6*d.* per dozen. What profit does he make on 1000 articles ?

8. Cloth was bought at 2*s.* 9½*d.* a yard, and sold at 3*s.* 3*d.* If the total profit came to £2. 4*s.* 0*d.*, how many yards were bought and sold ?

9. A wine merchant mixes 29 gallons of wine which cost him 15*s.* 8*d.* a gallon with 17 gallons at 23*s.* 4*d.* a gallon. What is the value of one gallon of the mixture ?

10. Coal is bought at 24*s.* per ton, and sold at 25*s.* 6*d.*; how many tons must be sold a year to clear an income of £350 ?

*11. How many coins, each containing 5 dwt. 6 gr. of standard gold, can be struck from a bar weighing 16 lbs. 10 oz. 7 dwt. 18 gr.?

*12. If 1869 sovereigns are coined out of 40 lbs. Troy of standard gold, what is the weight of a sovereign in grains ? And what is the value of an ounce of standard gold ?

*13. Coals at 32*s.* a ton are mixed with coke at 24*s.* a ton in the proportion of 5 tons of coal to 3 tons of coke ; find the money saved by using 11 tons of the mixture instead of 12 tons of coal.

*14. A man buys 45 bushels of apples at 2*s.* 6*d.* a bushel ; one-third of them are spoilt ; of the remainder, he sells one-third at 2*d.* a pound, and two-thirds at 4*s.* a bushel. If a bushel of apples weighs 40 lbs., what profit does he make ?

Third Week. *Prime Factors. Tests of Divisibility. Greatest Common Divisor. Least Common Multiple.*

Examples III.

1. Which of the following numbers are divisible by 8, 9, 11, 25? 15327, 13016, 43875, 361911, 81576, 4957425.

Find the remainders when each of the following numbers is divided by 8, 9, 11, and 25 : 46820379, 69014813, 916348512.

2. Obtain the remainders when (i) 468253 is divided by 385 (ii.) when 6324031 is divided by 792 by *short division*, carefully explaining the reason for the process.

3. From 98765 subtract any other number formed by the same digits in a different order, and divide the remainder by 9. Explain why in *all* such cases the remainder must be divisible by 9.

Break up the following numbers into their prime factors :

4. 385.	**5.** 702.	**6.** 1575.
7. 4410.	**8.** 19125.	**9.** 76725.

Find the Greatest Common Divisor (or Measure) of the following numbers :

10. 333 and 407.	**11.** 533 and 697.
12. 3451 and 7395.	**13.** 2604 and 3444.
14. 4131 and 11664.	**15.** 5544 and 6552.
16. 546, 624, and 676.	**17.** 663, 1547, and 1989.
18. 1458, 4131, and 1215.	**19.** 116039, 122067, 137137.

Find the Least Common Multiple of

20. 8, 14, 36, 108.	**21.** 42, 55, 70, 77.
22. 7, 3, 56, 21, 49.	**23.** 72, 81, 9, 8, 12.
24. 25, 27, 9, 11, 15.	**25.** 28, 52, 65, 70, 91.
26. 1430 and 2145.	**27.** 5292 and 8316.

Find the G.C.M. and L.C.M. of

28. 72, 24, 56, 120.	**29.** 51, 170, 153, 187.
30. 161, 253, 299, 322.	**31.** 3024, 4752, 7488.

32. *The Least Common Multiple of two numbers may be obtained by dividing their product by their Greatest Common Measure.* Explain by an example the reason for this, and adapt the theorem to finding the L.C.M. of three numbers.

33. What is the least sum of money that can be distributed exactly either in half-crowns or half-guineas?

34. What is the least sum of money that can be exactly divided into equal shares, each of 2s. 4d., or 2s. 9d., or 3s. 8d., or 4s. 8d.?

Fourth Week.   *Fractions.   Primary Rules.*

Examples IV.

1. Explain why the value of a fraction is not changed when both numerator and denominator are multiplied (or divided) by the same number; *e.g.* explain why $\frac{3}{5} = \frac{21}{35}$.

Reduce the following fractions to their lowest terms :

2. $\frac{3538}{3782}$.   **3.** $\frac{5115}{8184}$.   **4.** $\frac{25194}{89170}$.

5. $\frac{8679}{44973}$.   **6.** $\frac{63657}{123456}$.   **7.** $\frac{124321}{197451}$.

8. Explain carefully the reason of the rule for finding the sum (or difference) of two fractions : *e.g.* explain why $\frac{3}{5} - \frac{4}{9} = \frac{7}{45}$.

Add together the following fractions :

9. $\frac{9}{77}$, $2\frac{7}{33}$, $\frac{22}{63}$, and $7\frac{10}{11}$.   **10.** $2\frac{7}{46}$, $\frac{38}{45}$, $\frac{57}{230}$, and $9\frac{5}{9}$.

11. $7\frac{11}{18}$, $\frac{17}{24}$, $5\frac{4}{9}$, and $11\frac{11}{36}$.   **12.** $20\frac{11}{21}$, $3\frac{3}{16}$, $11\frac{13}{84}$, and $15\frac{57}{112}$.

13. What fraction must be subtracted from the sum of $16\frac{4}{7}$, $2\frac{15}{56}$, $\frac{11}{48}$, and $\frac{17}{35}$ to leave the remainder 19 ?

14. Find a fraction with 364 as denominator which is less than unity by $\frac{2}{13}$.

15. What fraction must be added to the sum of $\frac{9}{16}$, $\frac{1}{2}$, $\frac{3}{5}$ and $\frac{5}{8}$ to give the result $3\frac{11}{240}$?

16. Find the difference between $\frac{1}{12} + \frac{2}{3} + \frac{1}{10}$ and $\frac{1}{15} + \frac{3}{4} + \frac{1}{30}$.

17. What fraction must be added to the sum of $3\frac{3}{8}$ and $4\frac{1}{15}$ to make up the difference between $12\frac{1}{8}$ and $1\frac{17}{20}$?

18. Explain carefully what is meant by multiplying one fraction by another : *e.g.* explain why $\frac{2}{5} \times \frac{3}{7} = \frac{6}{35}$.

Multiply together the following fractions :

19. $1\frac{67}{186}$, $\frac{144}{161}$, $2\frac{55}{112}$, and $1\frac{97}{90}$.   **20.** $\frac{51}{91}$, $1\frac{27}{68}$, $2\frac{46}{55}$, and $\frac{77}{111}$.

21. Explain carefully why in order to divide a quantity by a fraction we multiply by its reciprocal : *e.g.* explain why $\frac{5}{9} \div \frac{2}{7} = \frac{35}{18}$.

22. Divide (i.) the product of $5\frac{4}{9}$, $\frac{12}{143}$, $2\frac{1}{7}$, $8\frac{99}{50}$ by $8\frac{2}{5}$;

(ii.) $\left(\frac{3}{4}+\frac{1}{3}-\frac{1}{6}\right)$ of $\frac{6}{25}$ by $\frac{11}{10}$;

(iii.) $\left(\frac{3}{7}+\frac{1}{3}+\frac{1}{2}-\frac{11}{42}\right)$ by $\left(\frac{7}{9}+\frac{5}{8}-\frac{3}{4}\right)$.

23. What fraction must be taken from the sum of $\dfrac{2\frac{5}{7}}{5}$ and $6\frac{3}{23}$ that the result may be equal to $\dfrac{10\frac{5}{6}}{1\frac{12}{13}}$?

Simplify the following complex fractions:

24. $\frac{2}{5}$ of $\frac{3}{7}$ of $\dfrac{9+\frac{12}{7}}{5+\frac{2}{1\frac{2}{3}}}$.

25. $\dfrac{\frac{7}{8}+\frac{1}{2}-\frac{8}{21}}{10\frac{1}{12}\text{ of }\frac{4}{3}\text{ of }\frac{15}{77}\text{ of }\frac{7}{11}}$.

26. $\dfrac{\frac{1}{2}+\frac{3}{5}+\frac{5}{7}}{\frac{5}{2}+\frac{4}{3}+\frac{1}{7}}$.

27. $\frac{1}{30}$ of $4\frac{1}{3}$ of $\dfrac{\frac{1}{2}+\frac{1}{3}+\frac{1}{4}}{\frac{1}{4}+\frac{1}{5}+\frac{1}{6}}$.

28. $\dfrac{1-\frac{1}{2}+\frac{1}{3}-\frac{1}{4}}{1+\frac{1}{2}-\frac{1}{3}+\frac{1}{4}}\div\frac{2}{17}$.

29. $\left(\frac{3}{4}+\frac{9}{11}-\frac{7}{8}\right)\div\dfrac{\frac{5}{8}+\frac{6}{5}}{3-\frac{1}{5}}$.

30. $\dfrac{2}{1+\dfrac{3}{1+\frac{2}{1\frac{3}{5}}}}$.

31. $\dfrac{6+\dfrac{1}{6-\frac{1}{8}}}{4-\dfrac{1}{4-\frac{1}{4}}}$ of $10\frac{8}{9}$.

32. By what fraction must $\dfrac{1}{2+\dfrac{1}{3+\dfrac{1}{1+\frac{1}{4}}}}$ be multiplied that the product may be unity?

33. By what fraction must unity be divided to give a quotient equivalent to the product of $\dfrac{2}{3+\frac{2}{3}}$ and $\dfrac{2}{3+\dfrac{2}{3+\frac{2}{3}}}$?

34. Simplify $\dfrac{4\frac{1}{5}-\frac{2}{3}\text{ of }\frac{5}{6}}{1\frac{2}{3}+2\frac{3}{7}-\frac{2}{5}}+\frac{4}{291}$; and from the result subtract $\dfrac{1}{\frac{1}{2}+\frac{1}{3}+\dfrac{\frac{1}{6}}{\frac{3}{4}-\frac{1}{3}}}$.

Reduce the following fractions to their simplest forms :

*35. $$\dfrac{\frac{1}{9}\text{ of }3\frac{6}{13}+\dfrac{2}{3\frac14}-\dfrac{1}{6\frac12}}{5+\dfrac{1}{2\frac16}+\frac{3}{13}\text{ of }2\frac16}.$$

*36. $$\dfrac{9\frac13\text{ of }4\frac57\div8\frac14}{2\frac{7}{9}-\frac{5}{12}+1\frac{7}{36}}.$$

*37. $$\dfrac{6\frac14-3\frac{1}{16}}{3\frac25\text{ of }1\frac14}\div\dfrac{12\frac{4}{15}-11\frac23}{3\frac13\text{ of }\dfrac{1}{4\frac16}}.$$

*38. $$\left\{1-\dfrac{-\frac{3}{10}}{(\frac12+\frac15)^2}\right\}\times\left\{1-\dfrac{\frac{1}{10}}{\frac14-\frac{1}{10}+\frac{1}{2\,5}}\right\}.$$

*39. $$\dfrac{2\frac19}{5\frac37}\text{ of }1\frac18-\dfrac{15\frac34-4\frac37\text{ of }3\frac12}{3\frac59\text{ of }\frac38+2\frac23}.$$

*40. $$\dfrac{3\frac37}{\frac45\left(\frac{3}{3\frac12}\text{ of }7\right)}+\dfrac{5\frac58}{4\frac16}\text{ of }\dfrac{4-(3\frac13-1\frac15)}{3\frac{4}{15}-2\frac{1}{60}+\frac{11}{20}}.$$

*41. $$\dfrac{100\frac{1}{13}-91\frac{89}{91}}{27\frac{2}{11}+2\frac{117}{143}}-\dfrac{3\frac37+6\frac35\text{ of }1\frac{11}{19}}{1\frac{7}{11}\text{ of }25\frac23+48\frac18}.$$

*42. $$\dfrac{44\frac23-3\frac{3}{11}\text{ of }10\frac12}{3\frac12+\frac{7}{85}\text{ of }13\frac{13}{14}}+\dfrac{\left(5\frac34-3\frac23\right)\text{ of }2\frac{5}{9}}{1\frac14\text{ of }\left(2\frac12+1\frac13\right)}.$$

*43. $$\dfrac{5+2\frac{7}{9}\text{ of }\frac{27}{50}}{30-6\frac{6}{11}\text{ of }2\frac15}+\dfrac{6\frac56}{23\frac37}\text{ of }\left(6\frac{1}{21}-5\frac{9}{14}\right).$$

*44. $$\dfrac{13\frac23}{11\frac{4}{11}}\text{ of }1\frac{49}{451}-\dfrac{6\frac12-5\frac14\text{ of }\frac67}{4\frac13\text{ of }\frac{7}{26}+1\frac{1}{66}}.$$

Fifth Week.   *Fractions. Compound Quantities.*

Examples V.

Find the value of

1. $\frac{4}{15}$ of £14. 10s. 7½d.

2. $\frac{24}{40}$ of £1. 15s. 8¾d.

3. $\frac{6}{11}$ of 14 fur. 13 p. 3 yds. 2 ft.

4. $1\frac{2}{7}$ of 2 tons 10 cwt. 2 qrs. 14 lbs.

5. Add together $1\frac{3}{11}$ of £15. 8s. 11d. and $\frac{4}{21}$ of £3. 1s. 3d.

6. From $\frac{11}{27}$ of £3079. 16s. subtract $\frac{1}{117}$ of £85. 16s.

7. Express (i.) £2. 19s. 9½d. as the fraction of £5. 2s. 6d.; and (ii.) 2 tons 3 qrs. 3 lbs. 8 oz. as the fraction of 3 tons 5 cwt. 1 qr.

8. Find the difference between $\frac{6}{7}$ of 15 ac. 3 r. 35 p. and $\frac{7}{9}$ of 31 ac. 2 r. 27 p.

9. By what fraction must £10. 8s. 8d. be multiplied to give the result £5. 17s. 4½d.?

10. By what fraction must £600. 12s. 6d. be divided to give the quotient £250. 5s. 2½d.?

11. What fraction of one pound is the difference between $\frac{3}{14}$ of half-a-guinea and $\frac{11}{18}$ of half-a-sovereign?

12. After $\frac{13}{21}$ of a plank had been cut off, the length of the remainder was 5 ft. 2$\frac{6}{7}$ in.: what was the length of the whole plank?

13. What fraction of 35 shillings is the sum of $2\frac{3}{4}$ of 7s. 6d. and $\frac{3}{2}$ of $\frac{5}{4}$ of 3 shillings?

14. If $\frac{5}{24}$ of a field is worth £1065, find the value of $\frac{3}{9}$ of it.

15. If one quarter is taken as the unit of weight, by what fraction will 10 lbs. 8 oz. be expressed?

16. If 1 mile 3 f. 10 p. 2 yds. 1 ft. 10 in. is taken as the unit of distance, what length will be represented by the fraction $\frac{2}{7}$?

17. Add together $\frac{7}{24}$ of £6. 5s. 4d. and $\frac{1}{6}$ of £12. 7s. 2d., and express $\frac{2}{9}$ of the sum as the fraction of £2. 11s. 10d.

18. A partner receives £219. 3s. 9d. as $\frac{3}{16}$ of the annual profits of a business. What are the whole profits, and what fraction of the whole is £250. 10s.?

19. If $\frac{25}{37}$ of a field is worth £925, and if the land is worth £40 an acre, find the area of the field.

20. What sum of money is the same fraction of £11. 2s. that 1 ton 12 cwt. 3 qrs. 7 lbs. is of 2 tons 19 cwt. 7 lbs. ?

21. What weight is the same fraction of 26 tons 18 cwt. 2 qrs. 7 lbs. that £6. 8s. 10d. is of £22. 10s. 11d. ?

22. If $\frac{7}{15}$ of one guinea be taken from $\frac{3}{12}$ of $\frac{5}{9}$ of £5, what fraction of £3. 9s. will remain ?

23. Add together $\frac{4}{15}$ of £1, $\frac{5}{8}$ of 1s., $\frac{5}{28}$ of 1 guinea, and $\frac{3}{5}$ of £1. 6s. 8d.

24. Find the value of $\frac{5}{6}$ of half-a-guinea $+ \frac{7}{15}$ of half-a-crown $+ \frac{7}{8}$ of a florin $- \frac{3}{4}$ of 6s. 8d.

25. The net profits of a business for a quarter of a year are £582. 14s. 11d. Of this sum the senior partner takes $\frac{7}{12}$, and the junior partner the remainder; but the latter has to pay $\frac{3}{19}$ of his share to an assistant. How much will the assistant receive at this rate per annum ?

26. A, B, C, and D are partners in a business; A takes $\frac{4}{9}$ of the profits, B takes $\frac{11}{45}$, and C takes $\frac{4}{15}$. There remains £202. 12s. 2d. for D. What were the total profits ?

27. The total area of three farms is 1768 acres. If the areas of the two smaller farms are respectively $\frac{3}{5}$ and $\frac{2}{3}$ that of the largest, find the acreage of each.

28. Of the annual profits of a firm the first partner takes $\frac{1}{3}$, and the second $\frac{1}{4}$, the third $\frac{2}{5}$ of the remainder. If £1025. 5s. is left to carry forward to next year's account, what are the total profits ?

29. A company regularly adds to its reserve fund in such a way that in each year the fund is $\frac{5}{4}$ of its amount in the preceding year. If it stood at £3276. 16s. in 1890, what was its amount in 1893 ? And if it reached £19,531. 5s. in 1898, how did it stand in 1895 ?

*30. Add together $\frac{5}{9}$ of a guinea, $\frac{17}{20}$ of half-a-crown, $1\frac{7}{38}$ of a shilling, and $\frac{1}{6}$ of a penny ; and reduce the sum to the fraction of one pound.

*31. A man pays one-tenth of his income in rates and taxes, and one-twelfth in insurances ; he has left £492. 13s. 1d. What is his income ?

*32. A man pays away half his money to A, a third of what he has left to B, and a fifth part of what he still has left to C. If after these payments he has 12s. 8d. left, how much had he at first ?

*33. Find the rent of an estate whereof $\frac{5}{12}$ is let at £2. 3s. 0d. an acre, and the remainder, consisting of 20 ac. 3 r. 12 p., at £1. 16s. 8d. an acre.

*34. A warehouse of five storeys is let in flats. Each flat (except the top one) lets for $\frac{3}{4}$ of the rent charged for all the flats above it, and the rent of the whole warehouse is £2401. What is the rent of the middle flat ?

Sixth Week. *Decimals. Primary Rules.*

Note on Approximate Results and Contracted Multiplication and Division. A quantity is said to be expressed as a decimal *correct to three places* when its value is given as nearly as it is possible to give it if we use only three decimal figures.

For instance $\frac{6}{17} = \cdot352941...$, from which it is seen that the value of $\frac{6}{17}$ lies between ·352 and ·353.

Now ·352 is *too small* by more than ·0009, while ·353 is *too great* by less than ·0001.

Thus ·353 is the nearer value ; and we say that $\frac{6}{17} = \cdot353$ *correct to three places of decimals.*

Example. Express $\frac{4}{19} + \frac{6}{17} + \frac{5}{13}$ as a decimal correct to *four* places.

$$\frac{4}{19} = \cdot2105 \,\big|\, 26...$$
$$\frac{6}{17} = \cdot3529 \,\big|\, 41...$$
$$\frac{5}{13} = \cdot3846 \,\big|\, 15...$$
$$\overline{\cdot9480 \,\big|\, 8} \quad = \cdot9481 \text{ correct to 4 places.}$$

Here we write down the first *six* figures of each decimal, adding in the usual manner. The *sixth* figure of the sum is not written down ; it would probably be wrong for want of a "carrying" figure. The *fifth* figure is retained in order to *correct* the fourth.

Notice that this result correct to *one* decimal place is ·9 ;

,,	,,	,,	*two*	,,	places is ·95 ;
,,	,,	,,	*three*	,,	places is ·948.

Contracted Multiplication.

The student is strongly recommended to master *contracted* methods of Multiplication and Division of decimals. These are of great practical utility, since they enable us to obtain without unnecessary labour products and quotients true to any assigned number of decimal places.

With a view to facilitating contracted multiplication, ordinary multiplication should be arranged as in the following example.

Example. Multiply 3·6804 by 25·13.

Arrange the decimals one under the other, so that the decimal points may be in the same column.

Begin the multiplication with the *left*-hand figure of the multi-

```
   3·6804
  25·13
  ───────
  73·608
  18·4020
    ·36804
    ·110412
  ─────────
  92·488452
```

plier, that is, with the 2 in the *tens* place; but since, in doing this, we are really multiplying by *twenty*, the result must be written one place to the *left*, the decimal point being kept in the same column as those above it. Then multiply in succession by 5, 1, and 3; at each stage ranging the result one place to the right of that last obtained, and preserving the column of decimal points. Lastly add the partial products as in ordinary addition of decimals.

We will now examine how this work might be abridged, if we needed a result *true only to two decimal places.*

As the result is to be true to *two* places of decimals, it is necessary at each stage to retain *three* decimal figures. We give below the full and contracted work for comparison, that the student may see what figures are superfluous, and how they may be discarded.

Full Working.		Contracted Working.	
3·68	04	3·6,8	0,4
25·13		25·1 3	
73·60	8	73·6 0	8
18·40	20	18·4 0	2
·36	804	·3 6	8
·11	0412	·1 1	0
92·48	8452	92·4 8	8

On comparison it will be seen that in the full working none of the figures printed in italics are needed; we may therefore proceed as follows:

Draw a vertical line after the second decimal figures; and begin, as before, by multiplying by the 2 (*i.e.* by 20), ranging the result accordingly one place to the left.

At the next stage the work is contracted *by marking off* and rejecting *the last figure* (*i.e.* 4) *on the right of the multiplicand*, before multiplying the figures that remain by the next figure (*i.e.* 5) of the multiplier. We must, however, carry 2 to the first place, as we should have done if the figure 4 had not been rejected.

Again, mark off the next figure 0 on the right of the multiplicand, and multiply the figures that remain by the next figure 1 of the multiplier. Continue this process till all the figures in the multiplicand are marked off, or all the figures in the multiplier are exhausted. Then, adding, we have 92·48 | 8 ; which corrected to two decimal places, gives 92·49.

Note. If there had been more than four decimal figures in the multiplicand, it will be seen that, in the above instance, those beyond the fourth might have been removed, as not affecting that part of the work which is to be retained.

For instance, if asked to multiply 3·6804182 by 25·13 *true to two decimal places*, we should begin by erasing the last three figures, 182.

We proceed to work out two more examples.

Examples. Multiply

(1) 3·260428103 by 501·0206 correct to 3 decimal places.

(2) 106·391684 by ·028563 correct to 3 decimal places.

(1)
```
      3·2 6 0|4 2 8 103
    501·0 2 0|6
   ─────────────
   1630·2 1 4|0
      3·2 6 0|4
       ·0 6 5|2
       ·0 0 1|0
   ─────────────
   1633·5 4 1|5
   ═════════════
```

(2)
```
    1 0 6·3 9 1|6 8 4...
         ·0 2 8|5 6 3
    ──────────────
       2·1 2 7|8
        ·8 5 1|1
        ·0 5 3|2
        ·0 0 6|4
        ·0 0 0|3
    ──────────────
       3·0 3 8|8
    ══════════════
```

Required product = 1633·542.   Required product = 3·039.

Note. If the number of partial products to be added is large, *two* figures should be retained to the right of the vertical line. One figure is, however, usually sufficient.

Contracted Division.

DEFINITION. The figures of a number or decimal, *other than* 0's *standing at the beginning or end*, are called **significant** figures.

For example, of the numbers 20708, 2070800, 20·708, and ·00020708, the significant figures in each case are 2, 0, 7, 0, 8.

We shall now show by examples how the work of long division may be contracted when only an approximate result is required.

Example 1. Divide 79·43806245 by 3·206402, giving only the *first two decimal figures* of the quotient.

First determine by inspection how many *integral* figures there will be in the quotient. In this case there will clearly be *two* integral figures. And as two decimal figures are required, it follows that we have to find the first *four* significant figures of the quotient.

We give below (on the left) the full division, printing in italics all the figures that would be superfluous, if *four* figures only are required in the quotient. On the right we give the contracted working for comparison. In each case the decimal points are omitted.

FULL WORKING.		CONTRACTED WORKING.	
3206402) 79438	*06245* (2477	32,0,6,4) 79438	(2477
64128	*04*	64128	
15310	*022*	15310	
12825	*608*	12826	
2484	*4144*	2484	
2244	*4814*	2244	
239	*93305*	240	
224	*44814*	224	
15	*48491*	16	

To perform the contracted working we retain in the divisor *five* figures, that is to say, *one more than the number of significant figures required in the quotient* : and in the dividend we retain as many figures as are needed to take the first step in the division, in this case also five.

We then proceed with the division in the ordinary way, except that at each stage, instead of bringing down a new figure from the dividend, *we mark off and reject a figure from the right of the divisor*, taking care, however, on multiplying, to make use of the figure last marked off for the purpose of obtaining a carrying number.

Thus we obtain as the quotient the figures 2477 : but since it has been already determined that there will be *two integral* figures, the required result is 24·77.

Example 2.   Divide ·02628947597 by 3·0685, the result to be given to *eight decimal figures*.

30,6,8,5,0,0) 26289475 (856753
     24548000
      ─────────
      1741475
      1534250
      ─────────
       207225
       184110
       ─────────
        23115
        21479
        ─────────
         1636
         1534
         ─────────
          102
           92

Here ·0262 ... is to be divided by a quantity intermediate between 3 and 4 ; therefore there will be *two* 0's in the quotient before the first significant figure. Hence to make up the required *eight* places of decimals, we have to find *six* significant figures.

This makes it necessary to retain *seven* figures in the divisor, which is done in this case by adding two 0's. It will be noticed that the first step in the division requires eight significant figures in the dividend.

The decimal points as before are omitted in the numerical work.

Thus the required result is ·00856753.

Examples VI.

1. What is meant by a *decimal fraction*? *e.g.*, explain the meaning of 207·30206.   Write down the following fractions in a decimal form :

$$\frac{27}{10}, \quad \frac{27}{100}, \quad \frac{27}{100000}, \quad \frac{207}{1000}, \quad \frac{2007}{100}.$$

What fractions (in their lowest terms) are equivalent to the following decimals ?

2. ·04.   3. ·028.   4. 2·015.   5. ·00375.

6. ·7625.   7. ·001375.   8. ·00009375.   9. ·078125.

Convert the following fractions into equivalent decimals :

10. $\frac{3}{400}$.   11. $2\frac{5}{8}$.   12. $\frac{29}{5000}$.   13. $\frac{49}{800}$.

14. $\frac{3}{192}$.   15. $\frac{1}{840}$.   16. $\frac{742}{875}$.   17. $\frac{7}{256}$.

Find (by decimals) the value of the following :

18. $\cdot1 + \cdot006 + \frac{4}{5} + \frac{3}{8} + \frac{1}{10000}$.

19. $53\frac{1}{2} + 36\cdot875 + 4\frac{5}{8} + \frac{2}{3}$ of $7\frac{1}{2}$.

20. $13\frac{1}{4} + 1\frac{2}{5} + 8\frac{3}{125}$ $(4\frac{1}{8} + \cdot549)$.

21. What decimal must be added to the sum of ·0023, 2·36, 250, and ·527 to give a total of 253 ?

22. Find the difference between 203·66519 and the sum of 181·5276, 10·0085, ·16709.

23. From the sum of ·90807, 6·05, ·0043, 22, and ·00068 take the difference between 30·101 and 1·14795.

State and explain a rule for multiplying one decimal by another; and find the value of

24. 41·2 × 3·9.   **25.** ·011 × 1100.

26. ·015 × ·273.   **27.** ·83 × ·073.

28. 200·002 × ·303.   **29.** ·0000152 × 87·5.

30. 28·395 × ·00114.   **31.** ·04705 × 24·0604.

32. [17·453]².   **33.** 3 − (1·732)².

State and explain a rule for dividing one decimal by another; and find the value of

34. 3·3252 ÷ 3·26.   **35.** 3·24 ÷ 11·25.

36. 42·5 ÷ ·017.   **37.** 157·311 ÷ 3405.

38. 128·36826 ÷ 3·204.   **39.** 13·5 ÷ ·001125.

40. ·045375 ÷ ·000015.   **41.** ·00371 ÷ 1·28.

42. Find the continued product of 2·05, ·0024 and 250; by what decimal must the result be multiplied to give ·06765?

Simplify the following (without converting decimals into vulgar fractions):

43. $\dfrac{6·1275 \times ·032}{·00024}$.

44. $\dfrac{·416 \times ·025}{3·25}$.

45. $\dfrac{·004 \times 32·4}{6·4 \times ·0045}$.

46. $\dfrac{·00281 \times ·0625}{1·405 \times ·00125}$.

47. $\dfrac{2·25}{·25} + \dfrac{2}{1·25} + \dfrac{·09}{·00625}$.

48. $\dfrac{·20705}{·0101} + \dfrac{8·32}{·512} + \dfrac{1·326}{·408}$.

49. $\dfrac{·02 − ·002 + ·302}{(·016)^2}$.

50. $\dfrac{25 − ·0025 + 3·82 ÷ ·25}{805·5 + ·05}$.

(*Contracted Multiplication and Division.*)

51. Multiply ·0914 by 32·56, giving the product correct to *three* places of decimals.

52. Multiply ·0731 by 163·2, giving the product true to the *nearest thousandth.*

53. Multiply 59·6159 by 3·0807, giving the result correct to *four* places of decimals.

54. Multiply 3·73205 by ·26795, giving the product correct to *five* places of decimals.

55. Show by contracted multiplication that the square of 1·73205 differs from 3 by less than one hundred-thousandth of unity.

56. Prove that the product of 1·231056 and ·81231056 differs from unity by less than ·000002.

57. Divide 43·7246 by 10·84589, giving the quotient true to *three* places of decimals.

58. Divide ·0492653 by ·020476, giving the quotient true to *four* significant figures.

59. Find the value of (i.) $2 \div 1\cdot41421$, (ii.) $1 \div 3\cdot14159$, in each case to *five* significant figures.

60. Find the value of

(i.) $\dfrac{\cdot000725 \times 31\cdot2501}{\cdot0625}$, (ii.) $\dfrac{236\cdot405 \times \cdot0026054}{4\cdot6082}$,

in each case to *four* places of decimals.

(*Miscellaneous Examples.*)

Reduce to their simplest forms :

*61. $\dfrac{\frac{3}{4} \text{ of } 0\cdot0603}{\frac{5}{4} \text{ of } 0\cdot00594}$.

*62. $\dfrac{\frac{3}{11} \times 25\cdot15}{4\frac{6}{11} \times 0\cdot4}$

*63. $(0\cdot1 \times 0\cdot01 \div 0\cdot0002) - \left(0\cdot6375 \times \dfrac{1}{1\cdot7} \div 0\cdot125\right)$.

*64. $\dfrac{0\cdot2 \times 0\cdot3}{0\cdot00012} \div \left\{\dfrac{5\cdot719}{1\cdot9} + 0\cdot3 \times 1\cdot2 \times 9\right\}$.

*65. $\dfrac{0\cdot002 \times 36\cdot25}{0\cdot029} - \dfrac{102\cdot85 \times 0\cdot04}{1\cdot7}$.

*66. $(0\cdot0057 \times 2\cdot09 \div 0\cdot361) - (0\cdot00165 \times 0\cdot077 \div 0\cdot0105)$

*67. $\dfrac{1\cdot61 \times 0\cdot0209}{0\cdot00253} - \dfrac{2\cdot03 \times 0\cdot336}{32\cdot48}$.

*68. $(18\cdot7 \times 0\cdot0039 \div 2\cdot21) - (0\cdot441 \times 0\cdot0091 \div 1\cdot911)$.

*69. $(7\cdot35 \times 0\cdot0143 \div 15\cdot015) - (0\cdot152 \times 0\cdot033 \div 2\cdot09)$.

*70. $(21\cdot7 \times 0\cdot087 \div 2\cdot03) + (102\cdot01 \times 0\cdot319 \div 2\cdot639)$.

Seventh Week.   *Recurring Decimals.*

Examples VII.

1. Before converting a vulgar fraction into a decimal, what means is there of determining whether the decimal will terminate or recur?

Will the decimals equivalent to the following fractions terminate or recur?

2. $\frac{19}{160}$ **3.** $\frac{157}{640}$.   **4.** $\frac{293}{960}$.

5. $\frac{1001}{625}$.   **6.** $\frac{563}{22400}$.   **7.** $\frac{1111}{13750}$.

Convert the following fractions into decimal form:

8. $\frac{5}{9}$.   **9.** $\frac{6}{35}$.   **10.** $\frac{1}{33}$.   **11.** $\frac{7}{240}$.

12. $\frac{259}{1100}$.   **13.** $\frac{149}{495}$.   **14.** $\frac{5}{28}$.

15. $\frac{1009}{1998}$.   **16.** $10\frac{10}{999}$.   **17.** $\frac{47}{468}$.

18. *Prove* that $\cdot\dot{9}=1$; and find terminating decimals equivalent to the following:

(i.) $9\cdot\dot{9}$.   (ii.) $\cdot34\dot{9}$.   (iii.) $7\cdot4\dot{9}\dot{9}$.   (iv.) $\cdot0099\dot{9}$.

19. Explain, by taking an example, the reason for the rule for converting a recurring decimal into a vulgar fraction: *e.g.* shew by strict reasoning that $\cdot725\dot{8}\dot{1}=\dfrac{72581-72}{99900}$.

Convert the following recurring decimals into equivalent vulgar fractions:

20. $\cdot\dot{0}\dot{1}$.   **21.** $\cdot32\dot{7}$.   **22.** $\cdot123\dot{7}$.

23. $\cdot013\dot{2}$.   **24.** $1\cdot0462\dot{9}$.   **25.** $\cdot4843\dot{2}$.

26. $55\cdot001\dot{8}$.   **27.** $\cdot\dot{4}2857\dot{1}$.   **28.** $\cdot0571428\dot{5}$.

Find the lowest multipliers which will convert the following recurring decimals into terminating decimals:

29. $\cdot0340\dot{9}$.   **30.** $\cdot\dot{2}42857\dot{1}$.   **31.** $\cdot115384\dot{6}$.

State the rule for the *exact* addition (or subtraction) of recurring decimals; and add together the following, giving the result true to *six* places of decimals:

32. $32\cdot1\dot{4}$, $\cdot3214$, $3\cdot21\dot{4}$, and $321\cdot\dot{4}$.

33. $101\cdot3\dot{0}\dot{1}$, $\cdot0283\dot{2}$, $45\cdot273$, $4\cdot\dot{3}$, and $\cdot42680\dot{1}$.

Add the following, giving the exact result as a recurring decimal:

34. $\cdot14$, $\cdot011\dot{6}$, and $\cdot3\dot{2}\dot{5}$.

35. $\cdot434\dot{8}$, $\cdot101$, and $\cdot455\dot{4}$

36. $\cdot3\dot{0}2\dot{5}$, $3\cdot02\dot{5}$, $302\cdot5$, and $30\cdot2\dot{5}$.

37. $\cdot101$, $\cdot000\dot{1}2\dot{6}$, $\cdot0\dot{2}\dot{0}$, and $\cdot00\dot{3}$.

Simplify the following :

38. $\cdot3\dot{0}2\dot{5} - \cdot3025$;

39. $\cdot3\dot{0}2\dot{5} - \cdot0000\dot{3}02\dot{5}$;

40. $8\cdot27 + 8\cdot2\dot{7} + 8\cdot\dot{2}\dot{7} - 24\cdot000\dot{1}$;

41. $\cdot\dot{4}2857\dot{1} - \cdot\dot{2}8571\dot{4}$.

and explain the last result by means of vulgar fractions.

Multiply

42. $\cdot1\dot{7}\dot{3}$ by 100.

43. $\cdot01368\dot{5}$ by 1000.

44. $\cdot4\dot{3}$ by 3.

45. $\cdot0\dot{3}\dot{6}$ by 11.

46. $\cdot15\dot{6}$ by 13.

47. $\cdot85714\dot{2}$ by $\cdot7$.

48. $\cdot4\dot{4}6153\dot{8}$ by $\cdot65$.

49. $\cdot01\dot{4}$ by $9\cdot9$.

Divide

50. $\cdot625$ by 9.

51. $9\cdot8$ by 11.

52. $406\cdot340\dot{6}$ by 1000.

53. $3\cdot683\dot{6}$ by 36.

54. $2\cdot61046\dot{2}$ by $5\cdot2$.

55. $2\cdot3\dot{4}\dot{5}$ by $7\cdot74$.

56. $1\cdot23\dot{5}$ by $\cdot32$.

57. $26\cdot33\dot{8}$ by $14\cdot14$.

Simplify the following, giving the results as decimals :

58. $\cdot2\dot{7} \times \cdot91\dot{6}$.

59. $\cdot53\dot{0} \times \cdot\dot{6}$.

60. $6060\cdot6\dot{0} \times \cdot01254$.

61. $4\cdot608\dot{1} \times \cdot3\dot{5}67$.

62. $\cdot03\dot{6} \div \cdot0\dot{3}$.

63. $\cdot3\dot{8} \div 12\cdot\dot{4}$.

64. $1 \div \cdot002\dot{5}$.

65. $\cdot4\dot{9}2\dot{5} \div \cdot70\dot{3}$.

66. $\dfrac{4 \times 1\cdot13\dot{6}}{4\cdot5\dot{4}}$.

67. $\dfrac{4\cdot997}{\cdot398\dot{4} \times 2\cdot09}$.

68. $\dfrac{1\cdot10\dot{2} + 2\cdot05\dot{9}}{3\cdot1\dot{6}}$.

69. $\dfrac{2\cdot791\dot{6} \times 3\cdot23\dot{7}}{1\cdot86\dot{1} \times \cdot8093\dot{4}}$.

Eighth Week. *Decimals of Compound Quantities.*

Note on Approximations. It should be carefully observed that a sum of money expressed in £s. and the decimal of a £., if correct to the *third* decimal place, will yield a result which differs from the true value by *less than one farthing.*

If, for instance, we are asked to find the value of £3·6818292... *to the nearest farthing*, we may begin by at once discarding all decimal figures beyond the third (using the fourth, if necessary, to correct the third): for £3·682 will give the same result (viz., £3. 13s. 7¾d.) as £3·6818292... to within less than a farthing.

Note. The reason of this is evident when we remember that 1 farthing = £$\frac{1}{960}$, which is *greater* than £$\frac{1}{1000}$ (or £·001). If, then, the decimal of £1 is corrected to the *third* place (*i.e.* made true to the nearest £·001) it cannot differ from the actual value by as much as £$\frac{1}{960}$, or 1 farthing. The student will find it convenient to acquire one of the various devices (explained in any text-book of Arithmetic) by which a sum of money can be expressed *at sight* as the decimal of £1 correct to the third place ; and also the converse process.

Example 1. Add together £17·0516, £8·197, and £12·31058, giving a result *true to the nearest farthing.*

```
17·051 | 65...
 8·197 |
12·310 | 58...
───────
£37·559 | 2
```

Here the addition (in decimals of £1) need only be performed *correct to the third decimal place.*

And £37·559 = £37. 11s. 2¼d. (to the nearest farthing).

Example 2. Find *to the nearest farthing* the rent and taxes on 306·0108 acres at £5. 17s. 5d. an acre.

```
   5·8 7 0, | 8 3,3
306·0 1 0  | S
─────────────────
1761·2 4 9 | 9
  35·2 2 5 | 0
     5 8   | 7
       4   | 6
─────────────────
£1796·5 3 8 | 2
```

£5. 17s. 5d. = £5·87083,
and we have to multiply 5·870833... by 306·0108, retaining a result *true* only to the third decimal place.

Now £1796·538 = £1796. 10s. 9d. (to the nearest farthing).

Similarly a *weight* expressed as the decimal of a *ton*, if correct to the *fourth* decimal place, will give a result true to the nearest lb. (since the number of lbs. in 1 ton is less than 10,000) ; and for a like reason a *length* expressed as a decimal of a *mile* need only be carried to the fourth place to give a result true to the nearest foot.

Example. Find *to the nearest lb.* the value of ·0165 of 8 tons 5 cwt. 37 lbs.

Now 8 tons 5 cwt. 37 lbs.
= 8·26651... tons.

$$
\begin{array}{r|l}
8\cdot2\,6\,6 & \text{tons.} \\
\cdot0\,1\,6\,5 & \\
\hline
\cdot0\,8\,2\,6 & 6 \\
4\,9\,5 & 9 \\
4\,1 & 3 \\
\hline
\cdot1\,3\,6\,3 & 8 \ \text{ton.}
\end{array}
$$

$$
112\left\{
\begin{array}{r|l}
16 & 37\cdot0 \ \text{lbs.} \\
7 & 2\cdot3125
\end{array}
\right.
$$

20 | 5·33035.. cwts.
8·26651.. tons.

·728 cwt.
112·
72·8
7·28
1·46
81·5 lbs.

And ·1364 ton = 2·728 cwt. = 2 cwt. 82 lbs. (to the nearest lb.).

Examples VIII.

Reduce to pence :

1. ·0375 of £1. 2. ·16875 of £1.
3. 3·4875 of 5s. 4. ·13375 of £5.
5. Reduce (i.) ·878125 of 1 ton to lbs.;
    (ii.) ·01875 of 1 mile to yards ;
    (iii.) ·4625 of 1 lb. Troy to dwt.

Find the value in pounds, shillings, and pence of

6. ·6375 of £1. 7. 3·88125 of £1.
8. 4·39375 of £1. 9. 3·15625 of £5.
10. ·0053125 of £20. 11. ·980078125 of £800.

Express as the decimal of £1 :

12. 17s. 10½d. 13. £3. 15s. 4½d.
14. 11s. 3¾d. 15. £4. 19s. 0¾d.
16. Reduce (i.) £13. 4s. 9d. to the decimal of £100 ;
    (ii.) £16. 13s. 9d. to the decimal of £12 ;
    (iii.) £10. 3s. 10½d. to the decimal of £8.

Find the value in compound quantities of

17. 1·475 acres. 18. ·28125 of a ton.
19. ·895 of a day. 20. 1·6425 of 10 miles.
21. ·430625 of 5 tons. 22. ·6383 of 125 acres,

23. Express (i.) 3 cwt. 3 qrs. 14 lbs. as the decimal of 1 ton ;

(ii.) 3 roods 13 poles as the decimal of 1 acre ;

(iii.) 2 weeks 4 days 6 hrs. as the decimal of 1 year ;

(iv.) 1 qr. 15 lbs. 5 oz. to the decimal of 1 cwt.

Find the value of

24. 2·125 of £3. 4*s.*   **25.** 3·625 of £4. 16*s.*

26. 8·24 of £2. 2*s.* 6*d.*   **27.** 5·84 of £2. 3*s.* 9*d.*

28. ·2272 of £14. 6*s.* 5½*d.*   **29.** ·8125 of 2 tons 4 cwt.

30. 3·875 of 11 ac. 2 r. 8 p.   **31.** 1·9375 of 4 bushels 2 pecks.

32. Multiply (i.) £1. 3*s.* 4*d.* by 3·15 ;

(ii.) £2. 13*s.* 4*d.* by 1·30625.

33. Divide   (i.) £10. 2*s.* 6*d.* by 4·05 ;

(ii.) £29. 12*s.* 6*d.* by 2·844.

34. By what decimal must £1. 6*s.* 8*d.* be multiplied to give £2. 18*s.* 4*d.*? And by what decimal must 9 cwt. 1 qr. be divided to give 1 ton 5 cwt.?

Find the value of

35. ·6416̇ of £1.   **36.** 2·183̇ of £1.   **37.** ·735416̇ of £100.

38. ·3416̇ of £5.   **39.** 4·3583̇ of £7.   **40.** ·83̇ of £2. 6*s.*

41. 2·309̇ of £1. 7*s.* 6*d.*   **42.** 3·0693̇ of 8*s.* 5*d.*

43. ·63663̇ of £45. 16*s.* 8*d.*   **44.** ·226851̇ of a cubic yard.

45. ·142857̇ of one cwt.   **46.** 3·6̇ of ·945̇ of ·428571̇ of 18*s.* 6*d.*

47. Express as decimals of £1 :

(i.) 1*s.* 5½*d.*;   (ii.) £2. 11*s.* 3½*d.*;   (iii.) £17. 16*s.* 8¾*d.*

Reduce

48. 4*s.* 2½*d.* to the decimal of 5*s.*

49. 3 lbs. 6 oz. to the decimal of 4 cwt. 2 lbs.

50. £19. 6*s.* 8*d.* to the decimal of £6.

51. ·056 of a sq. pole to the decimal of an acre.

52. ·0216 of £3. 8*s.* 4*d.* to the decimal of £123.

53. 3 yds. 2 ft. 2 in. to the decimal of 2½ poles.

54. 1 ac. 3 r. 8 p. to the decimal of 4 ac. 3 r. 32 p

55. 2 oz. 13 dwt. to the decimal of 1 lb. (Troy).

56. ·028 of 8*s.* 7½*d.* to the decimal of 11*s.* 6*d.*

57. ·0125 of 4 tons 19 cwt. 6 lbs. to the decimal of 3 cwt. 1 qr. 11 lbs.

58. ·159̇0̇ of one ton to the decimal of 5 cwt,

Add together

59 ·0015 of half-a-sovereign, 2·0615 of half-a-guinea, and 1·3357 of half-a-crown.

60. ·035 of £5, ·5 of a guinea, and ·5$\overset{\cdot}{3}$ of 7*s*. 6*d*.

61. ·35 of one ton, ·39 of one cwt., and ·44 of one qr.

62. 3·75 of half-a-crown, 4·$\frac{1}{4}$ of a guinea, and 1·0$\overset{..}{2}$4 of £22.

63. Subtract $\frac{15}{16}$ of £16. 2*s*. 4*d*. from ·0125 of £1626. 15*s*.; and find by what decimal the result must be multiplied to produce £1. 6*s*. 1½*d*.

Subtract

64. ·37$\overset{\cdot}{8}$ of 1*s*. 6½*d*. from ·3$\overset{..}{7}$8 of 2*s*. 9*d*.

65. the sum of ·3125 of 6 cwt., and ·032 of 3 cwt. 2 qrs. 14 lbs. 4 oz. from 2 cwt.

66. ·6$\overset{..}{2}$6 of £1. 4*s*. 9*d*. from the sum of 2·3125 of £12, and 4·825· of 5*s*.

Find the value *to the nearest farthing* of

67. ·7906 of £1.   **68.** 2·36842 of £1.

69. 3·701056 of £4. 15*s*.   **70.** 1·02 of £3. 5*s*. 2½*d*.

71. ·0914 of £32. 11*s*. 2½*d*.   **72.** 2·0172 of £500. 3*s*. 5½*d*.

73. £8. 6*s*. 11¼*d*. × 100·7.   **74.** £384. 7*s*. 5*d*. × ·008351.

75. £·561023 × 597·001.   **76.** £60. 12*s*. 0¼*d* ÷ 20·0002.

77. £191. 18*s*. 5¾*d*. ÷ 7·0071.   **78.** £10. 16*s*. 11*d*. ÷ ·24$\overset{..}{8}$05.

Find the value of

79. ·123725 of one ton *to the nearest ounce*.

80. ·97247 of one acre *to the nearest square yard*.

81. ·1225 of 1 mi. 6 fur. *to the nearest foot*.

82. ·3 of ·6$\overset{\cdot}{3}$ of £6. 2*s*. 3*d*. *to the nearest farthing*.

83. Find to the nearest inch the difference between ·437 of 1 fur. 3 p. and ·097 of a mile.

84. Find to the nearest ounce the sum of ·0237 of a ton and ·687 of a quarter.

*85. What decimal is 2¼*d*. of 7*s*. 6*d*.? And what decimal of 1 cwt. is 64 lbs.?

*86. How many pounds are there in 0·7086·4 of a ton?

*87. Find the value of 0·02545454... of £18. 19*s*. 6*d*.; and express the result as the decimal of £5. 10*s*. 0*d*.

*88. Divide £47. 17*s*. 6¼*d*. by 0·3727272...; and find what decimal of the quotient is £143.

*89. A sovereign consists of 22 parts by weight of pure gold to 2 parts of alloy, and it weighs 123·274 grains. Neglecting the value of the alloy, find (to the nearest penny) the value of pure gold per ounce Troy (480 grains).

Ninth Week. *Miscellaneous Examples on Fractions and Decimals.*

Examples IX.

Express as decimals :

1. $\cdot0003 + \frac{817}{3125} - \cdot00847 + \frac{361}{800}.$

2. $50\frac{21}{25} + 3\frac{1}{160} + 8\frac{43}{100} + 28\frac{32}{3125}.$

3. Simplify $\frac{442}{2431} + \frac{2151}{7887} + \frac{2094}{3830}$; and find what decimal must be multiplied by $\frac{1}{7}$ to give a result equal to the sum of $\frac{1}{11330}$ and $\frac{1}{16731}.$

4. Reduce $\dfrac{£127.\ 4s.\ 9d.}{£933.\ 1s.\ 6d.}$ to its simplest form ; and state what meanings may be assigned to this fraction.

5. Simplify $\dfrac{£4.\ 13s.\ 6d.}{£60.\ 15s.\ 6d.} + \dfrac{5\ \text{tons}\ 2\ \text{cwt.}}{16\ \text{tons}\ 11\ \text{cwt.}\ 2\ \text{qrs.}}$

6. Simplify $\dfrac{£1.\ 18s.\ 6d.}{1078d.} + \dfrac{\cdot35648\dot{1}\ \text{cub. ft.}}{1078\ \text{cub. in.}}$

Express as simple decimals :

7. $\dfrac{\cdot161}{1\cdot15} + \dfrac{\cdot00576}{\cdot12} + \dfrac{\cdot250625}{2\cdot5}.$

8. $\dfrac{1\cdot8}{6\times2\frac{2}{5}} + \dfrac{\cdot\dot{6}}{\cdot75} - \dfrac{36\cdot625+\frac{3}{8}}{\cdot024\ \text{of}\ 3000}.$

9. $\dfrac{\cdot875 \times \cdot27\dot{0}}{\cdot125 + \cdot125675} + \dfrac{3}{53}.$

10. Express $\frac{11}{17}$ of $1\cdot1\dot{3}$ of 2 cwt. 5 lbs. 8 oz. as the decimal of one ton.

11. A man by selling a horse for £12 more than he gave for it realises a profit equal to $\frac{2}{7}$ of the cost price. What was the cost price ?

Find the value of

12. $\dfrac{1}{1\frac{7}{20}}$ of $\cdot034\dot{8}$ of 18s. $6\frac{3}{4}d. + \frac{1}{7}$ of $\cdot2\dot{3}$ of 1s. 3d.

13. $\cdot6\dot{5}$ of $4\cdot\dot{1}\dot{1}$ of $\dfrac{3\frac{2}{3}}{13}$ of $2\cdot43\dot{2}$ of 13s. 6d.

14. $\cdot3375$ of £1 + $\cdot21\dot{6}$ of £15 + $1\cdot025$ of 6s. 8d.

15. A battalion lost $\frac{1}{30}$ of its numbers at its first engagement, $\frac{7}{20}$ of the remainder at the second, and $\frac{3}{11}$ of the remainder at the third. Then 512 men were left fit for duty. What was the original strength of the battalion ?

Simplify

16. $\dfrac{3}{2+\dfrac{2}{3+\dfrac{2}{3+\frac{2}{3}}}} \div \cdot 00039.$

17. $\dfrac{2}{3+\dfrac{\cdot 2}{3+\dfrac{\cdot 02}{3+\cdot 002}}}.$

18. A warehouse consists of seven floors, and the rent of each floor is ·875 times that of the floor below it. If the rent of the middle floor is £120. 1s., find that of the lowest.

19. Reduce ·055 of £2. 17s. 11d. to the decimal of the difference between $\frac{2}{3}$ of £5. 6s. 9d. and $\frac{3}{5}$ of £1. 2s. 1d.

20. A piece of copper weighing 3 lbs. 2 oz. is dropped gently into a vessel full of water. Find correct to the hundredth part of a cubic inch the volume of the water displaced, assuming that a cubic foot of water weighs 1000 oz., and that copper is 8·915 times as heavy as water.

Find, correct to four places of decimals, the values of the following series :

21. $1+\dfrac{1}{5}+\dfrac{1}{5^2}+\dfrac{1}{5^3}+\dfrac{1}{5^4}+\ldots.$

22. $\dfrac{1}{2}+\dfrac{1}{2\times3}+\dfrac{1}{2\times3\times4}+\dfrac{1}{2\times3\times4\times5}+\ldots.$

Find, correct to six places of decimals,

23. $\dfrac{1}{1\times9}+\dfrac{1}{3\times9^3}+\dfrac{1}{5\times9^5}+\dfrac{1}{7\times9^7}+\ldots.$

24. $\dfrac{1}{2}+\dfrac{1}{2\times4}+\dfrac{1}{2\times4\times6}+\dfrac{1}{2\times4\times6\times8}+\ldots.$

*25. Reduce to its simplest form

$$\dfrac{3\frac{4}{11}\times\left(1\frac{2}{9}\text{ of }1\cdot08\right)}{1\frac{11}{13}\times\left(0\cdot6+\frac{2}{9}\right)}.$$

*26. Find the sum of 2·25 of 2s. 1d. and 0·0̇3̇ of £6. 3s. 9d.; and express the result as the fraction of £2. 13s. 6d.

*27. Subtract 0·0625 of £113. 16s. 8d. from $\frac{23}{24}$ of £50. 5s. 6d.; and find the number by which the result must be multiplied to produce £985. 11s. 6d.

*28. An imperial gallon was declared by Act of Parliament in 1760 to contain 277·274 cubic inches ; a Winchester bushel contains 2150·42 cubic inches. How many Winchester bushels are equal to 100 imperial bushels ?

Tenth Week.　　　　*Practice.*

Notes and Hints for Solution. The method of Simple Practice is too familiar to need illustration here. We will work out one question in Compound Practice by the ordinary rule, and then give two examples to remind the student that in more complicated cases, where only an approximate result (true to the nearest penny or farthing) is required, a decimal process may often be employed with advantage.

Example 1. Find the value of 6 oz. 15 dwts. 15 grs. of standard gold at £3. 17s. 10½d. per ounce.

	£3. 17s. 10½d.	= value of 1 oz.
	6	
	£23. 7s. 3d.	= value of 6 oz.
10 dwts. = ½ of 1 oz.	1. 18s. 11¼d.	= ,, 10 dwts.
5 dwts. = ½ of 10 dwts.	19s. 5⅝d.	= ,, 5 dwts.
12 grs. = 1⁄10 of 5 dwts.	1s. 11$\frac{29}{80}$d.	= ,, 12 grs.
3 grs. = ¼ of 12 grs.	5$\frac{269}{320}$d. =	,, 3 grs.
	£26. 8s. 1$\frac{5}{64}$d.	= value of
		6 oz. 15 dwts. 15 grs.

Example 2. Find, *to the nearest farthing*, the value of a crop on 57 ac. 0 ro. 4 p. 18 sq. yds. at £2. 12s. 4d. an acre.

[The student will remember that a result expressed as a decimal of a £., if correct to the *third* place, will yield an answer true to the nearest farthing.]

Now 57 ac. 0 ro. 4 p. 18 sq. yds. = 57·0287... acres.

Required 57·0287 acres at £2. 12s. 4d. an acre.

	£57·028 7		18· sq. yds.
	2		4
	114·057 4	121{	11 | 72·
10s. = £½	28·514 3		11 | 6·54545..
2s. = ⅕ of 10s.	5·702 9		40 | 4·59504.. p.
4d. = ⅙ of 2s.	·950 5		4 | 0·11487.. r.
	£149·225 1 = £149. 4s. 6d.		·02871.. ac.

Example 3.   Find the value of 40 tons 2 cwt. 12 lbs. at £17. 5*s*. 7*d*. per ton to the nearest farthing.

£17. 5*s*. 7*d*. = £17·279166....
40 tons 2 cwt. 12 lbs. = 40·10535... tons.

£17·279	17			112 {	16	12·	lbs.
40·105	35				7	·75	
£691·166	8				20	2·10714..	cwt.
1·727	9					·10535..	ton.
86	4						
5	2						
	9						
£692·987 2	= £692. 19*s*. 9*d*.						

Examples X.

(*Simple Practice.*)

Find the cost of

1.   127 things at £9. 17*s*. 6*d*.   each.
2.   3451 things at £2. 18*s*. 9*d*.   each.
3.   6050 things at £2. 7*s*. 10*d*.   each.
4.   431 things at £5. 17*s*. 11½*d*.   each.
5.   735 things at   16*s*. 10½*d*. each.

6.   What sum will be required to enable a company to declare a dividend of 7*s*. 10½*d*. per share on 2135 shares?

7.   If a bankrupt pays 14*s*. 7¼*d*. in the pound, how much will a creditor receive to whom he owes £2510?

8.   Find the dividend on £1596 at 15*s*. 9¾*d*. in the pound.

9.   Find the total weight of 336 truck-loads of 3 tons 4 cwt. 1 qr. 17 lbs. each.

10.   What is the total area of 1536 allotments of land, each containing 4 acres 2 r. 36 p.?

(*Compound Practice.*)

Find by Practice the cost of

11.   2 cwt. 2 qrs. 17 lbs. at £1. 10*s*. 4*d*. per cwt.
12.   12 cwt. 3 qrs. 16 lbs. at £2. 17*s*. 2*d*. per ton.
13.   5 lbs. 10 oz. 17 dwt. of an alloy of gold at £2. 2*s*. 6*d*. per ounce.
14.   113 ac. 2 r. 13⅓ p. of land at £56. 5*s*. 0*d*. per acre.
15.   17 miles 7 fur. 60 yds. of iron rails at £17. 12*s*. per mile.

16. If a bankrupt pays 12*s.* 10*d.* in the pound, how much will he pay to a creditor to whom he owes £754. 10*s.*

17. Find the dividend on

(i.) £4287. 17*s.* 6*d.* at 14*s.* 4*d.* in the pound ;

(ii.) £2073. 6*s.* 3*d.* at 11*s.* 8*d.* in the pound.

18. Find the value of a nugget of gold weighing 3 lbs. 11 oz. 8 dwts. 4 grs. at £3. 17*s.* 6*d.* per ounce.

19. Find the cost of making a road 47 miles 3 fur. 5 yds. long at £38. 2*s.* 8*d.* a mile.

(*Approximations.*)

20. Find *to the nearest penny* the value of 3 ac. 1 r. $3\frac{1}{4}$ p. at £110 per acre.

21. Find *to the nearest penny* the value of $841\frac{1}{4}$ cwt. at £21. 13*s.* $7\frac{1}{4}d.$ per cwt.

22. Find *to the nearest farthing* the dividend on

(i.) £1483. 17*s.* at 8*s.* $11\frac{1}{4}d.$ in the pound ;

(ii.) £1710. 14*s.* 6*d.* at 13*s.* $4\frac{1}{2}d.$ in the pound.

23. Find *to the nearest penny* the rent of 315 ac. 3 r. 7 p. 11 sq. yds. at £1. 16*s.* 8*d.* per acre.

24. Find *to the nearest penny* the cost of 659 bales of cotton, each weighing 1 cwt. 1 qr. 21 lbs., at £2. 3*s.* $1\frac{1}{2}d.$ per cwt.

25. Find *to the nearest farthing* the value of 9 tons 4 cwt. 3 qrs. 21 lbs. of material at £14. 15*s.* 9*d.* per cwt.

*26. Find the value (*to the nearest farthing*) of 2 lbs. 7 oz. 15 dwt. 20 grs. of gold at £46. 14*s.* 6*d.* a pound.

*27. Find the value (*to the nearest farthing*) of 3 tons 17 cwts. 2 qrs. 8 lbs. at £32. 17*s.* 4*d.* a ton.

*28. Find (*to the nearest farthing*) the value of 3 lbs. 4 oz. 13 dwts. 19 grs. of gold at £3. 17*s.* $10\frac{1}{2}d.$ an ounce.

*29. Find the cost of 12 tons 2 cwts. 3 qrs. 11 lbs. of material at £17. 11*s.* 8*d.* per ton.

*30. Find the cost of 47 qrs. 4 bus. 2 pks. 1 gal. 3 qts. 1 pt. at £1. 17*s.* 4*d.* per bushel.

Eleventh Week. *Ratio and Proportion.*

Notes and Hints for Solution. The following examples illustrate typical questions to be solved by methods of ratio and proportion.

Example 1. Find the ratio which 7 days 3 hrs. 44 min. bears to 13 days 10 hrs.

The ratio which one quantity bears to another (of the same kind) is measured by the fraction that the first is of the second. This requires that the two quantities should be expressed in the same denomination.

Now 7 days 3 hrs. 44 min. = 10304 minutes ;

13 days 10 hrs. = 19320 minutes ;

∴ the required ratio $=\dfrac{10304}{19320}=\dfrac{1288\times 8}{1288\times 15}=\dfrac{8}{15}$ (or 8 : 15).

Example 2. What sum of money has to £11. 4s. the same ratio that 5 cwt. has to 1 ton 4 cwt. ?

Here the given ratio $=\dfrac{5\ \text{cwt.}}{1\ \text{ton 4 cwt.}}=\dfrac{5}{24}$;

∴ the required sum $=\frac{5}{24}$ of £11. 4s. = £2. 6s. 8d.

Example 3. If the carriage of 5 tons 2 cwt. of goods costs £2. 18s. 8d., what should be charged for carrying 7 tons 13 cwt. the same distance ?

Here 5 tons 2 cwt. = 102 cwt. ; and 7 tons 13 cwt. = 153 cwt. ; and we may reason thus :

To carry 102 cwt. costs £2. 18s. 8d. ;

∴ ,, 1 cwt. ,, $\frac{1}{102}$ of £2. 18s. 8d. ;

∴ ,, 153 cwt. ,, $\frac{153}{102}$ of £2. 18s. 8d.

$=\frac{3}{2}$ of £2. 18s. 8d. = £4. 8s.

After some practice the student may dispense with the middle step ; and, noticing that the required sum of money will be *greater* than £2. 18s. 8d. in the ratio of 153 : 102, he may multiply at once by the *increasing* ratio $\frac{153}{102}$ or $\frac{3}{2}$.

The work would then stand thus :

To carry 102 cwt. costs £2. 18s. 8d. ;

∴ ,, 153 cwt. ,, £2. 18s. 8d. $\times \frac{153}{102}$

$=$ £2. 18s. 8d. $\times \frac{3}{2}=$ £4. 8s.

Examples XI.

Express in simplest form

1. The ratio of 630 to 936.
2. The ratio of 2·375 to $\frac{19}{20}$ of 6·6̇.
3. The ratio of £1. 3*s*. to £4. 6*s*. 3*d*.
4. The ratio of 6 cwt. 2 qrs. 8 lbs. to 1 ton 3 cwt.
5. If four acres of land cost £121, at what rate is this per square yard?
6. If 8 cwt. 3 qrs. of material cost £40. 10*s*. 10*d*., at what rate is this per ton?
7. If a train travel 35 miles an hour, at what rate is this in feet per second?
8. What sum of money bears to £32 the same ratio that $85\frac{1}{2}$. bears to 114?
9. Find a sum of money which has to £244. 4*s*. 3¾*d*. the ratio of 4 : 51.
10. What weight bears to 47 tons the ratio which 11*s*. 8¼*d*. bears to £39. 19*s*.?
11. If a train takes 29 min. 45 secs. to run 10 miles 5 furlongs, how far will it run in 11 min. 40 secs.?
12. Find the cost of a foreign telegram of 425 words at the rate of £1. 12*s*. 6*d*. for 20 words.
13. If 17 men can do a piece of work in 68 days, how many must be employed to do the work in 4 days?
14. How long will it take 32 men to reap a field which 24 men can finish in 5 hours?
15. Steaming 18 knots an hour, a passage is made in 180 hours; how long would the same passage take at a speed of 10 knots an hour?
16. If 6 cwt. 2 qrs. 2 lbs. of sugar cost £9. 3*s*. 4*d*., how much will 4 cwt. 2 qrs. 7 lbs. cost?
17. If 5 ac. 3 r. 4 p. of land is worth £1125, what should be given for 44 ac. 3 r. 1 p.?
18. If from every 100 lbs. of sea-water $2\frac{1}{2}$ lbs. of salt may be obtained, what weight of water must be evaporated to yield half a ton of salt?
19. If 1 ton 16 cwt. 94 lbs. of coal cost £2. 7*s*. 6*d*., how much can be bought for £17. 16*s*. 3*d*.?
20. The debts of a bankrupt amount to £1792, and his whole property is worth £1344; how much can he pay in the pound?
21. The debts of a bankrupt amount to £563. 15*s*., and his assets to £371. 2*s*. 8½*d*.; how much can he pay in the pound?

22. After paying income-tax at the rate of 4d. in the pound, I have £578. 4s. per annum left. What is my gross income?

23. After paying income-tax at the rate of 5d. in the pound, a man has £576. 6s. 9d. a year. How much tax did he pay?

24. If one train runs 49 miles in 1 hour 10 minutes, and another runs 770 yards in $37\frac{1}{2}$ seconds, which has the higher speed?

25. Compare (as a ratio) the rates of travelling of a bicyclist who goes $34\frac{1}{8}$ miles in 2 hrs. 10 mins. and a train which travels $59\frac{1}{2}$ miles in 1 hr. 42 mins.

26. If 2 cwt. 3 qrs. 18 lbs. of tea cost as much as 27 cwt. 2 qrs. 17 lbs. of sugar, compare the values of equal weights of tea and sugar.

27. If 5 cwt. 3 qrs. 8 lbs. of tea cost as much as 55 cwt. 1 qr. 6 lbs. of sugar, what quantity of sugar should be given in exchange for 15 lbs. of tea?

28. A man drove from A to B, a distance of 54 miles, at an average rate of 8 miles an hour. Another man, starting half-an-hour after the first, arrived at B 15 minutes before him; find the ratio of their speeds.

29. A farmer bought 4 horses and 7 cows for £238, the prices of a horse and a cow being in the ratio of 5 : 2. How much did he give for each?

30. If it costs as much to feed 3 men as to feed 4 boys, and if for 3 boys the cost is 19s. $2\frac{1}{4}d$. per week, what will it cost per week to feed 51 men?

31. If 5 men *or* 7 women can do a piece of work in 37 days, in what time will 7 men *and* 5 women do a piece twice as great?

Express in simplest form

*32. The ratio of $2\frac{1}{4}$ to $7\frac{1}{3}$.

*33. The ratio of 4·333... to $10\frac{1}{11} + 3·0333....$

*34. Find the number that is to $7\frac{2}{3}$ in the ratio of £3. 1s. 3d. to £4. 13s. 11d.

Express in simplest form the ratio which

*35. $\frac{2}{5}$ of £27. 1s. $5\frac{3}{4}d$. bears to 0·6 of £42. 10s. $10\frac{3}{4}d$.

*36. $\frac{3}{4}$ of 53 cwt. 3 qrs. 3 lbs. bears to 0·4 of 65 cwt. 0 qrs. 11 lbs.

*37. $\frac{5}{8}$ of £5. 9s. 8d. bears to 1·4242... of £4. 16s. 3d.

*38. $\frac{3}{7}$ of $\frac{2}{9}$ of £5. 15s. 6d. bears to $\frac{11}{122}$ of £16. 0s. 3d.

*39. If either 5 men or 9 boys can do a certain piece of work in 19 days, in how many days can 13 men and 7 boys, working together, do a piece of work twice as great?

Twelfth Week.   *Compound Proportion.*

Notes and Hints for Solution. The student who has made himself familiar with the principles of Simple Proportion will have no difficulty in applying them to more intricate cases, where the required result is obtained by considering the combined effect of two or more ratios.

Example 1. If 27 men mow a field of 90 acres in 7 days, working 8 hours a day, how many men will be required to mow 200 acres in 16 days, if they work 10 hours a day ?

The question is *how many men* ? State the question so as to place last the term corresponding to the answer. Thus

90 acres mown in 7 days of 8 hours each by 27 men,

200 acres   ,,   16 days ,, 10 hours   ,,   *how many men*?

Now, other conditions remaining the same, if the number of acres were changed from 90 to 200, *more* men would be needed ;

∴   we multiply by the *increasing* ratio $\frac{200}{90}$ or $\frac{20}{9}$.

Again, if the number of days allowed were changed from 7 to 16, *fewer* men would be needed for the work ;

∴   we multiply by the *diminishing* ratio $\frac{7}{16}$.

Lastly, if the number of working hours per day were changed from 8 to 10, *fewer* men would do the work ;

∴   we multiply by the *diminishing* ratio $\frac{8}{10}$ or $\frac{4}{5}$.

Hence the required number of men $= 27 \times \frac{20}{9} \times \frac{7}{16} \times \frac{4}{5} = 21$.

Note. The reasoning, given in full in this example, should be done mentally.

Example 2. A man can read a book of 220 pages, each containing 28 lines, with an average of 12 words to the line, in $5\frac{1}{2}$ hours ; how long will it take him to read a book of 400 pages, each of 36 lines, with an average of 14 words to a line ?

220 pages   28 lines   12 words $5\frac{1}{2}$ hours

400   ,,   36 ,,   14 ,,   *how many hours* ?

Required number of hours

$$= 5\frac{1}{2} \times \frac{400}{220} \times \frac{36}{28} \times \frac{14}{12} = \frac{11}{2} \times \frac{20}{11} \times \frac{9}{7} \times \frac{7}{6} = 15.$$

Examples XII.

1. How much will 42 men earn in 30 days, if 69 men can earn £368 in 35 days?

2. How many horses will be required to plough 936 acres in 39 days, if 26 horses can plough 1152 acres in 24 days?

3. If 49 men can empty a reservoir in 65 days by pumping 8 hours a day, how many hours a day must 196 men work to empty it in 26 days?

4. If $16\frac{1}{4}$ tons of provisions last a garrison of 2100 men for 13 days, what weight of provisions will be required for 4340 men for 42 days?

5. If 4 men mow 15 acres in 10 working days of 7 hours each, in how many days of $6\frac{1}{2}$ hours could 7 men mow $19\frac{1}{2}$ acres?

6. If 285 men can dig a trench in 120 days, working 8 hours a day, how many men must be employed to do one-fifth of the work in 152 days, working 10 hours a day?

7. If 40 men can dig a trench in 4 days of 9 hours each, how many men must be employed to dig a trench twice as long, half as wide again, and three-quarters of the depth of the former in 5 days of 8 hours?

8. If 90 men can dig a ditch 50 yards long, 6 feet wide, and 12 feet deep in $4\frac{1}{2}$ days, how many men can dig a ditch 240 feet long, 2 yards wide, and 4 feet deep in 18 days?

9. If 24 men in 2 days of 12 hours each dig a trench 132 yards long, 4 yards wide, and 2 yards deep, how many hours a day must 90 men work to dig in 4 days a trench 3 times as long, the width being 5 yards, and the depth 3 yards?

10. If goods weighing 1 ton 5 cwt. $28\frac{1}{2}$ lbs. can be carried 100 miles for £23. 11s. 5d., how many miles can goods weighing 5 tons 42 lbs. be carried for £210. 15s. 9d.?

11. If 8 men can dig a field of $9\frac{1}{2}$ acres in 19 days, how long will it take 5 men to dig one containing 5 ac. 1 r. 10 p.?

12. If an express train travelling at the rate of 55 miles an hour can accomplish a journey in $3\frac{1}{2}$ hours, how long will it take a slow train to travel two-thirds of the distance, its rate being to that of the express train as 4 to 9?

13. The travelling expenses of 7 tourists for 5 weeks amounted to £75. 5s.; a second party of 18 made the same tour in 6 weeks, their average weekly expenditure per man being $\frac{4}{9}$ of that of the first party. What were the total expenses of the second party?

14. If 10 sheep *or* 15 lambs require 40 bushels of turnips for 7 days, how long should 36 bushels last 6 sheep *and* 18 lambs?

15. If a family by using 6 gas-burners for 5 hours a day pay £1. 5*s*. per quarter, when gas is at 5 shillings per 1000 cubic feet, what should be paid per quarter when 8 burners are used for 3 hours a day, gas being at 3*s*. 9*d*. per 1000 cubic feet?

16. Two horses can plough in a given time as much as 3 oxen, and the daily cost of 4 oxen is equal to that of 3 horses. A certain field can be ploughed by 3 horses in 8 days; find the cost of ploughing it by oxen in 6 days, if the daily cost of a horse is 3 shillings?

17. If 6 compositors, in 16 working days of $10\frac{1}{2}$ hours each, can set in type 720 pages, each of 60 lines with 40 letters to a line, in how many days of 7 hours each will 9 compositors set 960 pages, each of 45 lines with 50 letters to a line?

18. If 75 men can perform a piece of work in 12 days of 10 hours each, how many men will perform a piece of work twice as great in one-tenth of the time if they work the same number of hours a day, supposing that two of the second gang can do as much work in an hour as three of the first gang?

19. A piece of work may be done in 6 days by 10 men *or* by 18 boys in 8 days. If 5 men and 9 boys are employed on it, how long will they take? And if £60 is given to them in wages, how much does each man and each boy earn per day?

20. If a certain amount of work is done by 9 men, 12 women, and 13 boys in 11 days, how long will the same work take if 18 men, 3 women, and 5 boys are set to do it, assuming that the ratio of a man's work to a woman's is 5 : 3, and a woman's work to a boy's is 4 : 3?

21. If 175 men and 240 boys do in 1330 days the same amount of work as 603 men and 1005 boys do in 350 days; compare the average daily work done by each man with that done by each boy.

22. If 4 men and 6 boys working 9 hours a day mow 69·3 acres of corn in 10 days, in how many days will 5 men and 4 boys working 10 hours and 40 minutes a day mow 67·76 acres, it being assumed that 2 men mow as much as 5 boys?

23. If 20 English navvies, each earning 3*s*. 6*d*. a day, can do the same piece of work in 15 days that it takes 28 foreign workmen, each earning 3 francs a day, to complete in 20 days; determine which class of workmen it is more profitable to employ (1 franc = 10*d*.). If a piece of work done by the navvies cost £3000, what would be the cost of the same work done by the foreign workmen?

Thirteenth Week. *Proportional Parts.*

Notes and Hints for Solution. It is often required to break up a quantity into parts which bear to one another a given relation. For instance, we may be asked to distribute a sum of money in shares proportional to certain numbers. The method of effecting this will be seen from the following examples.

Example 1. Divide £79. 1s. 9d. among three persons, A, B, and C, so that the sums received may be proportional to the numbers 5, 6, 8.

Here the given sum is to be divided into $5+6+8$, or 19, equal shares, of which A is to have 5, B 6, and C 8.

Now *one* such share $= \frac{1}{19}$ of £79. 1s. 9d. $=$ £4. 3s. 3d.

$\therefore$ A gets *five* times £4. 3s. 3d. $=$ £20. 16s. 3d.

B „ *six* „ £4. 3s. 3d. $=$ £24. 19s. 6d.

C „ *eight* „ £4. 3s. 3d. $=$ £33. 6s.

Example 2. Divide £43. 12s. among A, B, and C, so that A gets $2\frac{1}{3}$ as much as B, and B $1\frac{1}{2}$ as much as C.

Suppose C gets 1 share; then B will get $1\frac{1}{2}$ shares, and A will get $2\frac{1}{3} \times 1\frac{1}{2} = 3\frac{1}{2}$ shares.

Now $3\frac{1}{2}$, $1\frac{1}{2}$, 1 are respectively equal to $\frac{7}{2}$, $\frac{3}{2}$, 1; and are therefore proportional to the numbers 7, 3, 2.

Thus proceeding as before,

A gets $\frac{7}{12}$ of £43. 12s. $=$ £25. 8s. 8d.

B gets $\frac{3}{12}$ $\left(i.e.\ \frac{1}{4}\right)$ of £43. 12s. $=$ £10. 18s. 0d.

C gets $\frac{2}{12}$ $\left(i.e.\ \frac{1}{6}\right)$ of £43. 12s. $=$ £7. 5s. 4d.

Example 3. A and B become partners in a business, A contributing £500 towards the capital, and B £600. After 5 months they are joined by C, with a capital of £300; but 9 months after starting B retires, taking out all his money. At the end of a year the profits are found to be £384. How should they be divided?

Here the claim of each partner depends partly on the *amount* of capital advanced, and partly on the *time* during which it was employed.

Suppose the employment of £100 for 1 month to constitute 1 share.

Then A may claim 5×12, or 60 shares;

B may claim 6×9, or 54 shares;

and C may claim 3×7, or 21 shares.

Thus we have simply to divide £384 into parts proportional to 60, 54, and 21; *i.e.* to 20, 18, and 7.

34

ARITHMETIC.

Examples XIII.

1. Shew how a sum of £5880 may be divided into three shares proportional to the numbers 3, 5, 7.

2. Divide £27,200 among three persons in shares proportional to the numbers 8, 5, 3.

3. Divide £2. 16s. 3d. into shares proportional to the numbers 13, 9, 3.

4. Divide £16. 17s. 6d. into shares proportional to the fractions $\frac{1}{4}, \frac{2}{5}, \frac{3}{5}$.

5. The sum of £81. 10s. is to be divided into four parts proportional to the fractions $\frac{1}{2}, \frac{2}{3}, \frac{3}{4}, \frac{4}{5}$; find the value of each part.

6. The sum of £32,818 is left to four persons to be divided into shares proportional to the fractions $\frac{2}{3}, \frac{3}{4}, \frac{4}{5}, \frac{5}{6}$. How much will each person receive?

7. Three persons A, B, and C join in a speculation, A contributing £500 to the capital, B £300, and C £200; if the total profits amount to £80, how much should each contributor receive?

8. A, B, and C enter into partnership, A contributing £1500, B £1650, and C £2100 to the capital. How ought an annual profit of £700 to be divided among them?

9. The sum of £5100 is to be raised jointly by three towns, whose populations are respectively 4250, 5250, and 7500. If the towns contribute proportionally to their population, for what payment is each town responsible?

10. A bankrupt's estate is sold for £4320 to meet the claims of four creditors to whom he owes respectively £1440, £1680, £2400, and £3120. How much should each creditor receive?

11. The debts of a bankrupt are £65, £75. 3s. 4d., and £108. 8s. 4d.; and his assets are £136. 14s. 5d. What will each creditor receive?

12. A field containing 13 ac. 3 r. 20 p. is divided into three allotments, whose areas are proportional to the numbers 9, 10, 11. If each allotment is subdivided into 37 cottage gardens, find the size of the gardens in each set.

13. Divide £1375 among three persons, A, B, and C, so that A may receive three times as much as B, and C half as much again as A and B together.

14. Divide £60. 2s. 1¾d. between A, B, and C, so that A may have half as much again as B, and B half as much again as C.

15. The sum of £9. 12s. 6d. is to be divided between 16 men, 18 women and 36 children; a woman is to have three times as much as a child, and a man as much as a woman and child together. What will be the share of each?

16. Divide £146 between A, B, and C, so that as often as A gets 4*s.*, B may get 5*s.* 4*d.*, and as often as B gets 8*s.* 9*d.*, C may get 7*s.* 6*d.*

17. Three men A, B, and C join in a speculation, for which A furnishes £100 for 3 months, B £80 for 5 months, and C £250 for 2 months. The result is a profit of £100: how should this be divided among them?

18. A commences business with a capital of £4000, and after 4 months takes B into partnership with a capital of £300. Two months later C joins the firm with a capital of £5000. At the end of the year the profits are found to be £1557. 15*s.*; how much of this sum should each partner receive?

19. A, B, and C enter into partnership on January 1st with capitals of £2200, £2600, and £3000 respectively. At the end of half-a-year B withdraws half his capital. At the end of the year how much should each partner receive of the total profits, viz., £789. 9*s.* 7*d.*?

20. Of four cisterns, the second contains half as much again as the first; the third contains one-third as much again as the first two together; the fourth contains one-fourth as much again as the first three together. The weight of water in the third cistern is 7 tons 4 cwt. What weight of water is contained in the fourth cistern?

*21. Divide 204 into three parts proportional to the numbers 7, 8, 9.

*22. Divide £4. 2*s.* 6*d.* amongst three persons A, B, and C; so that A's share may be $\frac{5}{8}$ of B's, and B's share $1\frac{1}{2}$ of C's.

*23. Divide £90. 6*s.* 0*d.* amongst A, B, and C, so that A may receive $\frac{3}{4}$ of what B receives, and C $\frac{1}{5}$ of what A and B receive together.

*24. The area of a country is 32,300,000 acres: it consists of three kinds of land, viz. *first*, arable; *secondly*, meadow; *thirdly*, waste. The areas of these kinds are proportional to the numbers 2, 3, $\frac{2}{5}$: how many acres are there of each kind?

*25. A sum of money is divided between A, B, and C: if C gets twice as much as A, and A and B together get £50, and B and C together get £60; find how much each person gets.

*26. A man owes three bills, one of which could be paid by a certain number of florins, another by twice that number of half-crowns, and the third by six times that number of shillings. The bills amount in all to £7. 3*s.* 0*d.* What are the several sums?

*27. A and B are partners in a business, in which A has invested £6750 and B £1500. B receives 15 per cent. of all the profits for acting as manager, and the remainder is divided between the partners in proportion to the sums invested. What will each receive out of profits amounting to £726. 9*s.* 2*d.*?

Fourteenth Week. *Square Measure. Rectangular Areas.*

Notes and Hints for Solution. In applying the funda-mental rule of Rectangular Areas, viz.

$$Area = Length \times Breadth,$$

care must be taken to express the length and breadth in units of the *same denomination*; the area will then be obtained in corresponding square units.

Similarly, the rule *Length = Area ÷ Breadth* requires that the linear and square units should correspond. Thus, if the area is expressed in *square feet*, the breadth must be expressed in *linear feet*; the length will then be obtained also in *linear feet*.

In addition to the ordinary Table of Square Measure, the student should observe the following:

The **chain**, used in land-surveying, is 22 yards in length, and is divided into 100 links (7·92 inches). Hence we have the following table:

$$22 \text{ yards} = 1 \text{ chain};$$
$$\therefore (22)^2, \text{ or } 484, square \text{ yards} = 1 \; square \text{ chain};$$

Now   4840 square yards = 1 acre;
$$\therefore 10 \text{ square chains} = 1 \text{ acre.}$$

Similarly, since   100 links = 1 chain;
$$\therefore (100)^2, \text{ or } 10,000 \; square \text{ links} = 1 \; square \text{ chain};$$
$$\therefore 100,000 \text{ square links} = 1 \text{ acre.}$$

Thus, *square chains* are converted into *acres* by dividing by 10; and *square links* are converted into acres by dividing by 100,000.

For instance,   987·5 square chains = 98·75 acres;
807,204 square links = 8·07204 acres.

Example 1. A rectangular allotment of ground is 24 chains 25 links long by 16 chains 80 links broad. Find to the nearest penny its rent at the rate of £5. 15*s.* 6*d.* an acre.

Here the length = 24 chains 25 links = 2425 links;
    the breadth = 16 chains 80 links = 1680 links;
$$\therefore \quad \text{the area} \quad = 2425 \times 1680 \text{ sq. links} = 4,074,000 \text{ sq. links}$$
$$= 40·74 \text{ acres.}$$

Required the rent of 40·74 acres at £5. 15s. 6d. an acre.

$$
\begin{array}{ll}
& \text{£40·74} \\
& \quad\ \ 5 \\ \hline
& 203·70 \\
10s. = \tfrac{1}{2}\ \text{£} & \quad 20·37 \\
5s. = \tfrac{1}{2}\ \text{of } 10s. & \quad 10·185 \\
6d. = \tfrac{1}{10}\ \text{of } 5s. & \quad\ \ 1·018\ |\ 5 \\ \hline
& 235·273 \quad =\text{£235. 5s. 6d. to the nearest penny.}
\end{array}
$$

Example 2. The area of a rectangular field is 4 ac. 2 r. 9 p., and its length is 12 chains 15 links : find its breadth.

$$
\begin{array}{r|l}
40 & 9·0 \quad \text{poles} \\ \hline
4 & 2·225 \ \text{roods}
\end{array}
$$

∴ 4 ac. 2 r. 9 p. = 4·55625 acres = 45·5625 sq. chains.

And $$breadth = \frac{area}{length} = \frac{45·5625}{12·15} \text{ chains}$$

$$= 3·75 \text{ chains} = 3 \text{ chains } 75 \text{ links.}$$

Example 3. A court-yard, 50 feet long by 42 feet broad, contains a rectangular lawn surrounded by a gravel path of uniform width. If the width of the path is 6 feet, find the dimensions of the lawn, and the area of the path.

Let *ABCD* represent the rectangular court-yard, and *EFGH* the lawn. Then *EF* = 38 ft., and *EH* = 30 ft.

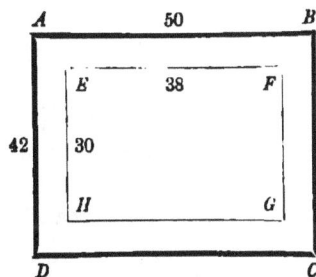

And the area of the path

= area *ABCD* – area *EFGH*

= 50 × 42 sq. ft. – 38 × 30 sq. ft.

= 960 sq. ft.

Example 4. Find the area of the four walls of a room, whose length is 18 ft. 7 in., breadth 14 ft. 5 in., and height 11 ft. 4 in.

The *area of four walls = twice (length + breadth) × height*

$$= 2 \times 33 \times 11\tfrac{1}{3} \text{ sq. ft.}$$

$$= 748 \text{ sq. ft.} = 83 \text{ sq. yds. 1 sq. ft.}$$

Examples XIV.

1. Find the areas of the rectangles in which

   (i.)   The length is 3 yds. 2 ft., and the breadth 1 yd. 1 ft.;

   (ii.) The length is 5 yds. 2 ft. 3 in., the breadth 2 yds. 1 ft. 6 in.

2. Find the areas of the following rectangles, giving the result in acres :

   (i.)   The length is 363 yds., the breadth 240 yds. ;

   (ii.) The length is 25 chains, the breadth 5 chains ;

   (iii.) The length is 20 chains, the breadth 8 chains 25 links.

3. Find the areas of the squares in which

   (i.)   Each side is 880 yards (result in acres);

   (ii.) Each side is 21 chains 75 links.

4. What will be the cost of paving a passage 47 ft. 6 in. long, and 9 ft. 6 in. wide, at the rate of 18s. per square yard ?

5. Find the cost of paving a rectangular area, whose length is 40 yds. and breadth 33 yds. 1 ft., at the rate of 3d. per square foot.

6. Find the rent of a rectangular field, 198 yards long by 165 yards broad, at £4. 6s. 8d. per acre.

7. A rectangular tennis-ground is 110 yds. long by 55 yds. wide. Find the expense of sowing it with grass-seed. at the rate of $3\frac{1}{2}$ bushels to the acre, the price of the seed being £1. 1s. 4d. per bushel.

8. Find the lengths of the following rectangles, having given that

   (i.)   The area is 3524 sq. yds. 8 sq. ft., and the breadth is 51 yds. 1 ft. ;

   (ii.) The area is $76\frac{1}{2}$ acres, and the breadth 36 chains;

   (iii.) The acreage is 37·6607, and the length 23 chains 45 links.

9. It cost £11. 15s. 9d. to floor a room with planking at 8d. per square foot ; if the breadth of the room is 17 ft. 3 in., find the length.

10. The rent of a rectangular plot of ground is £53. 7s. 6d. at the rate of 17s. 6d. per acre ; if the length is 40 chains find the breadth.

11. How many tiles, each 9 in. long by 6 in. wide, will be required to pave a court 24 ft. 6 in. in length, and 15 ft. 9 in. in breadth ?

12. A rectangular plot of ground, 50 yds. long and 40 yds. broad, is to be laid with turfs, $1\frac{1}{2}$ ft. long by 6 in. wide. Find the cost if the turfs are laid at the rate of 3s. per hundred.

13. How many planks, 2 ft. 6 in. long and 8 in. wide, will be needed to floor a room which measures 25 ft. by 16 ft.? Find the cost of supplying the planks at 5*d.* per foot.

14. How many yards of paper, 35 in. wide, are needed to cover a wall 22 ft. 6 in. long by 12 ft. 3 in. broad?

15. What will it cost to carpet a room 20 ft. 3 in. long by 16 ft. broad with carpet 2 ft. 3 in. wide at 3*s.* 4*d.* per yard?

16. How many yards of carpet $\frac{3}{4}$ yd. wide will be needed to cover the floor of a room 27 ft. long and 12 ft. 9 in. broad? And what will it cost at 3*s.* 10$\frac{1}{4}$*d.* per yard?

17. For a room, 21 ft. long by 15 ft. wide, the length of carpet required is 46 yds. 2 ft. What is the width of the carpet in inches?

18. If 64 planks, 6 ft. 3 in. long and 8$\frac{1}{4}$ in. wide, are required to floor a room 25 ft. long; what is its breadth?

19. It costs £12. 12*s.* to carpet a room with carpet 27 inches wide, at 3*s.* 6*d.* per yard. If the length of the room is 27 ft., find its breadth.

20. Find in square yards the area of a path 6 feet wide surrounding a lawn whose length is 30 yards and breadth 24 yards.

21. A court-yard, whose length is 55 yds. 1 ft. and breadth 33 yds. 1 ft., contains a rectangular lawn surrounded by a gravel path 8 ft. wide. Find the area of the lawn and of the path in square yards and feet.

22. A carpet is to be provided for a room 24 ft. 4 in. long and 17 ft. 6 in. wide, so as to leave a uniform margin 2 ft. wide. How many yards of carpet, 30 in. wide, will be required? And what will it cost to stain the margin at 6$\frac{3}{4}$*d.* per square yard?

23. What length of carpet 30 in. wide is required for a room 20 ft. 4 in. long by 18 ft. 8 in. broad, allowing for a margin of 2 ft. 8 in. all round? Find also the number of tiles, each 8 in. by 4 in., required to fill this margin.

24. How many square feet of gravel will be required to make a path, a foot wide, round the outside of a rectangular flower-bed, whose length and breadth are respectively 3 poles and 2$\frac{1}{2}$ yards?

25. A room 21 feet long by 14 ft. 3 in. broad is to be carpeted with Brussels carpet $\frac{3}{4}$ yd. wide at 3*s.* 10*d.* a yard, so as to leave a margin all round $\frac{1}{2}$ yard wide. This margin is to be covered with matting at 3*s.* the square yard. How much of each material will be required? And what will be the cost of the whole?

26. Find the area of the four walls of a room 27 ft. long, 23 ft. 6 in. wide, and 16 feet high.

27. A room, 16 ft. long, 9 ft. broad, and 10 ft. high, contains a door 8 ft. by 4 ft., two windows each 5 ft. by 4 ft., and a fire-place 6 ft. by 4 ft. 6 in. What area of the four walls remains to be papered ?

28. How many pieces of paper, 22 in. wide, are required for the walls of a room, 15 ft. 4 in. long, 14 ft. 8 in. wide, and 11 ft. high ? (*N.B.*—A *piece* of paper is 12 yds. long.)

29. A passage, 40 ft. long, 8 ft. wide, and 10 ft. high, has two doors 7 ft. by 4 ft., and a window 5 ft. 6 in. by 2 ft. Find the cost of paper $23\frac{1}{2}$ in. wide for the walls, at the rate of 3s. 6d. per piece of 12 yards.

30. For a room 16 ft. long, 12 ft. wide, and 10 ft. 6 in. high, the cost of paper is £1. 12s. 8d. If the width of the paper is 21 in., what is its price per piece (of 12 yds.)?

31. The total cost of papering a room 20 ft. long, 12 ft. broad, and 10 ft. high, with paper 32 in. wide is £1. 13s. 4d. If the paper cost 4s. 6d. per piece (of 12 yds.), how much per piece was charged for hanging ?

32. The walls of a room which measures 21 ft. long, 15 ft. 9 in. wide, and 11 ft. 8 in. high, are painted for £17. 17s. $3\frac{1}{2}d$. Find the additional cost of painting the ceiling at the same rate.

33. The cost of paper for a room, 15 ft. 7 in. long by 14 ft. 5 in. broad, is £1. 15s. 0d. If the paper is 30 in. wide, and costs 5s. per piece (of 12 yards), find the height of the room.

34. The breadth of a room is half as much again as its height, its length is twice its height, and it costs £5. 5s. 0d. to paint its walls at the rate of $1\frac{1}{4}d$. per square foot. What are its dimensions ?

35. Carpet 2 feet wide at 6s. a yard for a room 27 feet wide costs £40. 10s. ; and paper 1 ft. 6 in. wide at 6d. a yard for its walls costs £7. 12s. 0d. What is the height of the room ?

36. The length of a room is treble its breadth ; and the cost of carpet 36 in. wide at 7s. 6d. per yard is £28. 2s. 6d. ; also the cost of painting the four walls at $4\frac{1}{2}d$. per square foot is £28. 2s. 6d. What is the height of the room ?

*37. A room is 18 ft. long, 12 ft. wide, and 11 ft. high ; what length of paper, a yard wide, would be required to paper the four walls and ceiling ?

*38. All round the floor of a room, which is 28 ft. long and 22 ft. wide, there is a border 2 ft. wide which is left uncarpeted. Find the cost of staining the border at 1s. $1\frac{1}{2}d$. a square yard. Find also the number of yards of carpet, 27 in. wide, required for covering the rest of the floor, and the cost of the carpet at 3s. 9d. a yard.

Fifteenth Week. *Cubic Measure.* *Rectangular Solids.*

Notes and Hints for Solution. The following examples are intended to illustrate the fundamental rule connecting the volume of a rectangular solid with its linear dimensions, *i.e.*, its length, breadth, and thickness. The rule may be stated in two forms :

(i.) *Volume = Length × Breadth × Thickness.*

(ii.) *Volume = Area of Base × Thickness.*

Care must be taken to see that the *linear, square,* and *cubic* units correspond.

Example 1. Find the weight of a steel bar, 48 inches long, 4·5 inches broad, and 2·5 inches thick, taking steel to weigh 7·8 times its bulk of water, and 1 cubic foot of water to weigh 1000 ounces.

$Volume = length × breadth × thickness$

$= 48 × 4·5 × 2·5$ cubic inches $= 540$ cubic inches

$= \frac{5}{16}$ cubic foot ;

$Weight = \frac{5}{16}$ of 7800 oz. $= 2437\frac{1}{2}$ oz.

Example 2. It is found that a level seam of coal yields 3630 tons to the acre. If the coal is 1·28 times as heavy as water, and 1 cubic foot of water weighs 1000 oz., find the average thickness of the seam.

Weight of 1 cub. ft. of coal $= 1280$ oz. $= 80$ lbs. $= \frac{1}{28}$ ton.

Volume of 1 acre of coal $=$ total weight ÷ weight of 1 cub. ft.

$= (3630 ÷ \frac{1}{28})$ cub. ft. $= (3630 × 28)$ cub. ft.

Thickness of coal $=$ volume ÷ area of base

$= (3630 × 28)$ cub. ft. ÷ $(4840 × 9)$ sq. ft.

$= 2\frac{1}{3}$ linear feet $= 2$ ft. 4 in.

Example 3. Water passes from a reservoir into a canal by a channel 8 feet wide and $2\frac{1}{2}$ feet deep. If the water flows at a uniform rate of 4 miles an hour, find roughly how many gallons pass into the canal in ten minutes, given 1 cubic foot $= 6\frac{1}{4}$ gallons nearly.

In 10 minutes a column of water passes whose length is

$\frac{1}{6}$ of $4 × 1760 × 3$ feet $= 3520$ feet ;

∴ the volume of the column of water

$= 8 × 2\frac{1}{2} × 3520$ cub. ft. $= 70,400$ cub. ft.

∴ approx. number of gallons $= 70,400 × 6\frac{1}{4} = 440,000$.

Examples XV.

1. Find the volume of (i.) the rectangular solid, of which the length, breadth, and height are respectively 6 ft. 8 in., 5 ft. 3. in., and 2 ft.; (ii.) of the cube of which each edge measures 6 ft. 6 in.

2. What is the weight (in tons) of a rectangular block of granite whose dimensions are 7 ft., 5 ft., and 4 ft., if one cubic foot weighs 166 lbs.? And what will it cost to polish the whole surface at the rate of 1s. 6d. per square foot?

3. How many gallons are contained in a cubical vessel which measures internally 4 ft. in length, breadth, and depth, supposing that 1 cubic foot contains $6\frac{1}{4}$ gallons?

4. The internal measurements of a rectangular tank are 16 ft. long, 8 ft. wide, and 7 ft. deep; how many tons of water will it hold if 1 cubic foot of water weighs 1000 ounces? And what will be the cost of lining it with thin zinc at the rate of $6\frac{3}{4}d$. per square foot?

5. A cistern (without lid) 6 ft. long, 3 ft. broad, and $2\frac{1}{2}$ ft. deep, is to be lined with thin zinc at 5d. the square foot. What will be the cost? If the cistern is two-thirds full, find the weight of water, taking the weight of 1 cubic foot as 1000 ounces.

6. How long will it take to dig a trench 160 yards long, 16 ft. wide, and 14 ft. deep, if 30 tons of earth are removed in a day? [1 cubic foot of earth weighs $92\frac{1}{2}$ lbs.]

7. A brick (with mortar) measures 9 in. by $4\frac{1}{2}$ in. by 3 in. Find the number of bricks required for a wall 80 yds. long, 10 ft. high, and 9 inches thick; and find their cost at the rate of 35 shillings per thousand.

8. If a brick (with mortar) occupies a space 9 in. by $4\frac{1}{2}$ in. by 3 in., how many bricks would be required for a solid mass of masonry, 60 ft. long, 8 ft. 3 in. wide, and 9 ft. 3 in. high? And find to the nearest penny their cost at 30 shillings per thousand.

9. Four labourers undertake to excavate a pit measuring 24 ft. long, 20 feet wide, and 3 yards deep, at the rate of 1s. per cubic yard for the first yard in depth, 2s. for the second, and 4s. per cubic yard for the third yard in depth. What should each receive?

10. Find the cost of making a road half-a-mile long and 36 feet wide, the soil being first excavated to a depth of 1 foot at a cost of 1s. per cubic yard, rubble being then laid in 9 inches deep at a cost of 1s. 6d. per cubic yard, and 3 inches of gravel at 3s. 3d. per cubic yard being laid on the top, and the whole consolidated by a steam roller at a cost of 2d. per square yard.

11. Find the weight of an oak beam 12 ft. long, 18 in. wide, and 16 in. thick; it being given that seasoned oak weighs ·85 times its bulk of water, and that 1 cubic foot of water weighs 1000 oz.

12. Find the weight per square yard of sheet zinc $\frac{1}{9}$ in. thick; it being given that zinc weighs 7·2 times its bulk of water, and that 1 cubic foot of water weighs 1000 oz.

13. The external length, breadth, and thickness of a closed box are 4 ft., 2 ft., and 16 inches respectively, and the wood of which it is made is one inch thick. Find the number of cubic inches of wood used in making the box, and the cost of lining it with thin metal at $4\frac{1}{2}d.$ per square foot.

14. A closed box, whose external dimensions are 4 ft. 8 in., 4 ft. 2 in., and 2 ft. 6 in., is made of deal one inch thick. Find the weight of the box, given that one cubic foot of deal weighs 912 ounces.

15. A rectangular zinc cistern (open at the top) measures externally 2 ft. 8 in. long, 1 ft. 9 in. broad, and 1 ft. $4\frac{1}{2}$ in. deep. If the metal is $\frac{1}{2}$ inch thick, find (to the nearest lb.) the weight of the cistern, and also its total weight when filled with water. [1 cubic foot of water weighs 1000 oz., and zinc is 7·215 times as heavy as water.]

16. The capacity of a cistern is $67\frac{7}{24}$ cu. ft., and its length and breadth are respectively 6 ft. 4 in. and 4 ft. 3 in. Find its depth.

17. A cistern when full contains $187\frac{1}{2}$ gallons, and its length and breadth are respectively 4 ft. and $2\frac{1}{2}$ ft. If a cubic foot contains 6·25 gallons, find the depth of the cistern.

18. A tank, the area of whose base is 56 square feet, contains $6\frac{1}{4}$ tons of water. Find the depth of the water, given that one cubic foot of water weighs 1000 oz.

19. If 9 cwt. 57 lbs. of lead were rolled into a sheet 12 ft. long by 6 ft. wide, what would be its thickness? And if 1 ton of lead were rolled into a sheet of uniform thickness ·42 inch, find approximately its area in square yards. [Suppose 1 cubic foot of lead to weigh 710 lbs.]

20. A brick (without mortar) measures $8\frac{1}{2}$ inches in length, 4 inches in breadth, and $2\frac{1}{2}$ inches in thickness. How many of such bricks are contained in a stack 85 ft. long, 15 ft. wide, and 8 ft. 4 in. high? And if, allowing for mortar, each brick when laid occupies a space 9 in. by $4\frac{1}{2}$ in. by 3 in., what length of wall 6 feet high, and $13\frac{1}{2}$ in. wide, could be built from a stack of the above dimensions?

21. A level tract of land 21 miles long and $\frac{3}{4}$ mile wide is flooded to a depth of 5 feet. If a cubic foot of water weighs $62\frac{1}{2}$ lbs., find in tons the weight of water on the land.

22. Find to the nearest ton the weight of water which falls on an acre of ground during a rainfall of one inch, given that one cubic foot of water weighs 1000 oz. Also estimate the rainfall in gallons, given that 1 gallon of water weighs 10 lbs.

23. Water is drawn off from a reservoir through a channel 6 ft. wide and $2\frac{1}{2}$ ft. deep. If the water flows at the rate of 3 miles an hour, how many gallons pass out of the reservoir in 10 minutes? [Given that 1 cubic foot contains $6\frac{1}{4}$ gallons.]

24. Water flows into a rectangular tank through a pipe which admits 15 gallons per minute. Find approximately (in inches per hour) at what rate the water will rise in the tank, if the dimensions of the base are 24 ft. by 18 ft. [Given that 1 cubic ft. $= 6\frac{1}{4}$ galls.].

25. Supposing the capacity of a gallon to be nearly 277·25 cubic inches, find approximately the number of gallons per square mile in one inch of rainfall.

*26. The rainfall at a particular place is 29·2 inches in a year. To how many gallons per acre is this equivalent, taking a gallon as equivalent to ·16 of a cubic foot.

*27. A cubic foot of copper weighs 560 lbs. It is rolled into a square bar 40 ft. long. An exact cube is cut from the bar. What is its weight to the nearest thousandth of a pound?

*28. Find the cost of the bricks needed for building a wall 30 yards long, 6 feet high, and $13\frac{1}{2}$ inches thick, having given that bricks cost 25s. per 1000, and that each brick when laid fills up a space 9 inches long, $4\frac{1}{2}$ inches wide, and 3 inches deep.

*29. A pond whose area is half-an-acre is frozen over with ice 2 inches thick; find in tons to the nearest integer the weight of all the ice, if a cubic foot of it weighs $57\frac{1}{4}$ lbs.

*30. A section of a stream is 10 feet wide and 10 inches deep; the mean flow of the water through the section is at the rate of 3 miles an hour. Taking 25 gallons as equivalent to 4 cubic feet, find how many gallons of water flow through the section in 24 hours.

*31. A rectangular cistern, 8 ft. long, 7 ft. wide, holds when full $6\frac{1}{4}$ tons of water; find its depth, assuming that a cubic foot of water weighs 1000 oz. And find also within one-hundredth of a foot the depth of a cistern of the same capacity, having its length, width, and depth all equal.

Sixteenth Week. *Square Root.*

Notes and Hints for Solution. Examples of the method of extracting the square root of numerical quantities are given below for reference. The method is based on the following algebraical equivalents :

$$(a+b)^2 = a^2 + 2ab + b^2,$$
$$(a+b+c)^2 = (a+b)^2 + 2(a+b)c + c^2.$$

Example 1.   Extract the square root of 1371·9616.

```
            |13,71.,96,16  |37·04
            |  9           |‾‾‾‾
        67  |  4 71
            |  4 69
      7404  |    2 96 16
            |    2 96 16
            |‾‾‾‾‾‾‾‾‾‾‾‾
```

The square root required = 37·04.

Note. The periods, each consisting of two digits, are to be marked off from the *decimal point* each way. Thus to find the square roots of 518·076325, 7·08916, ·00000144, 3, we begin by marking off the periods as follows :

5,18,·07,63,25,   7,·08,91,60,   ·00,00,01,44,   3,·00,00,00,....

Observe that each period furnishes one digit to the square root.

Example 2.   Find the square root of (1) ·0007 correct to four places of decimals ; (2) $\frac{8}{13}$ correct to three places of decimals.

(1)   ·00,07,00,00,00,... | ·02645...   (2) $\frac{8}{13}$ = ·615384...
            4 |‾‾‾‾‾‾   ·61,53,84,... | ·7844...
      46 | 3 00   49 |‾‾‾‾‾‾
         | 2 76   148 | 12 53
     524 | 24 00 | 11 84
         | 20 96   1564 | 69 84
    5285 | 3 04 00 | 62 56
                                               15684 | 7 28...

∴ Square root required = ·0265   ∴ Square root = ·784 correct
    correct to 4 places.   to 3 places.

Examples XVI.

Find the square root of

1. 576.	2. 3249.	3. 7921.
4. 9409.	5. 17424.	6. 37249.
7. 299209.	8. 167281.	9. 1006009.
10. 18671041.	11. 13704804.	12. 1157428441.

[*To take the square root of a mixed number*, if the denominator is a perfect square, *reduce to an improper fraction, and take the square root of the numerator and denominator separately.*]

Find the square root of

13. $3\frac{1}{10}$ 14. $1\frac{25}{144}$.   15. $8\frac{17}{169}$.

16. $101\frac{92}{109}$.   17. $2406\frac{25}{9604}$.   18. $\frac{7}{8} + \frac{3}{4} + \frac{15}{16} + \frac{5}{64}$.

Find the square root of

19. 998·56.	20. 99·8001.	21. 1383·0961.
22. 3263·8369.	23. 2704·416016.	24. ·0001595169.
25. 8·027.	26. 2·00694.	27. 253089·4864.

[*To take the square root of a fraction whose denominator is* not *a perfect square, reduce the fraction to a decimal, and then take the square root.*] .

Find the value of the following, correct to three places of decimals :

28. $\sqrt{2}$.   29. $\sqrt{3}$.   30. $\sqrt{10}$.

31. $\sqrt{2·5}$.   32. $\sqrt{·07}$.   33. $\sqrt{\frac{7}{8}}$.

34. $\sqrt{\frac{5}{7}}$.   35. $\sqrt{\frac{8}{11}}$.   36. $\sqrt{4\frac{3}{13}}$.

37. Evaluate to three places of decimals : (i) $\dfrac{1}{\sqrt{8}}$; (ii) $\dfrac{1}{\sqrt{·4}}$.

[*To find the length of the side of a square, take the square root of the number of square units in the area.*]

38. The area of a square field falls short of 10 acres by 439 square yards ; find the length of each side.

39. Find the number of yards in the side of a square field whose area is 1 ac. 2 r. 4 p. 15 sq. yds.

40. The rent of a square allotment is £8. 3s. 4d., at the rate of 8s. 4d. per acre ; find in yards the length of each side.

41. The cost of paving a square court with stone slabs, each 18 inches square, at 2s. 3d. a slab, is £16. 4s. What is the length of each side of the court ?

42. The area of a square field is 12 ac. 3 r. 9 p. $18\frac{3}{4}$ sq. yds.; find in yards the length of each side.

43. Find (to the nearest yard) the length of each side of a square field whose area is 439 ac. 33 p.

44. Find (to the nearest penny) the cost of running a fence round a square field whose area is 4 acres at 1s. 6d. per yard.

[*To find the hypotenuse of a right-angled triangle, add the squares of the lengths of the sides containing the right angle, and take the square root of the sum. See Euc. I. 47.*]

45. The sides of a right-angled triangle (containing the right angle) are respectively 88 ft. and 105 ft. ; find the length of its hypotenuse.

46. The side of a square is 22 p. 1 yd. 2 ft. ; find to the nearest foot the length of its diagonal.

[*To find the* mean proportional *between two quantities, take the square root of their product.*]

Find the mean proportional between the following numbers :

47. 1692, 2303.   **48.** 7056, 105625.   **49.** 22·09, 6·6049.

Extract the square root of

*50. 0·00137641.   ***51.** 0·08450649.   ***52.** 4774671801.

Extract (correct to three decimal places) the square root of

*53. $8\frac{1}{3}$.   ***54.** 0·144.   ***55.** 0·51.

Extract (correct to four decimal places) the square root of

*56. 0·051.   ***57.** 0·571.   ***58.** 5·2.

*59. Arrange $\dfrac{1}{0\cdot742}$, $\sqrt{1\cdot81}$, and 1·346 in order of magnitude ; and find correct to four places of decimals by how much the sum of the largest and smallest of these numbers differs from twice the other.

*60. Find to the nearest integer how many inches there are in the length of one of the sides of a square field whose area is 2 acres.

*61. A square field contains 6 ac. 2 r. $6\frac{1}{4}$ sq. yds. ; find the length of a side in yards correct to the nearest hundredth.

*62. The sides of a rectangle are 16 feet and 10 feet respectively ; find in feet, to four places of decimals, the length of the diagonal of a square of equal area.

*63. There is a square enclosure of 10 acres ; a man walks at the rate of 3 miles an hour along one side, along a diagonal, along another side, and so returns along the other diagonal to the starting point. How many minutes does this take him ?

Seventeenth Week.   *Cube Root.*

Notes and Hints for Solution. The example worked out below exhibits a method of extracting the cube root of a number.

It is based upon the following algebraical equivalents :

$$\text{(i)} \qquad (a+b)^3 = a^3 + 3a^2b + 3ab^2 + b^3 \ ;$$
$$\text{(ii)} \quad (a+b+c)^3 = (a+b)^3 + 3(a+b)^2c + 3(a+b)c^2 + c^3.$$

Example.   Find the cube root of 425259008.

$$
\begin{array}{rr|l}
 & 425{,}259{,}008 & |\ 752 \\
 & 343 & \\
\text{three times } (70)^2 = 14700 & 82\,259 & \\
\text{three times } 70 \times 5 = 1050 & & \\
5^2 = 25 & & \\
\hline
15775 & 78\,875 & \\
\text{three times } (750)^2 = 1687500 & 3\,384\,008 & \\
\text{three times } 750 \times 2 = 4500 & & \\
2^2 = 4 & & \\
\hline
1692004 & 3\,384\,008 & \\
\end{array}
$$

Note. The cube root of a decimal is obtained in a similar manner, care being taken to mark off the periods of three figures each from the *decimal point* each way. Thus to find the cube root of 37·1594 and ·00007 we must mark off as follows :

$$37{,}159{,}400{,}\ldots, \quad 000{,}070{,}000{,}\ldots \ ;$$

in these cases the cube root can of course only be found approximately. Observe that each period furnishes one digit of the root.

Examples XVII.

Find the cube root of

1. 512.	**2.** 1331.	**3.** 3375.
4. 15625.	**5.** 103823.	**6.** 195112.
7. 438976.	**8.** 857375.	**9.** 160103007.
10. 8615125.	**11.** 193100552.	

Find the cube root of

12. ·729.	**13.** 4·913.	**14.** 50·653.
15. 614·125.	**16.** 64481·201.	**17.** 16·974593.
18. ·000012167.	**19.** ·000000125.	**20.** 2815166·528.

Find the cube root of

21. $166\frac{3}{8}$.　　　　**22.** $11\frac{323}{343}$.　　　　**23.** $190\frac{7}{64}$.

Extract the cube root of the following, giving the results true to two places of decimals :

24. $3\cdot00415$.　　　**25.** 5.　　　　**26.** 7.

27. $\frac{4}{9}$.　　　**28.** $3\frac{3}{4}$.　　　**29.** $\frac{2}{7}$.

30. $18\frac{2}{11}$.　　　**31.** $150\sqrt{2}$.

[*To find the length of the edge of a cube, take the cube root of the number of cubic units in its volume.*]

32. Find the edge of a cube whose volume is equal to that of a rectangular solid, of which the dimensions are 6 ft. 3 in., 2 ft. 6 in., and 1 foot.

33. Find the edge and surface of a cube whose volume is 4 cu. ft. 1088 cu. in.

34. Find the length of the edge of a cube of metal which cost £5407. 8*s.* 11½*d.*, one cubic inch being valued at 8*s.* 4*d.*

35. Find the number of square feet in the surface of a cube whose volume contains 2460375 cubic inches.

36. Find (to the nearest tenth of an inch) the dimensions of a cubical cistern capable of containing 800 gallons. [Suppose 1 gallon $= 277\cdot25$ cubic inches.]

37. Find (to the nearest hundredth of an inch) the edge of a cubical block of lead weighing one ton, having given that 1 cubic foot of lead weighs $709\frac{1}{2}$ lbs.

38. Find the edge of a cubical mass of brass whose weight is equal to that of a rectangular block of iron measuring 2 ft. 3 in. long, 1 ft. 4 in. broad, and 12 inches thick ; it being given that 1 cubic foot of brass weighs 8000 ounces, and 1 cubic foot of iron weighs 7788 ounces.

Extract the cube root of

*39.** $6434\cdot856$.　　　**40.** $768575\cdot296$.

*41.** Extract to the third decimal place the cube root of

(i) $0\cdot27$.　　(ii) $0\cdot9$.　　(iii) $15\frac{2}{3}$.

*42.** Shew that the square root of $0\cdot37$ exceeds the cube root of $0\cdot217$ by a difference which very nearly equals $\frac{1}{130}$.

*43.** A plate of metal is 106·58 inches long, 14·6 inches wide, and 2 inches thick ; supposing it to be melted and cast into an exact cube, what would be the edge of the cube ?

Eighteenth Week. *The Metric System of Measures and Weight.*

Notes and Hints for Solution. The Metric System is so constructed as to follow the arrangement of decimal notation. This implies that, a fundamental unit having been chosen, the higher denominations are obtained by taking 10 times, 100 times, 1000 times, ... that unit ; while the lower denominations are respectively *one-tenth, one-hundredth, one-thousandth,* ... of the unit. The same principle is carried out as far as possible through all the Tables of the System.

The whole Metric System of Measures and Weight rests on the unit of length, namely, the **metre.***

N.B. The metre = 39·37079... inches = 39⅜ inches (*nearly*).

The higher denominations (or multiples) are the **Decametre** (Dm.), the **Hectometre** (Hm.), the **Kilometre** (Km.), containing respectively *ten, one hundred,* and *one thousand* metres. The lower denominations (or submultiples) are the **decimetre** (dm.), the **centimetre** (cm.), and the **millimetre** (mm.), containing respectively *one-tenth, one-hundredth,* and *one-thousandth* of a metre.†

In practice the terms Hectometre, Decametre, decimetre are little used.

Hence a length given in terms of metric measurements can be at once expressed in any metric denomination by use of the decimal point.

For instance :—7 Km. 9 Dm. 4 cm. 3 mm. = 7090·043 metres

= 7·090043 Kilometres = 7090043 millimetres.

N.B. 1 Kilometre = 1093·633... yards = ⅝ mile (*nearly*).

---

*The metre was originally chosen, because it was believed to be one ten-millionth part of the distance from the North Pole to the Equator : but this estimate was based on an imperfect calculation of the Earth's magnitude.

†*Multiples* in the Metric System are distinguished by the *Greek* prefixes *Deca-, Hecto-, Kilo-* ; and for these it is convenient to use capital initial letters.

Submultiples are marked by the *Latin* prefixes *deci-, centi-, milli-* ; and may further be distinguished by small initials.

SQUARE MEASURE. Taking 1 square metre as the unit, it follows that

1 square Decametre $= 10^2$ (or 100) sq. metres,

1 square Hectometre $= 100^2$ (or 10,000) sq. metres, and so on ; while a sq. decimetre and a sq. centimetre are respectively $\frac{1}{100}$ and $\frac{1}{10,000}$ of a square metre.

For purposes of land measurement a square Decametre is called an **Are** ; it follows that a **Hectare** (or 100 Ares) is a square Hectometre. A square metre, being $\frac{1}{100}$ Are, is called a **centiare.**

Thus 19 Ha. 4 Ar. 3 sq. metres (or centiares)

$= 190403$ square metres

$= 1904·03$ Ares ; and so on.

N.B.   1 sq. metre $= 1550·059$ sq. inches $= 1\frac{1}{3}$ sq. yds. (*nearly*).

1 Hectare $= 2·4711...$ acres $= 2\frac{1}{2}$ acres (*nearly*).

CUBIC MEASURE. This follows the same general principle. The cubic metre (sometimes called a **stere**) is the unit. The cubic Decametre is therefore 10^3 (or 1000) cubic metres ; while the cubic decimetre and cubic centimetre are respectively $\frac{1}{1000}$ and $\frac{1}{1,000,000}$ of the cubic metre.

. CAPACITY. The unit of capacity is the **litre**, namely, the capacity of *one cubic decimetre.* To this the prefixes *Kilo-, Hecto-, Deca-, deci-, centi-, milli-* are attached with their usual meaning.

N.B.   1 cubic metre $= 1000$ cubic decimetres $= 1$ Kilolitre.

1 litre $= 61·027...$ cub. in. $= 1·760...$ pint $= 1\frac{3}{4}$ pint (*nearly*).

WEIGHT. The unit of weight is the **gramme**, namely, the weight of *one cubic centimetre of water* (at a temperature of 4° C.).

1 Kilogramme $= 1000$ grammes.

100 Kilogrammes $= 1$ Quintal.

1000 Kilogrammes $= 1$ Tonneau.

It follows that 1 litre (*i.e.* 1000 cubic centimetres) of water weighs 1 Kilogramme ;

also 1 cubic metre of water weighs 1000 Kilogrammes (or 1 tonneau).

N.B.   1 gramme $= 15·432...$ grains $= 15\frac{1}{2}$ grains (*nearly*).

1 Kilogramme $= 2·2046...$ lbs. $= 2\frac{1}{5}$ lbs. (*nearly*). .

Examples XVIII.

1. Express 5 Dm. 8 m. 7 cm. 3 mm. in terms of (i.) metres, (ii.) centimetres, (iii.) as the decimal of a Kilometre.

2. Express 3 Km. 4 Hm. 9 dm. 2 mm. in terms of (i.) metres, (ii.) Decametres, (iii.) decimetres.

3. In ·07801 of a Kilometre, how many metres are there? and how many millimetres? Express this length in multiples and sub-multiples of a metre, affixing the proper names to each denomination.

4. In 80460·2 centimetres, how many Kilometres are there? and how many metres? Express this length in multiples and sub-multiples of the metre.

5. (i.) Add together 400·82 metres, 5028 centimetres, and ·5489 of a Kilometre, expressing the result in the last denomination.

(ii.) From the sum of 902·03 metres, one hundred thousand millimetres, and 8 m. 1 dm. 9 cm., subtract 1022 centimetres. Express the result in terms of the Kilometre.

6. Multiply (i.) 2 Km. 5 Hm. 1 dm. 2 cm. 5 mm. by 8.
(ii.) 5 Dm. 2 cm. by 42.

7. Divide   (i.) 45 Km. 7 Dm. 4 m. 7 dm. by 9.
(ii.) 173 Km. 3 Hm. 6 Dm. 7 m. 2 dm. 7 cm. by 369.

8. Find the cost of cutting a road 12 Km. 800 m. long at £51. 12s. 6d. per Kilometre.

9. Find the cost of 16 Km. 64 m. of telegraph wire at £3. 2s. 6d. per Hectometre.

10. Estimate the cost of 136·75 metres of silk at 8 fr. 40 c. per metre, and give the result in terms of English money (to the nearest penny), supposing £1 sterling = 25·25 francs.

11. Five Kilogrammes of gold are to be divided among 110 persons; find (to the nearest hundredth of a gramme) what weight each person should receive.

12. Express (i.) $32\frac{1}{2}$ milligrammes as the decimal of a Kilogramme, and (ii.) $3\frac{1}{2}$ centilitres as the decimal of 7 Hectolitres.

13. Express (i.) 748 square metres as the decimal of a square Kilometre, and (ii.) ·748 Hectare in terms of square metres.

14. Express 690,420,489 sq. cm. as (i.) square metres, (ii.) as Hectares; and (iii.) multiply 17 Hectares 85 centiares by 110, giving the result in square metres.

15. The area of an allotment of ground is 4 Hectares 25 sq. metres, and its rent is 16,010 francs; at what rate is this per square metre?

16. Express in Hectares (i.) the area of a rectangular plot of ground, of which the length and breadth are respectively 805 metres and 74 metres ;

(ii.) the area of a square each side of which measures 523 m. 75 cm.

17. Find the rent of a rectangular plot of building ground measuring 750 metres by 88 metres at 1250 francs per Hectare.

18. (i.) A strip of planking is 12 m. 34 cm. in length, and its area is 2 sq. m. 96 sq. dm. 16 sq. cm. ; find its breadth in centimetres.

(ii.) The area of a plot of land is 25 Hectares 96·5 Ares and its length is 3462 metres ; find its breadth.

19. How many tiles, each measuring 15 cm. by 12 cm., are required to pave the floor of a hall 34·8 metres long and 14·52 metres wide ?

20. How many metres of paper 75 cm. wide will be required for the walls of a room whose length, breadth, and height are respectively 5 m. 65 cm., 4 m. 35 cm., and 4·5 metres ?

21. How many litres are contained in cisterns whose length, breadth, and depth are respectively :

(i.) 1 m. 25 cm.,   80 cm.,   70 cm. ;

(ii.) 4 m. 10 cm.,   3 m. 10 cm.,   51 cm. ?

22. Express in Kilogrammes the weight of water contained in a trench whose length is 26 m. 75 cm., and breadth 1 m. 10 cm., the depth of the water being 35 cm.

23. Find the weight in Kilogrammes of a beam of fir 4 m. 45 cm. long, 24 cm. broad, and 20 cm. thick ; it being given that fir weighs ·55 times as much as an equal bulk of water.

24. How many tonneaux of earth must be excavated in digging a trench 230 metres long, 4·5 metres wide, and 80 cm. deep ; it being given that the soil in question is 1·5 times as heavy as an equal volume of water ?

25. A bar of silver, 25 cm. long, measures 8·2 cm. and 5·4 cm. in width and thickness respectively. Find (i.) its weight in Kilogrammes, having given that silver weighs 10·5 times its bulk of water ; (ii.) its value, supposing 1 gramme of silver to be worth 1*d.* ; (iii.) the price of silver per ounce troy. [Take 1 gramme = 15½ grains.]

26. Find the weight per square metre of a sheet of zinc 5 mm. thick, having given that zinc weighs 7·14 times its bulk of water. Express the result in Kilogrammes.

Nineteenth Week. *The Metric System Continued. Foreign Coinage.*

Notes and Hints for Solution. We shall now give examples to illustrate methods of converting Metric Weights and Measures into the British Standard, and vice versâ. Similar questions dealing with foreign and British coinage are also provided.

In France, Belgium, and Switzerland the standard coin is the **Franc**, subdivided into 100 **Centimes**.

Italy, Greece, and (in part) Spain have adopted under different names the same coinage.

The value of the franc expressed in British currency varies with the rate of exchange; but the following may be taken as average values :

$$1 \text{ franc} = 9\tfrac{1}{2}d. \text{ (nearly)}; \quad £1 = 25\cdot25 \text{ francs.}$$

Information as to the currencies of other countries will be given in connection with questions relating to them.

Example 1.   Find the equivalent in French money (to the nearest centime) of £173. 12s. 3d.; the rate of exchange being £1 = 25·33 francs.

$$£173. \ 12s. \ 3d. = £173\cdot6125,$$
$$\text{and } £1 \qquad\qquad = 25\cdot33 \text{ francs.}$$
$$\therefore \quad £173. \ 12s. \ 3d. = 173\cdot6125 \times 25\cdot33 \text{ fr.}$$
$$\qquad\qquad\qquad = 4397\cdot60 \text{ francs.}$$

```
          173·6,1 |,2,5
           25·3 3 |
         3472·2 5 | 0
          868·0 6 | 2
           52·0 8 | 4
            5·2 0 | 8
         _____
         4397·6 0
```

Example 2.   Find (to the nearest farthing) the equivalent in British money of 286·14 German marks; it being given that £1 = 25·28 francs, and 100 francs = 80·85 marks.

[The standard coin of Germany is the **mark** (= 100 pfennige); 1 mark = 11¾d. nearly.]

Here

$$286\cdot14 \text{ marks} = 100 \times \frac{286\cdot14}{80\cdot85} \text{ francs}$$

$$= £\frac{100 \times 286\cdot14}{80\cdot85 \times 25\cdot28} = £\frac{28614}{2043\cdot888}$$

$$= £13\cdot9998\ldots$$

$$= £14 \text{ (to the nearest farthing).}$$

Example 3. Express 63 sq. metres 3 sq. dm. in square yards, square feet, and square inches (to the nearest sq. in.), having given 1 metre = 39·37079 inches.

Now 63 sq. metres 3 sq. dm. = 63·03 sq. metres,

and   1 sq. metre = $(39 \cdot 37079)^2$ sq. inches.

∴   63·03 sq. metres = $(39 \cdot 37079)^2 \times 63 \cdot 03$ sq. inches ....(i.)

$$= 1550 \cdot 06 \times 63 \cdot 03 \text{ sq. inches } ....(ii.)$$

$$= 97700 \text{ sq. in.}$$

$$= 75 \text{ sq. yds. 3 sq. ft. 68 sq. in.}$$

Note. The necessary multiplication is given below. It should be noted that since $(39 \cdot 37079)^2$ is to be multiplied by 63·03, a number *less than 100*, it is sufficient to work out $(39 \cdot 37079)^2$ correct to the second decimal place, in order to get a final result true to the nearest unit.

(i.)
```
 3 9·3 7|0 8
 3 9·3 7|0 8
118 1·1 2|4
35 4·3 3|7
 1 1·8 1|1
   2·7 5|6
    ·0 3|1
155 0·0 5|9
```

(ii.)
```
1550·0|6
  63·0|3
93003·6
 4650·1|8
   46·5|0
97700·3
```

Examples XIX.

(*Foreign Coinage.*)

1. Exchange into French coinage (to the nearest centime),
(i.) £46. 9s. 4½d. ;   (ii.) £101. 15s. 9¾d. ;
the rate of exchange being £1 = 25·28 francs.

2. Find the equivalent in British currency (to the nearest farthing) at the following rates of exchange, of
(i.)   837·63 francs :   [£1 = 25·36 francs.]
(ii.)   1098·44 marks.   [£1 = 20·44 marks.]

3. When the American dollar can exchange at 4s. 3·8d. in London, and at 5·44 francs in Paris, what is the rate of exchange between London and Paris? that is, what is the value of £1 sterling in francs?

4. Find in British money (to the nearest farthing) the equivalent of 60·95 Austrian florins ; it being given that £1 = 20·30 francs, and 100 francs = 50·25 florins. [The standard coin of Austria is the **florin** = 100 kreutzers ; 1 florin = 1*s.* 11½*d.* nearly.]

5. Find in English money (to the nearest farthing) the price per lb. of tea which costs 3 fr. 50 c. per Kilogramme. [Given £1 = 25·25 fr. ; 1 Kilogr. = 2·204 lbs.]

(Comparison of the Metric and Standard Systems.)

[*In the following examples, unless otherwise stated, the requisite data are to be taken from this table:*

1 *metre* = 39·37079... *inches.*
1 *litre* = *the capacity of* 1 *cubic decimetre.*
1 *gramme* = *the weight of* 1 *cubic centimetre of water at* 4°C.
1 *gallon* = 277·274... *cubic inches.*
1 *gallon of water weighs* 10 *lbs. at* 62° F. (10·016... *at* 4° C.)]

6. Verify carefully the following equivalents, using, where necessary, *contracted* multiplication and division, and giving results correct to the second place of decimals :

(i.) 1 Kilometre = 1093·63 yds.   (ii.) 1 sq. metre = 1550·06 sq. in.
(iii.) 1 Are   = 119·60 sq. yds. (iv.) 1 Hectare = 2·47 acres.
(v.) 1 cu. metre = 35·32 cu. ft.   (vi.) 1 litre   = 61·03 cu. in.⎫
                                                    = 1·76 pints. ⎭
(vii.) 1 gramme = 15·43 grains. (viii.) 1 Kilogramme = 2·21 lbs.

7. Verify the following equivalents (correct to two places of decimals) :

(i.) 1 yard = ·91 metre.   (ii.) 1 mile   = 1609·31 metres.
(iii.) 1 acre = ·40 Hectare.   (iv.) 1 gallon = 4·54 litres.
(v.) 1 oz. Av. = 28·35 grammes. (vi.) 1 ton   = 1016·05 Kilogrs.

8. Find to the nearest hundredth of an inch the difference between 35 yards and 32 metres.

9. Find to the nearest yard the difference between 5 miles and 8 Kilometres.

10. By how many grains does 5 Kilogrammes differ from 11 lbs. ? [1 gramme = 15·43235 grains.]

11. Find (to the nearest tenth of a yard) how many yards there are in 9·9 Kilometres ; and express 7½ miles in metric measurement to the nearest centimetre.

12. Reduce 8 m. 7 dm. 5 cm. to yards, feet, and inches.

13. Express in square decimetres 84 sq. yds., giving your result true to the nearest unit.

14. Express the velocity *five Kilometres per hour* in *feet per second.* •

15. Express the price 2 francs 80 centimes per Kilogramme in shillings per lb., having given 1 Kilogramme = 2·205 lbs. and £1 = 25·40 francs.

16. Find a multiplier that will convert a number expressing *metres per second* into *miles per hour.*

17. A cubic foot of water weighs nearly 1000 oz., and a cubic metre of water weighs 1000 Kilogrammes, also five furlongs are nearly equal to 1000 metres. Find approximately the equivalent in ounces of one Kilogramme.

18. If a Kilogramme = 2·204 lbs. Av., and a Hectolitre = 3·531 cub. ft. ; find to the nearest gramme the weight of 5·51 litres of a substance of which 10 cub. in. weigh 1 oz. ?

19. Find in inches (to the nearest hundredth) the edge of a cubical block of granite, weighing 1200·5 Kilogrammes, the specific gravity of granite being 3·5 times that of water.

*20. An acre is 0·40467 Hectares, and a pound sterling is taken as 25·25 francs. An estate measuring 1927 Hectares is sold for 10,100,000 francs. What is the selling price per acre in English money ?

*21. Given that a metre = 3·2809 ft., find how many square metres there are in 1000 square yards.

*22. Given that a gallon measures 277·274 cubic inches, and that a litre may be represented by a cube whose edge is 3·937 inches ; find correct to two places of decimals how many pints there are in a litre.

*23. How many Kilogrammes are there in 0·708624 of one ton, if 100 Kilos. are equivalent to 1·9684 cwts. ?

*24. Find how many metres there are in 0·581257 of one mile, having given that one metre = 1·093638 yards.

*25. Given that a metre is 3·3708 inches longer than a yard, find which is greater, 10 square metres or 12 square yards ; and express the difference as a decimal of a square metre correct to three places.

*26. A Kilometre being 1093·633 yards, find correct to four places of decimals how many Kilometres there are in 100 English miles.

*27. Taking a metre as 39·37 inches, and a gramme as 15·43 grains ; find the weight in grammes of a cubic metre of air, when 100 cubic inches of the air weigh 31 grains.

Twentieth Week.　　*Percentages.　Commission.*

Notes and Hints for Solution. A **percentage** is *a fraction of one hundred*; thus the term **5 per cent.** means 5 *in every* 100, and is equivalent to the fraction $\frac{5}{100}$. In such a case the number 5 is called the **rate** per cent. The words "per cent." are briefly expressed by the symbol % or by the letters "p.c."

Any fraction may be expressed as a percentage, and conversely.

Example 1.　Express (i.) $\frac{3}{20}$ as a percentage; (ii.) $8\frac{1}{3}$ p.c. as a fraction in lowest terms.

(i.) $\dfrac{3}{20} = \dfrac{\frac{3}{20}\text{ of }100}{100} = \dfrac{15}{100}$; *i.e.* 15 in every 100, or 15 p.c.

(ii.) $8\frac{1}{3}$ p.c. is equivalent to the fraction $\dfrac{8\frac{1}{3}}{100} = \dfrac{25}{300} = \dfrac{1}{12}$.

Example 2.　The population of a certain district is found to have decreased by 4 per cent. ; if it was originally 4325, what is it now ?

Every 100 has become $100-4$, or 96 ;

∴ the present population is 96 p.c. of the former population ;

that is, the present population $= \frac{96}{100}$ of $4325 = 4152$.

Commission is a *charge* due to an *agent* for conducting a business transaction, and is calculated as a percentage on the money spent in the transaction. Another common form of percentage is the **premium** paid on a sum of money for insurance of life or property.

Example 3.　For what sum should goods worth £573 be insured at $4\frac{1}{2}$ p.c., so that in case of loss the owner may recover the value of the goods and premium ?

If £100 were insured it would cover the value of goods worth £$95\frac{1}{2}$, together with the premium £$4\frac{1}{2}$;

∴ the sum to insure goods worth £573 $= £100 \times \dfrac{573}{95\frac{1}{2}} = £600$.

Examples XX.

Find, in lowest terms, the fraction equivalent to

1. $3\frac{1}{2}$ per cent.　　2. $4\frac{4}{9}$ per cent.　　3. $5\frac{3}{8}$ per cent.

4. 6·25 per cent.　　5. 5·625 per cent.　　6. $5\frac{1}{7}$ per cent.

Find the percentage expressed by the following fractions :

7. $\frac{2}{5}$.　　8. ·3.　　9. $\frac{27}{200}$.　　10. $\frac{3}{40}$.　　11. ·375.

12. Increase £66. 13s. 4d. by (i.) 5 % ; (ii.) 4 % ; (iii.) $7\frac{1}{2}$ %.

13. What is a man's income if after losing $22\frac{1}{2}$ % of it he has £186 left ?

14. The population of a certain district increased from 23,000 to 24,380 ; what was the rate per cent. of increase ?

15. A man whose annual income is £1275 spends 68 per cent. of it ; what does he save ?

16. What must be paid to insure a cargo valued at £8350 at $4\frac{1}{5}$ per cent.?

17. What is an agent's commission on £6. 13s. 4d. at $3\frac{1}{8}$ per cent. ?

18. Of the trees in an orchard 50 per cent. are apples, 25 per cent. pear trees, $16\frac{2}{3}$ per cent. plum trees, and there are fifty cherry trees ; how many trees are there altogether ?

19. For what sum should a cargo worth £7400 be insured at $7\frac{1}{2}$ per cent., so that in case of loss the owner may recover both cargo and premium ?

20. A man receives 4 per cent. on one-third of his capital, $4\frac{1}{2}$ per cent. on one-sixth, and 5 per cent. on the remainder ; what percentage does he receive on the whole ?

21. A merchant gains 30 per cent. on one-third of his capital, 45 per cent. on one-fourth, and loses 15 per cent. on the remainder ; what is his gain per cent. on the whole capital ?

22. A man embarks his whole capital in four successive ventures ; in the first he clears 100 per cent., and in each of the others loses 20 per cent ; shew that he has gained 2·4 per cent. on his original capital.

23. Find the amount of a bill for 864 yds. of linen at 2s. 1d. per yd., 520 yds. of cloth at 1s. 3d. per yd., and 280 yds. of silk at 16s. 9d. per yd., deducting $7\frac{1}{2}$ per cent. for ready money.

24. In an examination in which the full marks were 6000, A's marks exceeded B's by 12 per cent., B obtained 16 per cent. more than C, and C 20 per cent. more than D ; if A got 4872, find what percentage of the maximum were obtained by D.

*25. A certain number of persons ventured equal shares in a business so as to make up a total capital of £1719. 8s. 4d. The first dividend was $7\frac{1}{2}$ per cent. upon the capital, and amounted to £2. 14s. $10\frac{1}{2}d$. per share. What was the number of shares ?

*26. Seventy-five per cent. of the area of a farm is arable ; of the remainder eighty-five per cent. is pasture, and the rest is waste : the area of the waste is 3 a. 0 r. 20 p. What is the area of the farm?

Twenty-First Week.   *Profit and Loss.*

Notes and Hints for Solution. In working examples under this head it must be remembered that Profit or Loss, expressed as a percentage, always means a percentage reckoned on the *Cost Price.*

Example 1.   An article bought for 25*s*. is sold for £1. 6*s*. 10$\frac{1}{2}$*d*. ; what is the gain per cent. ?

Here 1*s*. 10$\frac{1}{2}$*d*., or 1$\frac{7}{8}$*s*., is the gain on the *cost price*, 25*s*. ;

we have therefore to express $\dfrac{1\frac{7}{8}}{25}$ as a percentage.

Now $$\frac{1\frac{7}{8}}{25} = \frac{1\frac{7}{8} \times 4}{100} = \frac{7\frac{1}{2}}{100} ;$$

∴   the gain is 7$\frac{1}{2}$ p.c.

Example 2.   If an article which cost 12*s*. 6*d*. is sold at a loss of 4 per cent., what is the selling price ?

Since the loss is 4 p.c. of the *cost price,*
∴   the selling price is 96 p.c. of the cost price ;
that is, the required selling price = $\frac{96}{100}$ of 12$\frac{1}{2}$*s*. = 12*s*.

Example 3.   By selling goods for £48 a profit of 20 per cent. is made.   What did the goods cost ?

Here the given *selling* price is 120 p.c. of the required *cost* price ;
that is,   £48 = $\frac{120}{100}$ of the cost price
                                  = $\frac{6}{5}$ of the cost price ;
∴   the required cost price = $\frac{5}{6}$ of £48 = £40.

Example 4.   By selling a horse for £68 I lose 15 per cent. ; at what price must it be sold to gain 10 per cent. ?

When the horse is sold for £68, the selling price is 85 p.c. of the cost price ; and to gain 10 p.c. the selling price must be 110 p.c. of the cost price ;
∴   the required price = £68 × $\frac{110}{85}$ = £88.

Example 5.   A man, having bought a cottage, sold it at a loss of 5 p.c.   Had he been able to sell it at a gain of 7 p.c., it would have fetched £24. 18*s*. more than it did.   What was the cost price ?

Here the 1st selling price = 95 p.c. of the cost price ;
and   the 2nd selling price = 107 p.c. of the cost price.
Thus the *difference* between the selling prices = 12 p.c. of the cost price ; that is,

£24. 18*s*. = $\frac{12}{100}$ of the cost price ;
∴   the cost price = $\frac{100}{12}$ of £24. 18*s*. = £207. 10*s*.

Examples XXI.

1. If I buy an article for 10s. 6d. and sell it for 14s., what is my gain per cent. ?

2. Goods are sold for £264. 5s. 3d., at a gain of £11. 2s. 9d. ; find the gain per cent.

3. What is the gain or loss per cent. on groceries retailed at 5¾d. per lb., and costing £2. 7s. 11d. per cwt. ?

4. If melons are bought at £1. 2s. 11d. a score, and sold at 17s. 5d. a dozen, what is the profit per cent. on the outlay ?

5. If a publisher gains £200 in selling 12000 shilling copies of a book, what is his gain per cent. ?

6. If an article which costs £25 is sold at a loss of $12\frac{1}{2}$ per cent., what does it sell for ?

7. At what price must an article costing £37. 10s. be sold so as to gain 8 per cent. ?

8. By selling goods for £17. 4s. a profit of $7\frac{1}{2}$ per cent. is made ; find the cost price.

9. A draper sells 150 yards of cloth at 8s. 9d. per yard, gaining thereby 25 p.c. What did he pay for the whole?

10. What was the cost price of goods which are sold for £95. 3s. 3¼d., at a profit of $5\frac{1}{4}$ p.c. ?

11. By selling apples at 3 a penny 5 per cent. is gained ; find the loss or gain per cent. when they are sold at 25 for sixpence.

12. By selling tobacco at 1s. 3d. per lb. a man gains 35 per cent. ; what does his profit amount to on the sale of 4 cwt. 11 lbs. ?

13. A bookseller's profit is $\frac{3}{13}$ of his *selling* price ; find his gain per cent. ; also his profit on 260 copies of a book sold for 7s. 6d.

14. If $17\frac{1}{2}$ per cent. be lost by selling an article at 8s. 3d., at what price should it have been sold to secure a gain of 40 per cent. ?

15. By disposing of goods for £364 a man loses 9 per cent., what should have been the selling price in order to have made a profit of 7 per cent. ?

16. If by selling an article at 5s. 6d. I gain $\frac{3}{8}$ of my outlay, what should I have gained per cent. if I had sold it for 6s. 6d. ?

17. By selling tea at 2s. 8d. per lb. a merchant gains $\frac{1}{8}$ of his outlay ; he then raises the price to 3s. 1d. per lb. What does he gain per cent. on the prime cost ?

18. What percentage of loss or gain will result from selling cloth at 7s. 9d. per yard, if 5 per cent. is gained by selling it at 8s. 9d. per yard ?

19. Sugar is bought at £25 per ton ; refining costs £1. 1s. 8d. per cwt. ; it is sold at 5½d. per lb.   What is the gain per cent. ?

20. A person bought a horse and sold it at a loss of 10 per cent. ; if he had received £9 more he would have gained 12½ per cent. ; what did the horse cost him ?

21. If 3 per cent. more be gained by selling a horse for £83. 5s. than by selling him for £81, what was the original price ?

22. An article of commerce passes successively through the hands of three dealers, each of whom in selling adds as his profit 10 per cent. of the price at which he bought it.   What did the first dealer pay for goods which the third dealer sells for £11. 1s. 10d. ?

23. A buys a cask of wine and sells it to B at a profit of 5 per cent., B sells it to C at a profit of 5 per cent., C sells it to D for £49. 12s. 3d., making a profit of 12½ per cent. ; what did the wine cost A ?

*24. A builder sold a house for £945, thereby gaining 8 per cent. on his outlay ; what did it cost him to build it ?

If the purchaser lets the house at £70 a year, find how much per cent. per annum he makes on the purchase money.

*25. Two-parts of chicory, costing £1. 9s. 9d. per cwt., are mixed with 5 parts of coffee, costing £8. 4s. 6d. per cwt. ; the mixture is sold at 1s. 4d. per pound ; find the profit per cent.

*26. One kind of tea is sold at 3s. a pound, and the profit is 20 per cent. ; another kind costs 2s. 8d. a pound.   If 4 pounds of the former are mixed with 5 pounds of the latter, and the mixture is sold at 3s. 4d. a pound, what is the profit per cent. ?

*27. A shopkeeper marks his goods with a price from which he can deduct 7½ per cent. for prompt payment, and still have a profit of 10 per cent. on what the goods cost him.   Find the cost price of an article which he marks at £2. 15s.

*28. A publisher sells books to a retail dealer at 5s. a copy, but allows 25 copies to count as 24 ; if the retailer sells each of the 25 copies at 6s. 9d., what profit per cent. does he make ?

*29. A quantity of ore containing 23 per cent. of copper is bought at 9s. per cwt. ; 95 per cent. of the copper is extracted at a cost of 2s. 10½d. per cwt. of ore ; find the price per ton at which the copper must be sold if a profit of 15 per cent. is to be made.

*30. A man sells 10 cwt. of sugar at 2¾d. per pound, thereby gaining 11s. 8d. ; what is his profit per cent. ?

Twenty-Second Week.   *Invoices and Estimates.*

Notes and Hints for Solution. The following example will explain the way in which Invoices are usually made out.

Princes Street,

EDINBURGH, *Dec. 30th, 1898.*

J. W. Clark, Esq.,

Bought of JAMES DUNCAN,

Draper and Hosier.

			£.	s.	d.
July 12th	9½ yds. flannel at 1/4 per yd., -	-		12	8
,, ,,	5½ doz. buttons at 3¼d. per doz.,	-		1	6
,, ,,	26 yds. calico at 1/8½ per yd., -	-	2	4	5
Sept. 10th	6 prs. socks at 1/9 per pr., -	-		10	6
Oct. 27th	23 yds. muslin at 3/4¾ per yd.,	-	3	18	1¼
,, ,,	18 yds. linen at 2/8 per yd., -	-	2	8	
Nov. 10th	2½ doz. collars at 7/6 per doz., -	-		18	9
			10	13	11¼

Each of the separate charges should be made out correct to the nearest farthing.

Thus, in the second line above, we write 1s. 6d. instead of the exact value 1s. 5⅞d.

Examples XXII.

Make out an invoice for each of the following sets of articles, supplying names and dates :

1. 12 yds. of muslin at 5½d. per yd.; 9 yds. of flannel at 1s. 11¾d. per yd. ; 3 pairs of gloves at 2s. 11½d. per pair ; 4 pairs of stockings at 2s. 3d. per pair ; 8 handkerchiefs at 16s. per dozen.

2. 10 lbs. of sugar at 2½d. per lb. ; 24 lbs. ditto at 3½d. per lb. ; 6 lbs. of cocoa at 1s. 2d. per lb. ; 5 lbs. of raisins at 7½d. per lb. ; 8 bananas at 1s. 3d. per doz. ; 5 lbs. of apples at 2½d. per lb. ; 30 oranges at 8d. per doz. ; 5 lbs. of biscuits at 7½d. per lb.

3. 20 tumblers at 16s. per dozen ; 2 dozen wine glasses at 13s. per dozen ; 6 water bottles at 4s. 3d. each ; set of 3 jugs at 1s. 3d., 1s. 6d., 1s. 9d. each ; 4 candlesticks at 11½d. each ; 4 dishes at 1s. 9d. each ; 4 ditto at 1s. 3d. each ; 1 tea service at 38s.

Deduct 5 per cent. for cash payment.

4. 3½ dozen stockings at 8s. 4d. per doz.; 2 gross of buttons at 3¼d. per doz.; 15½ yds. of velvet at 1s. 11½d. a yd.; 13½ yds. of flannel at 3s. 4d. a yd.; 12 pairs of gloves at 1s. 9d. per pair; 17½ yds. of silk at 11s. 7½d. a yd.; 1¼ yds. of ribbon at 8½d. a yd.

5. 37 yds. of Brussels carpet at 4s. 3d. a yd.; 18¾ sq. yds. of linoleum at 3s. a yd.; 7½ yds. of matting at 2s. 6d. a yd.; 6½ sq. yds. of floorcloth at 3s. 9d. a yd.; 10 yds. of binding at 1½d. a yd.; 24 yds. of Axminster carpet at 6s. 6d. a yd.

6. 3 Cashmere vests at 2s. 9d.; 8 shirts at 7s. 11d.; 4 Cardigan jackets at 5s. 4d.; 6 jerseys at 3s. 9d.; 8 pairs of golf hose at 2s. 7½d. a pair; 4 pairs of knitted gloves at 1s. 9d. a pair; 3 ditto at 2s. 3d.; 8 cravats at 2s. 6d. per doz.; 1 gross of buttons at 3¼d. a doz.; packing case, 10d.; carriage, 2s. 6d.

7. A courtyard 15 yds. by 12 yds. is to be paved with pebbles at 3s. per sq. yd., except two footpaths at right angles to the sides each 4 ft. broad, which meet in the centre, forming a cross. These are to be laid in paving stone at 3s. 3d. per sq. yd. Find the cost of the whole to the nearest penny.

8. A publisher in disposing of books to the retail bookseller charges for 25 copies as 24 and accepts 30 per cent. less than the selling price, and upon the whole receipts takes 10 per cent. commission for himself, it being agreed that the cost of production is to be borne by the author. If a book selling at 5s. costs 1s. for paper and printing, 5½d. for binding, and 2½d. for advertisements and other expenses; and if the bookseller allows a discount of 3d. in the 1s., what will be the profit made respectively by author, publisher, and bookseller on an edition of 10,000 copies?

*9. 1 cwt. of indigo at 14s. 6d. per lb.; 1 ton of cloves at 1s. 2d. per lb.; 5 cwt. 3 qrs. 18 lbs. spelter at 4½d. per lb.; 7 cwt. 1 qr. 14 lbs. block tin at £64 per ton.

Deduct 10 per cent. discount for cash.

*10. Make out the following contractor's bill :

300,000 bricks at 35s. per 1000.
240 tons lime at 25s. per ton.
670 yards gravel at 12s. 6d. per yard.
250 yards sand at 17s. 6d. per yard.
Cartage : lime 1s. 6d. per ton, gravel and sand 9d. per yd.

Deduct 5 per cent. discount.

*11. In building a wall 22,500 bricks are used at £1. 12s. a thousand, 135 bushels of lime at 1s. 4½d. a bushel, 16½ loads of sand at 3s. 6d. a load; the labour is reckoned at 9s. 6d. per thousand bricks laid; and 300 coping stones are used at 1s. 7½d. a piece, including cost of laying. Make out the above in the form of a bill, and find the amount after deducting 7½ per cent. for prompt payment.

Twenty-Third Week. *Simple Interest.*

Notes and Hints for Solution. Interest is the name given to money paid for the use of money lent. The sum lent is called the **Principal**, and the number of pounds paid in consideration of every £100 of the loan is called the **Rate per cent.** The principal together with its interest for any time is called the **Amount.** When the interest is paid on the original principal only and is not used to increase the principal, it is called **Simple Interest.**

Case I. *To find the* **simple interest** *when the principal, rate per cent., and time are given.*

Example 1. Find the Simple Interest on £350 for 4 years at 3 p.c. per annum.

The interest for 1 year $= 3$ p.c. of $£350 = £350 \times \frac{3}{100}$;

$\therefore$ the interest for 4 years $= £350 \times \frac{3}{100} \times 4 = £42.$

Thus we easily obtain the rule : *Multiply the principal by the rate per cent. and by the number of years, and divide the product by 100.* The application of this rule is illustrated in the following examples.

Example 2. Find to the nearest penny the Simple Interest on £227. 10*s*. 6*d*. (i.) for 5 years at $3\frac{1}{2}$ per cent.; (ii.) for 4 years 7 months at $2\frac{2}{5}$ per cent.

(i.) The required interest

$$= £227. \ 10s. \ 6d. \times 5 \times \frac{3\frac{1}{2}}{100}.$$

£.	*s.*	*d.*
227	10	6
		5
1137	12	6
		$3\frac{1}{2}$
3412	17	6
568	16	3
39,81	13	9
20		
16,33		
12		
4,05		

$\therefore$ Interest $= £39. \ 16s. \ 4d.$

(ii.) The required interest

$= £227 \cdot 525 \times 4\frac{7}{12} \times 2\frac{2}{5} \div 100$

$= £227 \cdot 525 \times \frac{55}{12} \times \frac{12}{5} \times \frac{1}{100}$

$= £2 \cdot 27525 \times 11$

$= £25 \cdot 028$ (correct to 3 decimal places)

$= £25. \ 0s. \ 7d.$
(correct to the nearest penny).

Example 3.   Find to the nearest penny the Amount of £512. 4s. 6d. at 3 per cent. Simple Interest from March 15th to Sept. 14th.

The number of days is 183, since it is customary to include *one only* of the dates mentioned ; and the interest on £512·225 for 183 days

$$= £512\cdot225 \times \tfrac{3}{100} \times \tfrac{183}{365}$$

$$= £512\cdot225 \times \frac{3 \times 366}{73000}$$

$$= \frac{£512\cdot225 \times 1\cdot098}{73}$$

$$= \frac{£562\cdot423}{73} = £7\cdot7044..$$

$$= £7. \ 14s. \ 1d.$$

```
512·2,2,5
  1·0 9 8
512·2 2 5
 46·1 0 0 | 2
  4·0 9 7 | 8
562·4 2 3 |
```

Thus the *Amount* is £512. 4s. 6d. + £7. 14s. 1d. = £519. 18s. 7d.

Examples XXIII.

Find the Simple Interest on

1. £750   for 3 yrs. at 5%.   2. £3375   for 4 yrs. at 3%.
3. £540   ,, $2\frac{1}{2}$,,   3%.   4. £1500   ,, $3\frac{1}{4}$,, $4\frac{1}{3}$%.
5. £1350   ,, $5\frac{3}{4}$,, $3\frac{1}{2}$%.   6. £750   ,, $3\frac{3}{4}$,, $4\frac{1}{2}$%.
7. £903. 15s. ,, $3\frac{5}{12}$,,   4%.   8. £1885. 15s.,, $7\frac{1}{4}$,,   5%.
9. £244. 1s. 8d. for $3\frac{3}{4}$ yrs. at 3%.
10. £200. 16s. 8d. for $3\frac{1}{2}$ yrs. at $2\frac{3}{4}$%.

Find, to the nearest penny, the Amount of

11. £257   for $5\frac{1}{2}$ years   at $2\frac{3}{4}$ per cent.
12. £63. 5s. 9d.   ,, $10\frac{1}{2}$,,   ,, $3\frac{1}{4}$,,
13. £305. 2s. 1d.   ,, $3\frac{1}{4}$,,   ,, 4   ,,
14. £146. 12s. 2d. ,, 225 days   .. $2\frac{1}{2}$,,
15. £30. 8s. 4d.   ,, 234 ,,   ,, 5   ,,
16. £820. 4s. 2d. ,, 2 years 146 days ,, $2\frac{1}{2}$,,
17. £219 from January 1st to July 16th at $3\frac{1}{3}$ per cent.
18. £252. 1s. 8d. from June 13th to August 25th at 3 per cent.
19. £1368. 15s. from May 17th to Dec. 15th at $4\frac{3}{4}$ per cent.

Twenty-Fourth Week. *Simple Interest Continued.*

Notes and Hints for Solution. In questions on Simple Interest we are concerned with four quantities, viz. principal, rate per cent., number of years, and interest. If any three of these are given we can find the fourth. Case I. has been discussed in the last section.

Case II. *To find the* **time** *when the principal, interest, and rate per cent. are given.*

Example 1. In what time will £1250. 12*s*. 6*d*. amount to £1375. 13*s*. 9*d*. at 4 p.c. per annum?

$$\text{The req. no. of yrs.} = \frac{\text{whole interest}}{\text{interest for } one \text{ year}}.$$

Now the whole interest $= $ £1375. 13*s*. 9*d*. $-$ £1250. 12*s*. 6*d*.

$$= \text{£}125. \; 1s. \; 3d. = \text{£}125\tfrac{5}{80} \; ;$$

and int. for 1 yr. $= \text{£}1250\tfrac{25}{40} \times \dfrac{4}{100} = \text{£}50\tfrac{1}{40} \; ;$

req. no. of yrs. $= 125\tfrac{5}{80} \div 50\tfrac{1}{40} = 2\tfrac{1}{2}.$

Case III. *To find the* **rate per cent.** *when the principal, interest, and time are given.*

Example 2. At what rate per cent. will the Simple Interest on £422. 10*s*. for 3 years be £47. 10*s*. 7$\frac{1}{2}$*d*. ?

$$\text{The req. rate per cent.} = \frac{\text{whole interest}}{\text{interest at } one \text{ per cent.}}.$$

Now the whole interest $= \text{£}47\tfrac{85}{160} = \text{£}47\tfrac{17}{32} \; ;$

and int. at 1 p.c. $= \text{£}422\tfrac{1}{2} \times \dfrac{1}{100} \times 3 \; ;$

$\therefore$ req. rate p.c. $= 47\tfrac{17}{32} \div \dfrac{845 \times 3}{2 \times 100}$

$$= \frac{1521}{32} \times \frac{2 \times 100}{3 \times 845} = 3\tfrac{3}{4}.$$

Case IV. *To find the* principal *when the interest, rate per cent., and time are given.*

Example 3. What principal will produce £114. 3*s*. 9*d*. as Simple Interest for $7\frac{1}{4}$ years at $3\frac{1}{2}$ per cent. ?

Under the given conditions,

$$£100 \text{ gives as interest } £\tfrac{29}{4} \times \tfrac{7}{2}.$$

Hence *interest* $£\tfrac{29}{4} \times \tfrac{7}{2}$ is derived from *principal* £100 ;

$$\therefore \quad \text{interest } £114\tfrac{3}{16} \quad ,, \qquad ,, \quad \text{principal } £100 \times \frac{114\tfrac{3}{16}}{\tfrac{29}{4} \times \tfrac{7}{2}} ;$$

that is, the required principal $= £100 \times \tfrac{1827}{16} \times \tfrac{4}{29} \times \tfrac{2}{7}$

$$= £450.$$

Example 4. What principal will amount to £1865. 12*s*. 6*d*. in $6\frac{1}{2}$ years at $3\frac{3}{4}$ per cent. ?

Under the given conditions

$$£100 \text{ amounts to } £100 + £\left(\frac{13}{2} \times \frac{15}{4}\right), \text{ or } £\frac{995}{8}.$$

Hence an *amount* $£\dfrac{995}{8}$ corresponds to a *principal* £100 ;

$$\text{amount } £1865\tfrac{5}{8} \quad ,, \qquad ,, \quad \text{principal } £100 \times \frac{1865\tfrac{5}{8}}{\frac{995}{8}} ;$$

that is, required principal $= £100 \times \dfrac{14925}{995} = £1500.$

Examples XXIV.

For what time would the simple interest on

1. £750   be £375   at 5 per cent. per annum ?
2. £333. 6*s*. 8*d*. ,, £67. 10*s*. ,, $2\frac{1}{4}$,,   ,,
3. £3756 ,, £633. 16*s*. 6*d*. ,, $4\frac{1}{2}$,,   ,,
4. £1500 ,, £365. 12*s*. 6*d*. ,, $3\frac{3}{4}$,,   ,,.

5. In what time will £3100 amount to £3384. 3*s*. 4*d*. at $3\frac{1}{3}$ per cent. per annum ?

6. In how many days will £2187. 10*s*. amount to £2243. 5*s*. $7\frac{1}{2}d.$ at $4\frac{1}{4}$ per cent. per annum?

7. In how many days will the interest on £1572. 17*s*. 6*d*. come to £23. 14*s*. $0\frac{1}{4}d.$ at $5\frac{1}{2}$ per cent. per annum?

8. In what time will a sum of money double itself at $4\frac{1}{6}$ per cent. simple interest?

At what rate per cent. would the simple interest on

9. £1225   be £192. 18*s*. 9*d*. for 3 years?

10. £3725. 15*s*. ,, £434. 13*s*. 5*d*. ,, $3\frac{1}{2}$ years?

11. £3643. 6*s*. 8*d*. ,, £605. 14*s*. 1*d*. ,, $4\frac{3}{4}$ years?

12. £980   ,, £178. 8*s*. 10*d*. ,, 3 years 2 months?

13. If the interest on £650 for 5 months is £12. 3*s*. 9*d*., what is the rate per cent.?

14. At what rate per cent. will £514. 7*s*. 6*d*. amount to £694. 8*s*. $1\frac{1}{2}d.$ in $7\frac{1}{2}$ years?

15. What is the rate of simple interest when £868. 0*s*. 3*d*. amounts to £1012. 13*s*. $7\frac{1}{2}d.$ in 5 years 4 months?

16. The sum of £437. 10*s*. was lent at simple interest, and at the end of two-thirds of a year the debt was cancelled by the payment of £449. 3*s*. 4*d*. What was the rate of interest?

17. At what rate per cent. will a sum of money treble itself at simple interest in 25 years?

18. What sum will amount to £470. 8*s*. $10\frac{3}{4}d.$ in 4 years at $2\frac{1}{2}$ per cent.?

19. On what principal will the interest at 4 per cent. in 3 years come to £33. 3*s*. 3*d*.?

20. What principal will amount to £1751. 16*s*.- $10\frac{1}{2}d.$ in $3\frac{1}{2}$ years at $4\frac{1}{4}$ per cent.?

21. What sum of money must be put out at $2\frac{2}{5}$ per cent. for $2\frac{1}{2}$ years to produce £98. 4*s*. 6*d*. interest?

22. What sum will amount to £393. 19*s*. 9*d*. at 4 per cent. in 292 days?

23. What sum of money put out at 3 per cent. from July 14th to Sept. 25th will amount to £253. 11*s*. 11*d*.?

24. What principal will give the same interest in 4 months at 3 per cent. per annum as £312. 10*s*. will give in 8 months at $4\frac{1}{2}$ per cent. per annum?

Twenty-fifth Week.   *Compound Interest.*

Notes and Hints for Solution. Money is said to be put out at **Compound Interest** when each instalment of interest as it becomes due is added to the principal instead of being paid over to the lender of the principal sum. In this case the principal is continually being increased, and the interest for each period is the interest on the *Amount* at the end of the preceding period. Thus, if £100 is put out at 5 per cent. compound interest, at the end of one year the amount is £105 ; this is the principal for the second year, and the interest on it is found to be £5. 5s. ; thus the principal for the third year is £110. 5s., and so on. It follows that under this system the interest for each period must be separately calculated, the complete Compound Interest being the sum of the interests for the several periods ; though this is most conveniently obtained by subtracting the original *principal* from the final *amount*. Unless otherwise stated, interest is supposed to be payable *yearly*.

Example L.   Find the Compound Interest on £273. 12s. for 3 years at 4 per cent. to the nearest penny.

£	
273·6	1st principal
10·944	1st year's interest
284·544	2nd principal
11·381 76	2nd year's interest
295·925 76	3rd principal
11·837 03	3rd year's interest
307·762 79	Amount in 3 years
273·6	Original principal
34·163	Interest required
20	
3·26	
12	
3·12	

Compound Interest = £34. 3s. 3d.

We first express the principal as the decimal of a pound. To multiply this by $\frac{4}{100}$, we multiply the principal by 4 and set down each figure *two places to the right*, the position of the decimal point remaining fixed. Each year's principal is treated in the same way. Lastly, the original principal is subtracted from the final amount.

To obtain an answer correct to the nearest penny it is only necessary to secure accuracy to three decimal places in the final result. Thus we retain *five* decimal figures throughout the work. [See page 9.]

Example 2. Find the amount at Compound Interest of £157. 16*s*. 6*d*. in 2 years at 5 per cent. per annum, interest payable half-yearly.

Here interest is to be allowed for 4 periods at $2\frac{1}{2}$ % per period.

£
157·825
3·156 50 ⎱ Interest for 1st
 ·789 12 ⎰ half-year.
————
161·770 62
3·235 41 ⎱ Interest for 2nd
 ·808 85 ⎰ half-year.
————
165·814 88
3·316 30 ⎱ Interest for 3rd
 ·829 07 ⎰ half-year.
————
169·960 25
3·399 20 ⎱ Interest for 4th
 ·849 80 ⎰ half-year.
————
174·209 25
Amount = £174·209 = £174. 4*s*. 2*d*.

We first decimalize the principal as in Example 1. To find the interest for the first period we have to multiply the principal by $\frac{2\frac{1}{2}}{100}$, that is, by $\frac{2}{100}+\frac{1}{200}$. For the first step of work we multiply by 2, setting down each figure two places to the right. For the second step we take $\frac{1}{5}$ of the principal after mental division by 100. Each half-year's principal is treated in the same way.

Example 3. What principal will amount to £171. 15*s*. 10*d*. in 3 years at 4 % Compound Interest?

Here *principal* £100 in 1 year at 4 % *amounts* to £104 ;

∴ the amount at the end of any year is found by multiplying the principal at the beginning of that year by $\frac{104}{100}$, or 1·04.

∴ the amount at the end of 3 years = principal × (1·04)^3.

Thus to find the principal we must divide the amount by (1·04)^3.

1·04
  416   $=\frac{4}{100}$ of 1·04
————
1·0816 $=1·04$ of $1·04=(1·01)^2$
 43264 $=\frac{4}{100}$ of 1·0816
————
1·124864 $=1·04$ of $(1·04)^2=(1·04)^3$

Thus the required principal

$$=£\frac{171·7917}{1·124864}$$

$$=£152·722...$$

$$=£152.\ 14s.\ 5d.$$

£171. 15*s*. 10*d*. = £171·7917.

1·1,2,4,8,6,4) 171·7917 (152·722
            112 4864
            ————
             59 3053
             56 2432
             ————
              3 0621
              2 2497
              ————
                8124
                7874
                ————
                 250
                 225
                 ————
                  25
                  22
                  ——

Examples XXV.

Find to the nearest penny the amount at Compound Interest on

1. £225 for 2 yrs. at 4%.   **2.** £3000. 15s. for 2 yrs. at 4%.

3. £1000 for 3 yrs. at 5%.   **4.** £415 for 3 yrs. at 6%.

5. £350. 12s. 6d. for 5 yrs. at 4%.   **6.** £3546 for 3 yrs. at 5%.

Find to the nearest penny the Compound Interest on

7. £1256. 10s. for 2 yrs. at $3\frac{1}{2}$%.   **8.** £4500 for 2 yrs. at $4\frac{1}{4}$%.

9. £745. 10s. for 2 yrs. at $3\frac{1}{3}$%.   **10.** £1485 for 3 yrs. at $5\frac{1}{4}$%.

11. £5016. 11s. 6d. for 2 yrs. at $4\frac{1}{2}$%.

12. £1601. 4s. 8d. for 3 yrs. at $3\frac{1}{4}$%.

[*To find the Compound Interest on a sum of money for $3\frac{1}{2}$ years at $2\frac{1}{2}$%, calculate the amount for the first three years as in former examples, then consider the $\frac{1}{2}$ year as a* **complete period** *for which the rate of interest is $\frac{1}{2}$ of $2\frac{1}{2}$%, or $1\frac{1}{4}$%.*]

Find to the nearest penny the Compound Interest on

13. £3600 for $3\frac{1}{2}$ yrs. at $2\frac{1}{2}$%.

14. £8457. 14s. 6d. for $2\frac{1}{4}$ yrs. at $1\frac{1}{4}$%.

15. £504. 13s. 9d. for 2 yrs. 4 mos. at $4\frac{1}{2}$%.

16. Find to the nearest penny the amount of £320. 15s. in $1\frac{1}{2}$ years at 4% Compound Interest, payable half-yearly.

17. Find to the nearest penny the Compound Interest on £425 for 2 years at 6%, payable half-yearly.

18. If interest be payable half-yearly, find the Compound Interest on £410 for $2\frac{1}{2}$ years at $4\frac{1}{2}$%.

19. What sum at 4% Compound Interest will amount to £3515. 4s. in 3 years?

20. What sum at 5% Compound Interest will amount to £264. 12s. in 2 years?

21. Find the principal which will amount to £810. 6s. 9d. in 4 years at 5% Compound Interest.

22. What sum must be put out at Compound Interest at 8 per cent. per annum, payable half-yearly, so as to amount to £3661. 13s. 4d. in $1\frac{1}{2}$ years?

23. What principal, lent out at Compound Interest for 2 years at $5\frac{1}{10}$%, will amount to £4602. 10s. 1d.?

Twenty-Sixth Week. *Present Worth and Discount.*

Notes and Hints for Solution. Suppose A owes B a sum of £102, payment being due in six months' time, and that money is worth 4 per cent. Since at this rate £100 would amount to £102 in six months, A can equitably discharge his debt to B by paying him £100 at once instead of £102 at the end of half-a-year. In this case £100 is called the **Present Worth** of £102, and £2 is called the **True Discount** on £102 for six months at 4 per cent. per annum.

Thus the Present Worth of a debt is that sum which with its interest for the given time would amount to the sum due, and the True Discount on the *debt* is the Simple Interest on the *Present Worth* ; or briefly

Present Worth + Interest on P.W. = Sum due,

Present Worth + Discount on Debt = Sum due.

In practice discount is often calculated as **interest on the sum due,** and is then known as **Banker's or Commercial Discount.**

A **Bill** is a written promise to pay a sum at a given date, and the amount due is called the **Face Value** of the Bill. A Bill nominally due on a certain date is not legally due until 3 days later. These days are called **days of grace ;** thus a bill dated May 20th and due in 3 months is legally due on Aug. 23rd.

Example 1. Find the true discount and present worth of £204. 16s. due in 146 days at 6 %.

Int. on £100 at 6 % for 146 days $= £6 \times \frac{146}{365} = £1\frac{2}{5}$,

that is, a *principal* £100 amounts to $£\left(100 + 1\frac{2}{5}\right)$ or $£\frac{512}{5}$.

Thus, on a *debt* $£\frac{512}{5}$, the true *discount* $= £\frac{12}{5}$,

,,   £204$\frac{4}{5}$ $= £\frac{12}{5} \times \frac{1024}{5} \div \frac{512}{5}$

$\qquad\qquad\qquad\qquad\qquad\qquad\qquad = £\frac{1024}{5} \times \frac{12}{512} = £4. \ 16s.$

And the present worth $= £204. \ 16s. - £4. \ 16s. = £200.$

Note. If present worth alone is required we use the method of Example 4, p. 68.

Example 2. Find the true present worth of a bill for £9397. 10s. drawn March 16th at 12 months and discounted on May 31st at $6\frac{1}{4}$ % per annum, allowing the usual 3 days of grace.

The bill becomes legally due on March 19th, and, reckoning from May 31st, has 292 days to run.

Int. on £100 at $6\frac{1}{4}$ % for 292 days $= £\frac{25}{4} \times \frac{292}{365} = £5$;

Thus, on a *bill* for £105 the *present worth* $= £100$;

$$,, \qquad £9397. 10s. \qquad ,, \qquad = £100 \times \frac{9397\frac{1}{2}}{105} = £8950.$$

When a debt is due after a number of years the discount must be calculated *at compound interest* on the present worth.

Example 3. Find the present worth of £171. 15s. 10d. due 3 years hence at 4 % compound interest.

Here the question is: What *principal* will *amount* to £171. 15s. 10d. in 3 yrs. at 4 % compound interest?

This example has already been solved [see Ex. 3, p. 71]. Other cases may be treated similarly.

It may be observed that the inverse questions on interest, discussed at pages 67, 68, furnish illustrations of discount and present worth. We have only to substitute the words *present worth* and *face value* (or *sum due*) for *principal* and *amount* respectively, and to remember that the *interest* on the present worth is the *discount* on the sum due.

Examples XXVI.

[*In the following examples true discount is always to be found unless the contrary is stated.*]

Find the present worth of

1.	£333	due 5 years	hence at	$7\frac{3}{4}$	per cent.
2.	£188. 2s. 6d.	,, 18 months	,,	5	,,
3.	£1336. 11s. 3d.	,, $3\frac{1}{2}$ years	,,	5	,,
4.	£279. 2s. 6d.	,, $4\frac{1}{2}$ months	,,	4	,,
5.	£353. 12s.	,, 2 yrs. 4 months	,,	$4\frac{1}{2}$	,,

Find the discount on

6.	£962	due $4\frac{1}{2}$ years	hence at	$4\frac{1}{2}$	per cent.
7.	£2160	,, 2 ,,	,,	4	,,

Find the discount on

8. £275. 12s. 6d. ,, 15 months ,, 4 per cent.

9. £76. 15s. ,, 8 ,, ,, $3\frac{1}{2}$,,

10. £328. 13s. 5d. ,, 3 ,, ,, 4 ,,

11. The true discount on a certain sum due $2\frac{1}{2}$ years hence at 3 % per annum is £17. 14s. 6d. What is the sum?

12. Shew that the interest on £266. 13s. 4d. for 3 months at $4\frac{1}{2}$ % per annum is equal to the true discount on £83 due 15 months hence at 3 % per annum.

13. At what rate per cent. will the present worth of £100. 10s. 3d. payable 2 years hence be £93. 10s.?

14. If the difference between the interest and the discount on a sum of money for 2 months at $4\frac{1}{2}$ per cent. is 2s. 3d., find the sum.

15. What rate per cent. per annum does a man get for his money when in discounting a bill due in 10 months he deducts as discount 4 per cent. of the total amount of the bill?

16. Find the true discount, allowing the usual 3 days' grace, upon a bill for £603, drawn Oct. 4th at 4 months, and discounted Nov. 26th at $2\frac{1}{2}$ per cent. per annum.

17. Find the present worth of a bill for £554. 17s. $11\frac{3}{4}d.$ drawn April 15th at 6 months, paid Aug. 6th at $3\frac{1}{2}$ per cent., days of grace being allowed.

18. A bill for £1812 drawn July 13th at 5 months is discounted on October 4th at $3\frac{1}{3}$ per cent. Find the true discount, allowing the usual three days of grace.

19. Find the present worth at $2\frac{1}{2}$ % compound interest on £143. 11s. $8\frac{1}{2}d.$ payable 3 years hence.

20. Find the present worth of £1447. 0s. $7\frac{1}{2}d.$ due 3 years hence at 5 % compound interest.

***21.** If a sum of £1000 becomes due three months hence, what is its present value as commonly calculated, and what as correctly calculated, interest being reckoned at 5 per cent.?

***22.** Find to the nearest penny the simple interest on £248. 18s. 9d. for $2\frac{1}{2}$ years at $3\frac{3}{4}$ per cent. per annum.

Find also the sum of money which must be invested at $3\frac{3}{4}$ per cent. per annum simple interest in order to amount to £248. 18s. 9d. in $2\frac{1}{2}$ years.

***23.** Find the difference between the simple interest and the compound interest on £2343. 15s. for 2 years at 4 per cent. per annum.

Find the present value of £84. 10s. due two years hence, compound interest being reckoned at 4 per cent. per annum.

Twenty-Seventh Week. *Stocks.*

Notes and Hints for Solution. Money borrowed by Governments, Town Corporations, Railways, and some other large undertakings is known as **Stock.** The public who lend the money are known as **Stockholders,** and receive interest periodically (usually half-yearly or quarterly) as a percentage on the amount of Stock held. This interest is sometimes at a fixed rate, and the percentage gives the name to the Stock, "$2\frac{1}{2}$ per cent. Stock," "4 per cent. Stock," and so on. Sometimes the periodical interest is made after a division of profits, and is called a **dividend.** Dividends vary from time to time according to the prosperity of the business. The *Stock* also varies in value owing to different causes. If £100 *Stock* is worth £100 *in cash*, it is said to be **at par**; if it is worth more than £100 cash it is said to be at a **premium**; if it is worth less than £100 cash it is said to be **at a discount.** The price of Stock is quoted at so much **per cent.**; thus a Stock is said to be "at 97" (or at a discount of 3 per cent.) when £100 Stock can be bought for £97; a Stock "at 105" is one in which the cash value of £100 Stock is £105, and so for other cases.

In working examples in Stocks it is important to distinguish clearly between *Stock* and *Cash,* and to remember that the dividends do not depend on the price at which the Stock is bought.

Example 1. How much 5 per cent. Stock at $112\frac{1}{2}$ can be bought for £3037. 10*s.*?

Here £$112\frac{1}{2}$ *cash* will buy £100 *Stock*;

∴ £$3037\frac{1}{2}$,, ,, £$100 \times \dfrac{3037\frac{1}{2}}{112\frac{1}{2}}$, or £2700.

[It should be noticed here that we are not concerned with any question of *income*, hence it is unnecessary to name the particular class of Stock so long as we know the market price.]

Example 2. What is the cash value of £5640 in $4\frac{1}{2}$ % Stock which is 6 per cent. below par?

Here £100 *Stock* sells for £94 *cash*;

∴ £5640 ,, ,, £$94 \times \frac{5640}{100}$;

∴ required cash value = £5301. 12*s.*

Example 3. A man *holds* £5640 in $4\frac{1}{2}$ % Stock at 94 ; find his income.

On £100 *Stock* he gets an *income* of $£4\frac{1}{2}$;

∴ on £5640 ,, ,, $£4\frac{1}{2} \times \frac{5640}{100} = £253. 16s.$

Example 4. A man *invests* £5640 in $4\frac{1}{2}$ % Stock at 94 ; what is his income ?

Here £94 *cash* will purchase an *income* of $£4\frac{1}{2}$;

∴ £5640 ,, ,, $£4\frac{1}{2} \times \frac{5640}{94}$, or £270.

Example 5. How much money must I invest in $4\frac{1}{2}$ % Stock at 94 in order to get an income of £45 ?

To purchase an *income* of $£4\frac{1}{2}$ I must invest £94 *cash* ;

∴ ,, ,, £45 ,, $£94 \times \dfrac{45}{4\frac{1}{2}}$, or £940.

Example 6. Which is the better investment, 3 % Stock at 87 or 4 % Stock at 115 ?

Suppose £87 × 115 is invested in each stock ; then, as in Example 4,

Income from first Stock $= £3 \times \dfrac{87 \times 115}{87} = £345.$

Income from second Stock $= £4 \times \dfrac{87 \times 115}{115} = £348.$

Thus the second is the better investment.

Examples XXVII.

1. How much stock at 124 can be bought for £496?

2. How much stock at $97\frac{1}{2}$ can be bought for £3900 ?

3. When railway stock is 5 per cent. below par, how much can be bought for £2527 ?

4. How much stock can be bought for £621 when the price quoted is at a premium of $3\frac{1}{2}$ per cent. ?

5. For how much will £3075 stock at 102 sell ?

6. What will be obtained by the sale of £2735 stock at £95 ?

7. Find to the nearest penny what the sale of £3572 Consols will amount to when they are $9\frac{1}{2}$ per cent. above par.

8. What is the cash value of £4715 stock which is quoted at $£87\frac{1}{2}$?

9. A man holds £1240 in $2\frac{3}{4}$ p.c. Consols, what is his annual income?

10. What is the annual income arising from £410 of $3\frac{1}{2}$ p.c. stock?

11. What is the half-yearly dividend due to a man who owns £860 railway $4\frac{1}{2}$ p.c. preference stock?

12. What income will be obtained by investing £1188 in 3 p.c. stock at 81?

13. What income will be derived from £967. 10s. laid out in the purchase of India 5 per cents. at $107\frac{1}{2}$?

14. If a 3 p.c. stock is at $94\frac{3}{8}$, what income will be derived by investing £7927. 10s.?

15. How much money must be invested in $5\frac{1}{2}$ p.c. stock at 110 so as to get an income of £60?

16. How much must I invest in railway stock at $139\frac{1}{2}$ paying $7\frac{3}{4}$ p.c. per annum so as to secure an income of £25?

17. What sum must be invested in 3 p.c. stock at $87\frac{5}{8}$ in order to obtain an income of £435? What income would be obtained from the same sum if the stock was at 87 at the time of purchasing?

18. Which is the better investment, 4 p.c. stock at 105 or $5\frac{1}{2}$ p.c. stock at 140?

19. One man invests a certain sum in $2\frac{3}{4}$ p.c. stock at $97\frac{1}{2}$, another invests an equal sum in $3\frac{1}{2}$ p.c. stock which is 20 p.c. above par; compare their incomes.

20. Which is the more profitable investment, a stock quoted at 152 paying $4\frac{1}{2}$ p.c. or a $2\frac{1}{2}$ p.c. stock at 85? If £3230 is invested in each, what is the difference in income?

21. How much must be invested in $4\frac{3}{4}$ p.c. stock at 95 to produce an income of £613. 12s. after paying an income tax of 4d. in the £?

22. An income of £1000 is made up of £240 from a 6 p.c. stock, £340 from an 8 p.c. stock, and the remainder from a $3\frac{1}{2}$ p.c. stock; how much of each stock is held?

23. What income would be obtained by investing the sum produced by selling $35\frac{1}{2}$ acres at £350 an acre in 3 per cents. at 71? How much would it differ from that obtained by letting the land at £14 an acre?

24. The difference between the incomes derived from investing a certain sum in $4\frac{1}{2}$ p.c. stock at 150, and in $3\frac{1}{2}$ p.c. stock at 125, is found to be £6. 10s. What was the sum invested?

Twenty-Eighth Week. *Stocks and Shares.*

Notes and Hints for Solution. The purchase and sale of stock is conducted through an agent called a **Broker**, who charges a commission known as **Brokerage** on every £100 bought or sold. The brokerage is usually 2s. 6d. on £100 stock, and is quoted as "$\frac{1}{8}$ per cent." Thus in investing money the *buyer* pays $\frac{1}{8}$ per cent. *more* than the quoted price of the stock, and the *seller* receives $\frac{1}{8}$ per cent. *less* than the quoted price. In the examples hitherto discussed the brokerage has been supposed to be included in the price of stock.

Sometimes the Capital of a Company is divided into **Shares** of a definite amount such as £1, £5, or £10. Thus, instead of speaking of £50,000 stock, we may have 50,000 shares of £1, or 10,000 shares of £5, or 5000 shares of £10. But, unless otherwise stated, it is to be understood that the dividends are quoted as so much *per cent.*

Example. 1. A person invests £9075 in 3 % stock at $90\frac{5}{8}$, and when it has risen to $91\frac{1}{8}$ he sells out and reinvests in $3\frac{1}{2}$ % Stock at $97\frac{3}{8}$; find the change in his income, brokerage at $\frac{1}{8}$ per cent. being charged on each transaction.

Here $£\left(90\frac{5}{8} + \frac{1}{8}\right)$ *cash* has to be paid for £100 *Stock*;

hence with $£90\frac{3}{4}$ *cash* he buys *income* £3;

$\therefore$ with £9075 ,, ,, $£3 \times \dfrac{9075}{90\frac{3}{4}} = £300.$

On selling, for £100 *Stock* he receives $£\left(91\frac{1}{8} - \frac{1}{8}\right)$ *cash*;

i.e. stock which was bought for $£90\frac{3}{4}$ *cash* sold for £91 *cash*;

$\therefore$,, ,, ,, £9075 ,, $£91 \times \dfrac{9075}{90\frac{3}{4}}.$

Again, for $£97\frac{1}{2}$ *cash* he buys *income* $£3\frac{1}{2}$;

$\therefore$ for $£91 \times \dfrac{9075}{90\frac{3}{4}}$,, $£3\frac{1}{2} \times \dfrac{91 \times 9075}{90\frac{3}{4} \times 97\frac{1}{2}} = £326\frac{2}{3}.$

Thus the gain in his income is £26. 13s. 4d.

Example 2. What rate per cent. on capital is obtained by investing in $4\frac{1}{2}$ % Stock at $134\frac{7}{8}$, brokerage being $\frac{1}{8}$ per cent.

Here $£\left(134\frac{7}{8} + \frac{1}{8}\right)$ *cash* has to be paid for £100 *Stock*.

So that £135 *cash* will buy an *income* of $£4\frac{1}{2}$;

$\therefore$ £100 ,, ,, $£4\frac{1}{2} \times \dfrac{100}{135} = £3\frac{1}{3}.$

Example 3. What price must be paid for $4\frac{1}{2}$ % Stock so that a man may get 6 % for his money ?

Here we have to consider what the price of $4\frac{1}{2}$ % Stock is when £100 cash invested in it produces an income of £6.

An *income* of £6 is bought for £100 *cash* ;

$$\therefore \qquad ,, \qquad £4\frac{1}{2} \qquad ,, \qquad £100 \times \frac{4\frac{1}{2}}{6}, \text{ or } £75.$$

Example 4. If £3 shares in a Company are sold at 7*s*. 6*d*. premium, find (i) how many shares can be bought for £540, (ii) the cost price of 40 shares, (iii) the dividend at 4 % on 60 shares.

(i) Since each share costs $£3\frac{3}{8}$,

the required number of shares $= £540 \div £3\frac{3}{8} = 160$.

(ii) 40 shares cost $£3\frac{3}{8} \times 40 = £135$.

(iii) 60 shares represent a capital of £3 × 60, or £180, the interest on which at 4 % is £7. 4*s*.

Example 5. A person invests half his capital in 3 per cent. debentures at $101\frac{1}{2}$ and the other half in 4 per cent. debentures at 135 ; his total income from both sources is £202. 15*s*. How much did he invest ?

Suppose he invests $£\left(135 \times 101\frac{1}{2}\right)$ in *each* stock.

$$\text{Income from 3 \% stock} = £135 \times 3 = £405,$$

$$,, \qquad 4\% \quad ,, \quad = £\frac{203}{2} \times 4 = £406.$$

Whole income £811 arises from £135 × 203 invested ;

$$\therefore \qquad ,, \qquad £202\frac{3}{4} \qquad ,, \qquad £135 \times 203 \times \frac{202\frac{3}{4}}{811}$$

$$= £\frac{135 \times 203}{4} = £6851. \ 5s.$$

Examples XXVIII.

[*Unless otherwise stated brokerage is included in the prices quoted.*]

1. What annual income is obtained by investing £9175 in 4 per cent. stock at $91\frac{5}{8}$, brokerage being $\frac{1}{8}$ per cent. ?

2. How much stock at $97\frac{3}{8}$ can be bought for £5850, brokerage being $\frac{1}{8}$ per cent. ?

3. If I buy £1400 stock in 3 per cents. at $92\frac{3}{8}$, paying brokerage $\frac{1}{8}$, what does it cost me ? What will be the annual interest on my outlay ?

4. What interest per cent. is derived from investing in 3 per cent. stock at $83\frac{3}{4}$, paying brokerage $\frac{1}{4}$ per cent.?

5. What sum must be invested in $4\frac{1}{2}$ per cent. stock at $102\frac{1}{4}$ to produce an income of £363, brokerage being $\frac{1}{4}$ per cent.?

6. Calculate the price of $2\frac{1}{2}$ per cent. Consols when £8670 can be bought for £8279. 17s. (brokerage $\frac{1}{8}$ per cent.).

7. A person invests £2852 in 5% stock at 115; he sells out at 125 and invests in 3% stock at 93; find the change in his income.

8. I invest £5187. 10s. in 3% stock at 83, but afterwards transfer three-fifths of the sum to the 4 per cents. at 96? How is my income affected?

9. What annual income is obtained by investing £1900 in $3\frac{1}{2}$% stock at 95? Find the change in income if the stock be sold at 90, and the proceeds re-invested in 3% stock at 81.

10. A man sells out £28500 $2\frac{1}{2}$% Consols at 92, and invests the proceeds partly in land which pays an annual rent of $1\frac{3}{4}$% on the purchase money, and partly in $4\frac{1}{2}$% debentures at 150, so as to derive the same income from each purchase. Find the change in his income.

11. Find the alteration in income occasioned by transferring £3200 from 3 per cent. stock at $86\frac{3}{8}$ to 4 per cent. stock at $114\frac{7}{8}$, the brokerage being $\frac{1}{8}$ per cent. in each case.

12. A certain amount of 3% stock at 84 is transferred to $3\frac{1}{2}$% stock without change of dividend. What is the price of the latter stock?

13. With 3% stock standing at 90 I sold out and bought railway shares at 180 which paid 10%, and my income was increased by £10; how much stock did I sell?

14. I derive an income of £126. 13s. 4d. from 3% stock; I sell out at 93 and buy Russian $3\frac{1}{2}$ per cents. at 95; what is the difference in my income?

15. My income from 3 per cents., after deducting 5d. in the pound income-tax, is £452. 7s. 6d. I sell out at $78\frac{5}{8}$ and buy 4% stock at $102\frac{3}{8}$; find to a penny the alteration in income, brokerage $\frac{1}{8}$ per cent. being charged for each transaction.

16. A person finds he can obtain £5 more per annum by investing in $3\frac{1}{2}$% stock at 96 than in 3% stock at 88; how much has he to invest?

17. A man sells out $2\frac{3}{4}$% Consols at $96\frac{1}{4}$, and by investing the proceeds in shares which pay an annual dividend of £4 per share raises his income 5 per cent. What was the price of each share?

18. If fifty £10 shares in a company paying a dividend of 8 % are sold for £18 each, and the proceeds invested in £5 shares in another company at £3. 10s. each, find what difference in income results if the second company pays a dividend of $3\frac{1}{2}$ %.

19. One company pays $5\frac{1}{2}$ % on shares of £100 each ; another pays at the rate of $3\frac{1}{2}$ % on shares of £10 each ; if the price of the former be £115. 10s., and of the latter £7. 15s., compare the rates of interest which the shares return to a purchaser.

*20. A person sells out £3965 from 3 per cent. stock at 74, and re-invests in $5\frac{1}{2}$ per cents. at 143. What is his gain or loss in annual income ?

*21. A man invests £3600 in 3 per cent. stock at 90. He sells out at 80 and lends $\frac{5}{8}$ths of his money at 4 per cent. and $\frac{3}{8}$ths at 5 per cent. How long must the loan last so that when he re-invests his money in 3 per cents. at 90 his gain on interest may exactly equal his loss upon principal ?

*22. If 3 per cent. stock is at $98\frac{1}{4}$, how much money must be invested in the stock to yield an income of £120 a year ? Find also to the nearest penny the annual income from the same sum of money invested in 4 per cent. stock at $127\frac{1}{8}$.

*23. What is the rate per cent. of the interest that a man gets on money invested in a 4 per cent. stock, the price of which is $119\frac{3}{8}$ (including brokerage)? What income would he get on £1500 so invested ?

If £100 of $2\frac{3}{4}$ per cent. consolidated stock cost $£97\frac{5}{8}$ of money, what quantity of the stock would cost £7497. 12s., and what annual income would be derived from it ?

*24. A man invests two equal sums of money ; the one bears interest at the rate of $2\frac{1}{2}$ per cent. per annum, the other at the rate of 3 per cent. per annum ; at the end of $5\frac{1}{2}$ years he has received £18. 7s. $1\frac{1}{2}d.$ more from the latter investment than from the former; find the sums of money invested.

*25. A man invests a fourth of his capital in $2\frac{1}{4}$ per cent. stock at 90, and the remainder in $3\frac{1}{2}$ per cent. stock at 105 ; find the average rate per cent. return on his capital.

*26. A man invested £2000 in a 4 per cent. guaranteed stock when the price of £100 of stock was $£97\frac{5}{8}$ in money; he sells out when the price of £100 has risen to £134, and invests the proceeds in a $2\frac{1}{2}$ per cent. stock, the price of £100 of which is £98 ; neglecting the commissions, find whether his income is increased or diminished, and to the nearest penny by how much.

Twenty-Ninth Week. *Miscellaneous Examples.*

Notes and Hints for Solution. The following examples illustrate some useful types of questions relating to Time, Work, and Velocity.

Example 1. A work can be done by A, B, and C working separately in 40, 60, and 120 days respectively. When A and B have been at work 4 days, B falls ill and his place is taken by C; if A leaves off ten days later, how much longer will C require to finish the work?

A, B, C respectively do $\frac{1}{40}$, $\frac{1}{60}$, $\frac{1}{120}$ of the work in 1 day;
$\therefore$ in 4 days A and B will do $\frac{4}{40} + \frac{4}{60}$, or $\frac{1}{6}$ of the work, leaving $\frac{5}{6}$ to be completed. In the next ten days A and C do $\frac{10}{40} + \frac{10}{120}$, or $\frac{1}{3}$ of the work, and thus leave $\frac{5}{6} - \frac{1}{3}$, or $\frac{1}{2}$ still to be done by C working alone. Hence the required number of days is $\frac{1}{2} \div \frac{1}{120} = 60$.

Example 2. If 4 men with 9 boys can do a piece of work in $1\frac{1}{5}$ day, and 3 men with 6 boys can do the same work in $1\frac{5}{7}$ day, how long would one boy or one man working alone take to do the work?

$$
\begin{cases}
4 \text{ men with } 9 \text{ boys do } \frac{5}{6} \text{ of the work in 1 day,} \\
3 \quad ,, \quad\quad 6 \quad ,, \quad\quad \frac{7}{12} \quad\quad ,, \quad\quad 1 \text{ ,,}
\end{cases}
$$

$$
\therefore \begin{cases}
12 \quad ,, \quad\quad 27 \quad ,, \quad\quad \frac{5}{2} \quad\quad ,, \quad\quad 1 \text{ ,,} \\
12 \quad ,, \quad\quad 24 \quad ,, \quad\quad \frac{7}{3} \quad\quad ,, \quad\quad 1 \text{ ,,}
\end{cases}
$$

$$
\therefore \; 3 \quad ,, \quad\quad \frac{5}{2} - \frac{7}{3}, \text{ or } \frac{1}{6} \quad ,, \quad\quad 1 \text{ ,,}
$$

$\therefore$ one boy would take 18 days working alone,
and it easily follows that one man would take 12 days.

Example 3. I row against a stream flowing $1\frac{1}{2}$ miles an hour to a certain point, and then turn back, stopping two miles short of the place whence I originally started. If the whole time occupied in rowing be 2 hrs. 10 mins., and my uniform speed in still water be $4\frac{1}{2}$ miles an hour, find how far upstream I went.

Upstream the rate of rowing is $4\frac{1}{2} - 1\frac{1}{2}$, or 3 miles per hour.
Downstream ,, ,, $4\frac{1}{2} + 1\frac{1}{2}$, or 6 ,, ,,
If I had returned to the starting point, the last 2 miles would have taken 20 minutes, and the whole time would have been $2\frac{1}{2}$ hours, two-thirds of which must have been occupied by the upstream journey, since the rates of rowing up and down are as 1 to 2.
Thus the distance upstream $= \frac{2}{3} \times \frac{5}{2} \times 3 = 5$ miles,

Examples XXIX.

1. If A can do a piece of work in 10 days, B in 15 days, and C in 30 days; how long will they take when they work together?

2. A and B together do a work in 15 days, A working alone could do it in 20 days; how long would B take?

3. Three persons, A, B, C, can complete a work in 10 days; if A takes 30 days and B 45 days to do the same work, how long will C take?

4. A cistern is filled in 3 hours when water flows in from two pipes; when one of these alone is open it takes 8 hours to fill the cistern, how long would be required for the other pipe alone?

5. A cistern has two pipes, one of which can fill it in 2 hrs., the other in 3 hrs.; a third pipe can empty it in 5 hrs.; if while the cistern is empty all these are opened, in what time will it be one-quarter filled?

6. A and B can do a piece of work in 12 days; after working 2 days they are assisted by C, who works at the same rate as A, and the work is finished in $6\frac{1}{4}$ days more; in how many days would B alone do the work?

7. A cistern contains 60 gallons; it has a tap which will fill it in 12 minutes, and a waste pipe which will empty it in 15 minutes. (1) The cistern being empty, both taps are turned on at once; how long will the cistern take to fill? (2) The supply tap is turned on for 5 minutes, then both for quarter of an hour, and then the waste-pipe is turned off; how long after this will it be before the cistern is filled?

8. In what time would a cistern be filled by three pipes whose diameters are $\frac{1}{2}$ in., $\frac{3}{4}$ in., and 1 in., running together, when the largest alone would fill it in 58 minutes; the amount of water flowing in by each pipe being proportional to the square of its diameter?

9. A, B, and C can do a piece of work together in 60 days; after they have worked together for 10 days A withdraws, and B and C work together at the same rate for 20 days more; B then withdraws, and C completes in 96 days more, working $\frac{1}{3}$ longer each day. Working at his former rate C alone could do the work in 222 days; find how long A and B would each take to do it separately.

10. A works for 6 days at the rate of 8 hours per day; B works for 5 hours on the first day, and on each of the subsequent days one hour longer than on the preceding day: A does as much in 4 hours as B does in 5 hours. If the total sum paid to A and B as wages for the week be £2. 2s., how much should each receive?

11. If 5 men or 7 women can do a piece of work in 37 days, in what time will 7 men and 5 women do a piece of work twice as great?

12. A piece of work can be done by 4 men in 6 days, or by 5 women in 8 days; 3 men and 3 women are employed; what is the total expense, if a man's daily wage is 2*s.* 8*d.*, and a woman's 1*s.* 8*d.*?

13. A piece of work can be done by 3 men and 4 boys in 6 days, by 3 men and 1 boy in 8 days, and by 4 women and 8 boys in 5 days. How long would a woman take to complete the work single-handed?

14. A farmer engaged a number of men and boys to reap corn; the men were to receive 4*s.* a day and the boys 2*s.* 6*d.*; if the work of 3 boys was equal to that of 2 men, and 4 men and 5 boys could together reap 22 acres in 4 days, what sum should 6 men and 7 boys receive for reaping 48 acres?

15. If a certain amount of work is done by 9 men, 12 women, and 13 boys in 11 days, how long will the same work take if 18 men, 3 women, and 5 boys are set to do it: assuming that the ratio of a man's work to a woman's is as 5 to 3, and a woman's work to a boy's as 4 to 3?

16. Two men and three boys can level and turf 352 yards of a cricket ground in 4 days, and three men and two boys can complete 276 yards in 3 days: compare the amount of work done by a man and a boy.

17. A man rode a bicycle from A to B, a distance of 54 miles, at an average rate of 8 miles an hour; another man started from A on horseback, half an hour after the bicyclist, and arrived at B 15 minutes before him; find the ratio of their speeds.

18. A and B start at the same time from London to Blisworth, A walking 4 miles an hour, B riding 9 miles an hour. B reaches Blisworth in 4 hours, and immediately rides back to London. After 3 hours' rest he starts again for Blisworth at the same rate. How far from London will he overtake A, who has in the meantime rested 7 hours?

19. At what distance from London will a train which leaves London for Rugby at 2.45 P.M., and goes at the rate of 41 miles an hour, meet a train which leaves Rugby for London at 1.45 P.M., and goes at the rate of 25 miles an hour, the distance between London and Rugby being 80 miles?

20. A and B start from the same point to run in opposite directions round a circular race-course 9755 feet in circumference, A not starting till B has run 105 feet. They pass each other when A has run 4850 feet. Which will first come round to the starting-point, and what distance will they then be apart?

86 ARITHMETIC.

Thirtieth Week. *Miscellaneous Examples.*

Notes and Hints for Solution. The following examples illustrate questions relating to Clocks, Chain Rule, Mixtures, and Races.

Example 1. At what time between 4 and 5 o'clock will the hands of a clock be (i) at right angles to each other, (ii) opposite to each other ?

(i) This will happen *twice*, namely when the minute-hand is 15 minute-spaces before and behind the hour-hand. Considering the former case alone : since at 4 o'clock the hour-hand is 20 minute-spaces ahead of the minute-hand, the latter must gain 35 minute-spaces.

Now the min.-hand passes over 60 minute-spaces while the hr.-hand passes over 5 ;

∴ the min.-hand gains 55 min.-spaces in every 60′,

that is   ,,   35   ,,   ,, $60' \times \frac{35}{55}$, or $38\frac{2}{11}'$.

(ii) This will happen when the min.-hand has gained $20+30$ min.-spaces on the hr.-hand.

That is, in $60' \times \frac{50}{55}$, or $54\frac{6}{11}'$.

Thus the required times are $38\frac{2}{11}'$ and $54\frac{6}{11}'$ past 4.

Example 2. If 48 lbs. of tea are worth 55 gals. of ale, and 63 gals. of ale 24 bottles of wine, and 11 bottles of wine 9 pairs of gloves, how many pounds of tea must be given for 20 pairs of gloves?

$$20 \text{ pairs of gloves} = 20 \times \tfrac{11}{9} \text{ bottles of wine}$$
$$= 20 \times \tfrac{11}{9} \times \tfrac{63}{24} \text{ gals. of ale}$$
$$= 20 \times \tfrac{11}{9} \times \tfrac{63}{24} \times \tfrac{48}{55} \text{ lbs. of tea.}$$

Thus the req. no. of lbs. of tea $= \dfrac{20 \times 11 \times 63 \times 48}{9 \times 24 \times 55} = 56.$

If we arrange the separate statements as follows, we have an example of **Chain Rule** :

Req. no. of lbs. of tea = 20 pairs of gloves,
9 pairs of gloves   = 11 bottles of wine,
24 bottles of wine   = 63 gals. of ale,
55 gals. of ale   = 48 lbs. of tea.

By multiplying together the numbers on each side and dividing the product on the right by the product on the left, we at once obtain the required result.

Example 3. In what proportion must coffee at 1*s*. 2*d*. per lb. be mixed with coffee at 2*s*. per lb. so that the mixture may be sold at 2*s*. 1*d*. per lb. at a profit of 25 per cent. ?

To sell *without gain or loss* the selling price per lb. must be

$$25d. \times \tfrac{100}{125}, \text{ or } 20d.$$

By selling at this price

On 1 lb. of 14*d*. coffee there is a gain of 6*d*.,

On 1 lb. of 24*d*. ,, loss of 4*d*. ;

∴ 2 lbs. of the former must be taken with 3 lbs. of the latter, so that the gain in the one case may balance the loss in the other.

Thus the required proportion is 2 : 3.

Examples XXX.

1. When are the hands of a clock together between the hours of 6 and 7 ?

2. At what times between 3 and 4 o'clock is the minute-hand of a watch (1) at right angles to the hour-hand, (2) one minute ahead of the hour-hand ?

3. When between 4 and 5 o'clock will there be 13 minutes between the two hands ?

4. A watch which gains 5″ in every 3′ was set right at 6 A.M. What was the true time in the afternoon of the same day when the watch indicated 3 hrs. 15′?

5. Two clocks are together at 12 o'clock ; one loses 7″ and the other gains 8″ in 12 hours ; when will one be half-an-hour before the other, and what o'clock will it then shew ?

6. A clock loses 5 seconds in every 24 minutes ; at 10 P.M. on Sunday it is 19 minutes fast : when will it shew the right time ?

7. Two clocks point to 2 o'clock at the same instant on the afternoon of Christmas day ; one loses 8 seconds and the other gains 7 seconds in 24 hours ; when will one be half-an-hour before the other, and what time will each clock then shew ?

8. If 5 fowls are worth 3 ducks, 14 ducks worth 5 geese, and 3 geese worth 2 turkeys, what is the price of a fowl when a turkey costs a guinea ?

9. If $1\tfrac{3}{5}$ yards of cloth are worth $\tfrac{5}{12}$ of a bushel of corn, and 12 yards of cloth cost $4\tfrac{1}{2}$ dollars, what is the value of 5 quarters of corn, a dollar being equal to 4*s*. 2*d*. ?

10. When 52 lbs. of coffee are worth as much as 12 lbs. of tea, 22 lbs. of tea are worth as much as 572 lbs. of sugar, a cask of sugar costs 2 guineas, and 1 cwt. of coffee costs 8 guineas, what is the weight of a cask of sugar ?

11. If £3=20 thalers, 25 thalers=93 francs, 27 francs=5 scudi, and 62 scudi=135 gulden, how many gulden=£1 ?

12. In what proportion must coffee at 1s. 6d. per lb. be mixed with coffee at 2s. per lb., so that the mixture may be sold without loss or profit at 1s. 8d. per lb. ?

13. Wine at 18s. a gallon is mixed with wine at 25s. a gallon ; in what proportion must the mixture be made so as to be worth 22s. a gallon ?

14. In what proportion must a merchant mix one kind of tea at 3s. per lb. with another at 1s. 6d. per lb., in order that by selling the mixture at 2s. 8d. per lb. he may make a profit of 25 per cent. ?

15. With a tea worth half-a-crown a pound a dealer mixes an inferior quality worth 1s. 6d. a pound. In what proportion must he mix them, so that by selling the mixture at the higher price he may gain 16 per cent. ?

16. A milkseller pays 1s. 1d. per gallon for his milk ; he adds water and sells the mixture at 2d. per pint, thereby making altogether 40 per cent. profit. Calculate the proportion of water to milk his customers receive ?

17. A can beat B by 5 yards in 100, and C can beat B by 14½ yards in the same distance ; by how much will C beat A in a mile race, the rates of running remaining uniform ?

18. A hare sees a hound 176 yards away from her, and scuds off in the opposite direction at a speed of 12 miles an hour ; thirty seconds later the hound perceives her, and gives chase at a speed of 18 miles an hour. How soon will he overtake the hare, and at what distance from the spot whence the hare took flight ?

19. A and B run a mile race ; A goes 5 feet at each step, and takes 3·3 steps per second all through the race ; B goes 6 feet at each step, and takes 3 steps per second for three-quarters of a mile, but in the last quarter he only goes 5·5 feet at each step, and takes 2·5 steps per second ; which won the race, and what time did each take?

*20. A and B together can do a piece of work in 5⅔ days, and A does twice as much work as B in a given time ; how long would it take A alone to do the work ?

*21. A, B, and C run a race for a mile. B has one minute start and C two minutes on A. A, B, C run respectively at the rate of 10, 8½, 7½ miles per hour. Who wins and who loses, and how long after the winner does the last man pass the winning post ?

*22. A, B, and C working together do a piece of work for £3. 7s. 6d. A, working alone, could do it in 10 days ; B, working alone, could do it in 12 days ; and C, working alone, could do it in 15 days. Divide the money between them in proportion to the quantity of work done by each.

ANSWERS TO ARITHMETIC.

I. Page 1. **1.** £113. 13s. 4d. **2.** £535. 2s. 11d.
3. £6338. 6s. 5d. **4.** £7. 9s. 1½d. **5.** 883⅓ ft.
6. 12 tons 1 cwt. 0 qr. 14 lbs. 13 oz. **7.** £28. 2s. 8½d.
8. 300,000. **9.** 397.
10. 15 ac. 1 r. 17 p. 2 sq. yds. 6 sq. ft. 108 sq. in. **11.** £248. 15s.
12. 10 pieces; rem., 5 ft. **13.** £232. 6s. 3½d. **14.** 8s. 2d.
15. £123. 4s. **16.** 4964. **17.** 510 lbs. **18.** 31,556,952.

II. Page 2. **1.** £430. 12s. 4½d.
2. 1 r. 23 p. 10 sq. yds. 3 sq. ft. 23 sq. in. **3.** £1. 11s. 9¾d.
4. 102. **5.** 30,000. **6.** 142,074. **7.** £8. 6s. 8d.
8. 96. **9.** 18s. 6d. **10.** 4666⅔ tons. **11.** 771.
12. 123¼ nearly; £3. 17s. 10½d. **13.** £3. 5s. **14.** £1. 14s. 2d.

III. Page 3. **1.** 9; 8; 9, 25; 11; 8, 9, 11; 9, 11, 25.
3, 3, 1, 4; 5, 5, 10, 13; 0, 3, 2, 12.
2. 93; 703. **4.** $7 \times 5 \times 11$. **5.** $3^3 \times 2 \times 13$. **6.** $5^2 \times 7 \times 9$.
7. $2 \times 5 \times 7^2 \times 3^2$. **8.** $5^3 \times 3^2 \times 17$. **9.** $5^2 \times 3^2 \times 11 \times 31$.
10. 37. **11.** 41. **12.** 493. **13.** 84. **14.** 243.
15. 504. **16.** 26. **17.** 221. **18.** 243. **19.** 1507.
20. 1512. **21.** 2310. **22.** 1176. **23.** 648. **24.** 7425.
25. 1820. **26.** 4290. **27.** 58,212. **28.** 8; 2520.
29. 17; 16,830. **30.** 23; 46,046. **31.** 144; 1,729,728.
33. £2. 12s. 6d. **34.** £7. 14s.

IV. Page 4. **2.** $\dfrac{29}{31}$. **3.** $\dfrac{5}{8}$. **4.** $\dfrac{2}{7}$.
5. $\dfrac{11}{57}$. **6.** $\dfrac{33}{64}$. **7.** $\dfrac{17}{27}$. **9.** $10\frac{37}{83}$. **10.** $12\frac{4}{5}$.
11. $25\frac{5}{72}$. **12.** $50\frac{3}{8}$. **13.** $\dfrac{133}{240}$. **14.** $\dfrac{308}{364}$. **15.** $\dfrac{11}{20}$.

16. 0.   **17.** $2\frac{5}{6}$.   **19.** 6.   **20.** 1.

22. (i) 1 ; (ii) $\frac{1}{5}$; (iii) $1\frac{2}{4}\frac{5}{7}$.   **23.** $\frac{61}{60}$.   **24.** $\frac{2}{7}$.

25. $\frac{167}{168}$.   **26.** $\frac{127}{241}$.   **27.** $\frac{65}{333}$.   **28.** $3\frac{1}{2}$.   **29.** $\frac{21}{22}$.

30. $\frac{6}{7}$.   **31.** 18.   **32.** $\frac{43}{19}$.   **33.** $\frac{13}{4}$.   **34.** $\frac{7}{37}$.

35. $\frac{22}{155}$.   **36.** $1\frac{1}{2}$.   **37.** 1.   **38.** $\frac{9}{49}$.   **39.** $\frac{3}{8}$.

40. $2\frac{4}{3\,5}$.   **41.** $\frac{29}{210}$.   **42.** $3\frac{4}{0}$.   **43.** $\frac{77}{144}$.   **44.** $\frac{5}{12}$.

V. Page 7.   **1.** £3. 17s. 6d.   **2.** 17s. 6d.
3. 7 fur. 32 p. 5 yds.   **4.** 3 tons 5 cwt. 10 lbs.   **5.** £20. 4s. 10d.
6. £1254.   **7.** (i) $\frac{7}{12}$; (ii) $\frac{5}{8}$.   **8.** 10 ac. 3 r. 31 p.

9. $\frac{9}{16}$.   **10.** $2\frac{2}{5}$.   **11.** $\frac{1}{480}$.   **12.** 13 ft. 9 in.   **13.** $\frac{3}{4}$.

14. £1917.   **15.** $\frac{3}{8}$.   **16.** 3 fur. 8 p. 3 yds. 2 ft. 8 in.

17. £3. 17s. 9d. ; $\frac{1}{3}$.   **18.** £1169 ; $\frac{3}{14}$.   **19.** 34 ac. 36 p.

20. £6. 3s. 4d.   **21.** 7 tons 13 cwt. 3 qrs. 14 lbs.   **22.** $\frac{8}{135}$.

23. £1. 5s. $8\frac{1}{2}d$.   **24.** 6s. 8d.   **25.** £153. 7s. 1d.
26. £4558. 13s. 9d.   **27.** 780 ; 468 ; 520 acres.   **28.** £4101.

29. £6400 ; £10,000.   **30.** $\frac{3}{4}$.   **31.** £603. 5s.

32. £2. 7s. 6d.   **33.** £70. 3s. $2\frac{1}{2}d$.   **34.** £336.

VI. Page 13.   **1.** 2·7 ; ·27 ; ·00027 ; ·207 ; 20·07.

2. $\frac{1}{25}$.   **3.** $\frac{7}{250}$.   **4.** $2\frac{3}{200}$.   **5.** $\frac{3}{800}$.   **6.** $\frac{61}{80}$.

7. $\frac{11}{8000}$.   **8.** $\frac{3}{32000}$.   **9.** $\frac{5}{64}$.   **10.** ·0075.   **11.** 2·625.

12. ·0058.   **13.** ·06125.   **14.** ·015625.   **15.** ·0015625.
16. ·848.   **17.** ·02734375.   **18.** 1·2811.   **19.** 100.
20. 18.   **21.** ·1107.   **22.** 11·962.   **23.** ·01.
24. 160·68.   **25.** 12·1.   **26.** ·004095.   **27.** ·06059.
28. 60·600606.   **29.** ·00133.   **30.** ·0323703.   **31.** 1·13204182.

32. 304·607209.   33. ·000176.   34. 1·02.   35. ·288.
36. 2500.   37. ·0462.   38. 40·065.   39. 12000.
40. 3025.   41. ·0028984375.   42. 1·23 ; ·055.
43. 817.   44. ·0032.   45. 4·5.   46. ·1.
47. 25.   48. 40.   49. 1250.   50. ·05.
51. 2·976.   52. 11·930.   53. 183·6587.   54. 1·0000.
57. 4·031.   58. 2·406.   59. (i) 1·4142 ; (ii) ·31831.
60. (i) ·3625 ; (ii) ·1336.   61. 6$\frac{1}{11}$.   62. 3·7725.
63. 2.   64. 80.   65. ·08.   66. ·0209.
67. 13·279.   68. ·0309.   69. ·0046.   70. 13·2608.

VII. Page 16. 2, 3, 5 terminate ; 4, 6, 7 recur.   8. ·5̇.
9. ·1̇0̇9̇.   10. ·0̇3̇.   11. ·02916̇.   12. ·235̇4̇.
13. ·3̇0̇1̇.   14. ·1̇785714̇2̇.   15. ·5̇05̇0̇.   16. 10·0100̇1̇.
17. ·1̇004273̇5̇.   18. (i) 10 ; (ii) ·35 ; (iii) 7·5 ; (iv) ·01.

20. $\dfrac{1}{99}$.   21. $\dfrac{59}{180}$.   22. $\dfrac{49}{396}$.   23. $\dfrac{22}{1665}$.   24. 1$\frac{5}{108}$.

25. $\dfrac{448}{925}$.   26. 55$\frac{1}{550}$.   27. $\dfrac{3}{7}$.   28. $\dfrac{2}{35}$.   29. 11.

30. 7.   31. 13.   32. 357·121473.   33. 151·362473.
34. ·47691̇.   35. ·9912940̇.   36. 336·08061̇1̇.
37 ·124661479̇.   38. ·00003025̇.   39. ·3025.
40. ·8201̇.   41. ·1̇42857̇.   42. 17·3̇7̇.   43. 13·68513̇.
44. 1·3.   45. ·4̇.   46. 2·036̇.   47. ·6̇.
48. ·29.   49. ·14.   50. ·069̇4̇.   51. ·890̇.
52. ·4063̇.   53. ·1023̇.   54. ·50201̇.   55. ·30̇.
56. 3·8601̇1̇.   57. 1·8626̇.   58. ·25̇.   59. ·35̇.
60. 76.   61. 1·644.   62. 1·1.   63. ·03125.   64. 396.
65. ·7̇.   66. 1.   67. 6.   68. 1.   69. 6.

VIII. Page 19.   1. 9d.   2. 40$\frac{1}{2}d$.   3. 209$\frac{1}{4}d$.   4. 160$\frac{1}{2}d$.
5. (i) 1967 lbs. ; (ii) 33 yds. ; (iii) 111 dwt.   6. 12s. 9d.
7. £3. 17s. 7$\frac{1}{2}d$.   8. £4. 7s. 10$\frac{1}{2}d$.   9. £15. 15s. 7$\frac{1}{2}d$.
10. 2s. 1$\frac{1}{2}d$.   11. £784. 1s. 3d.   12. ·89375.
13. 3·76875.   14. ·565625.   15. 4·953125.
16. (i) ·132375 ; (ii) 1·390025 ; (iii) 1·27421875.
17. 1 ac. 1 r. 36 p.   18. 5 cwt. 2 qrs. 14 lbs.
19. 21 hrs. 28 m. 48 sec.   20. 16 m. 3 fur. 16 p.
21. 2 tons 3 cwt. 7 lbs.   22. 79 ac. 3 r. 6 p.

23. (i) ·19375 ; (ii) ·83125 ; (iii) ·05 ; (iv) ·38671875.
24. £6. 16s.   25. £17. 8s.   26. £25. 15s.
27. £12. 15s. 6d.   28. £3. 5s. 1d.   29. 1 ton 15 cwt. 3 qrs.
30. 44 ac. 3 r. 1 p.   31. 8 bush. 2 pks. 1 gal. 3 qrts.
32. (i) £3. 13s. 6d. ; (ii) £3. 9s. 8d.
33. (i) £2. 10s. ; (ii) £10. 8s. 4d.   34. (i) 2·1875 ; (ii) ·37.
35. 12s. 10d.   36. £2. 3s. 8d.   37. £73. 10s. 10d.
38. £1. 14s. 2d.   39. £30. 10s. 2d.   40. £1. 18s. 4d.
41. £3. 3s. 6d.   42. £1. 5s. 10d.   43. £29. 3s. 7d.
44. 6 cub. ft. 216 cub. in.   45. 16 lbs.   46. £1. 7s. 6d.
47. (i) ·072916 ; (ii) 2·564583 ; (iii) 17·8364583.   48. ·8416.
49. ·0075.   50. 3·2.   51. ·00035.   52. ·0006.
53. ·270.   54. ·36.   55. ·22083.   56. ·021.
57. ·3698.   58. ·63.   59. £1. 5s.   60. 19s. 2d.
61. 7 cwt. 2 qrs.   62. £27. 13s. $4\frac{1}{2}d.$ 63. £5. 4s. 6d. ; ·25.
64. $5\frac{1}{2}d.$ 65. 1 lb.   66. £28. 3s. $7\frac{1}{4}d.$
67. 15s. $9\frac{3}{4}d.$ 68. £2. 7s. $4\frac{1}{2}d.$ 69. £17s. 11s. $7\frac{1}{4}d.$
70. £3. 6s. 6d.   71. £2. 19s. $6\frac{1}{4}d.$ 72. £1008. 18s. $11\frac{3}{4}d.$
73. £840. 10s. $7\frac{1}{4}d.$ 74. £3. 4s. $2\frac{1}{4}d.$ 75. £334. 18s. $7\frac{1}{2}d.$
76. £3. 0s. $7\frac{1}{4}d.$ 77. £27. 7s. $9\frac{1}{2}d.$ 78. £43. 14s. 6d.
79. 2 cwt. 1 qr. 25 lbs. 2 oz.   80. 3 r. 35 p. 18 sq. yds.
81. 1 fur. 28 p. 3 yds. 1 ft.   82. £1. 5s. 11d.
83. 67 yds. 1 ft. 1 in.   84. 72 lbs. 5 oz.
85. (i) ·025 ; (ii) ·571428.   86. 1587·31776 lbs.
87. (i) 9s. 7·92d. ; (ii) ·08781.
88. (i) £128. 8s. $11\frac{1}{2}d.$; (ii) 1·1132….   89. £4. 4s. 11d.

IX. Page 22.   1. ·70452.   2. 90·28649.   3. (i) 1 ; (ii) ·001.

4. $\dfrac{3}{22}.$ 5. $\dfrac{5}{13}.$ 6. 1.   7. ·28825.   8. $\dfrac{1}{2}.$ 9. 1.

10. ·075.   11. £42.   12. $6\frac{1}{4}d.$ 13. £1. 5s.   14. £3. 18s. 7d.

15. 960.   16. 3000.   17. $\dfrac{22565}{34598}$, or ·6537….

18. £179. 4s.   19. ·055.   20. 9·69 cub. in.   21. 1·2500.
22. ·7183.   23. ·111572.   24. ·648719.   25. $2\frac{3}{4}\frac{7}{0}.$
26. 8s. $5\frac{1}{4}d.$; $\dfrac{135}{856}.$ 27. 24.   28. 103·15 nearly.

X. Page 25.   1. £1254. 2s. 6d.   2. £10,137. 6s. 3d.
3. £14,469. 11s. 8d.   4. £2542. 0s. $0\frac{1}{2}d.$ 5. £620. 3s. $1\frac{1}{2}d.$

6. £840. 13s. 1½d.   7. £1832. 16s. 5½d.   8. £1261. 16s. 9d.
9. 1081 tons 19 cwt.   10. 7257 ac. 2 r. 16 p.
11. £4. 0s. 5¼d.   12. £1. 16s. 9¾d.   13. £150. 11s. 1½d.
14. £6389. 1s. 3d.   15. £315. 4s.   16. £484. 2s. 9d.
17. (i) £3072. 19s. 6½d.; (ii) £1209. 8s. 7¾d.   18. £183. 14s. 1¾d.
19. £1806. 13s. 6d.   20. £359. 14s. 8d.   21. £18,238. 9s. 6d.
22. (i) £663. 1s. 11d. ; (ii) £1144. 0s. 11¼d.
23. £578. 19s. 2d.   24. £2042. 12s. 10d.   25. £2734. 15s. 3¼d.
26. £123. 15s. 9¼d.   27. £127. 9s. 6¼d.   28. £158. 8s. 8½d.
29. £213. 10s. 1d.   30. £710. 14s. 1d.

XI. Page 28.   1. 35 : 52.   2. 3 : 8.   3. 4 : 15.
4. 2 : 7.   5. 1½d.   6. £92. 13s. 4d.   7. 51⅓ ft.
8. £24.   9. £19. 3s. 1d.   10. 13¾ cwt.   11. 4¹⁄₆ miles.
12. £34. 10s. 7½d.   13. 289.   14. 3 hours 45 min.
15. 324 hours.   16. £6. 8s. 4d.   17. £8718. 15s.
18. 20 tons.   19. 13 tons 16 cwt. 1 qr. 5 lbs.   20. 15s.
21. 13s. 2d.   22. £588.   23. £12. 5s. 3d.   24. Equal.
25. 9 : 20.   26. 19 : 2.   27. 142½ lbs.   28. 8 : 9.
29. £35, £14.   30. £21. 14s. 11d.   31. 35 days.
32. 27 : 88.   33. 30 : 91.   34. 5.   35. 70 : 99.
36. 285 : 184.   37. 2 : 3.   38. 8 : 21.   39. 11¼ days.

XII. Page 31.   1. £192.   2. 13.   3. 5.
4. 108½ tons.   5. 8.   6. 36.   7. 81.   8. 12.
9. 9.   10. 225.   11. 17 days.   12. 5¼ hours.
13. £103. 4s.   14. 3½ days.   15. 15s.   16. 13s. 6d.
17. 20.   18. 1000.   19. 6⁶⁄₇ days ; £1 ; 8s. 4d.
20. 11 days.   21. 3 : 2.   22. 8 days.   23. £4000.

XIII. Page 34.   1. £1176 ; £1960 ; £2744.
2. £13,600 ; £8500 ; £5100.   3. £1. 9s. 3d. ; £1. 0s. 3d. ; 0s. 9d.
4. £3. 7s. 6d. ; £5. 8s. ; £8. 2s.
5. £15 ; £20 ; £22. 10s. ; £24.
6. £7173. 6s. 8d. ; £8070 ; £8608 ; £8966. 13s. 4d.
7. £40 ; £24 ; £16.   8. £200 ; £220 ; £280.
9. £1275 ; £1575 ; £2250.   10. £720 ; £840 ; £1200 ; £1560.
11. £35. 15s. ; £41. 6s. 10d.; £59. 12s. 7d.   12. 18 p.; 20 p.; 22 p.
13. £412. 10s. ; £137. 10s. ; £825.
14. £28, 9s. 5¼d. ; £18. 19s. 7½d. ; £12. 13s. 1d.

15. 5*s.* ; 3*s.* 9*d.* ; 1*s.* 3*d.*   **16.** £42 ; £56 ; £48.

17. £25; £33. 6*s.* 8*d.*; £41. 13*s.* 4*d.*   **18.** £930; £46. 10*s.*; £581. 5*s.*

19. £242. 18*s.* 4*d.* ; £215. 6*s.* 3*d.* ; £331. 5*s.*

20. 15 tons 15 cwt.   **21.** $76\frac{1}{2}$; 68 ; $59\frac{1}{2}$.

22. £1. 2*s.* 6*d.* ; £1. 16*s.* ; £1. 4*s.*   **23.** £32. 5*s.*; £43 ; £15. 1*s.*

24. 11,400,000 ; 17,100,000 ; 3,800,000.   **25.** £10 ; £40 ; £20.

26. £1. 2*s.* ; £2. 15*s.* ; £3. 6*s.*   **27.** £505. 4*s.* $4\frac{1}{2}d.$; £221. 4*s.* $9\frac{1}{2}d.$

XIV. Page 38.

1. (i) 4 sq. yds. 8 sq. ft. ; (ii) 14 sq. yds. 3 sq. ft. 54 sq. in.

2. (i) 18 acres ; (ii) $12\frac{1}{2}$ acres ; (iii) $16\frac{1}{2}$ acres.

3. (i) 160 acres ; (ii) 47 ac. 1 r. 9 p.   **4.** £45. 2*s.* 6*d.*

5. £150.   **6.** £29. 5*s.*   **7.** £4. 13*s.* 4*d.*

8. (i) 68 yds. 2 ft. ; (ii) $21\frac{1}{4}$ chains ; (iii) 16 chains 6 links.

9. 20 ft. 6 in.   **10.** 15 chains 25 links.   **11.** 1029.

12. £36.   **13.** 240 ; £12. 10*s.*   **14.** $31\frac{1}{2}$ yds.

15. £8.   **16.** 51 ; £9. 16*s.* $6\frac{3}{4}d.$ **17.** 27 inches.

18. 11 feet.   **19.** 18 feet.   **20.** 232 sq. yds.

21. 1400 sq. yds. ; 444 sq. yds. 4 sq. ft.   **22.** 36·6 yds. ; 9*s.* $5\frac{1}{2}d.$

23. 26 yds. 2 ft.; 808.   **24.** 118 sq. ft.

25. 30 yds. ; $10\frac{3}{4}$ sq. yds. ; £7. 7*s.* 3*d.*   **26.** 179 sq. yds. 5 sq. ft.

27. 44 sq. yds. 5 sq. ft.   **28.** 10.   **29.** £2. 4*s.* 4*d.*

30. 3*s.* 6*d.*   **31.** 6*d.*   **32.** £6. 17*s.* $9\frac{3}{4}d.$ **33.** 10 ft. 6 in.

34. 24 ft. ; 18 ft. ; 12 ft.   **35.** 12 feet.   **36.** $12\frac{1}{2}$ feet.

37. 97 yds. 1 ft.   **38.** £1. 3*s.* ; 64 yds. ; £12.

XV. Page 42.   **1.** (i) 70 cu. ft.; (ii) 274 cu. ft. 1080 cu. in.

2. (i) $10\frac{3}{4}$ tons ; (ii) £12. 9*s.*   **3.** 400 gallons.

4. (i) 25 tons ; (ii) £13. 1*s.*   **5.** (i) £1. 6*s.* 3*d.* ; (ii) 1875 lbs.

6. 148 days.   **7.** 25,600 ; £44. 16*s.*   **8.** 65,120 ; £97. 13*s.* 7*d.*

9. £4. 13*s.* 4*d.*   **10.** £605.   **11.** 1275 lbs.   **12.** 42 lbs. 3 oz.

13. 4264 cu. in. ; 10*s.* $2\frac{3}{4}d.$ **14.** 376 lbs. $13\frac{1}{3}$ oz.

15. 305 lbs. ; 664 lbs.   **16.** 2 ft. 6 in.   **17.** 3 feet.

18. 4 feet.   **19.** $\frac{1}{4}$ inch ; 10 sq. yds.   **20.** 216,000 ; 750 yds.

21. 61,256,250 tons.   **22.** 101 tons ; $22,687\frac{1}{2}$ gallons.

23. 247,500.   **24.** 4 inches per hour.   **25.** 14,479,674.

26. 662,475 gallons.   **27.** 2·214 lbs.   **28.** £10. 16*s.*

29. 93 tons.   **30.** 19,800,000.   **31.** 4 ft. ; 6·07 ft.

XVI. Page 46.   **1.** 24.   **2.** 57.   **3.** 89.
4. 97.   **5.** 132.   **6.** 193.   **7.** 547.   **8.** 409.
9. 1003.   **10.** 4321.   **11.** 3702.   **12.** 34021.   **13.** $1\frac{3}{4}$.
14. $1\frac{1}{12}$.   **15.** $2\frac{11}{13}$.   **16.** $10\frac{1}{13}$.   **17.** $49\frac{5}{98}$.   **18.** $1\frac{5}{8}$.
19. 31·6.   **20.** 9·99.   **21.** 37·19.   **22.** 57·13.   **23.** 52·004.
24. ·01263.   **25.** 2·83.   **26.** 1·416.   **27.** 503·08.   **28.** 1·414.
29. 1·732.   **30.** 3·162.   **31.** 1·581.   **32.** ·265.   **33.** 1·183.
34. ·845.   **35.** ·853.   **36.** 2·057.   **37.** (i) 354 ; (ii) 1·5811.
38. 219 yds.   **39.** 86 yds.   **40.** 308 yds.   **41.** 18 feet.
42. 249 yds.   **43.** 1458 yds.   **44.** £41. 14s. 9d.   **45.** 137 feet.
46. 520 feet.   **47.** 1974.   **48.** 27300.   **49.** 12·079.
50. ·0371.   **51.** ·2907.   **52.** 69099.   **53.** 2·887.
54. ·379.   **55.** ·714.   **56.** ·2258.   **57.** ·7556.   **58.** 2·2804.
59. 1·3476; 1·3454; 1·3460; ·0010.   **60.** 98 yds. 1 ft. 2 in.
61. 177·39 yds.   **62.** 17·8885 feet.   **63.** 12 minutes (nearly).

XVII. Page 48.   **1.** 8.   **2.** 11.   **3.** 15.   **4.** 25.
5. 47.   **6.** 58.   **7.** 76.   **8.** 95.   **9.** 543.
10. 205.   **11.** 578.   **12.** ·9.   **13.** 1·7.   **14.** 3·7.
15. 8·5.   **16.** 40·1.   **17.** 2·57.   **18.** ·023.   **19.** ·005.
20. 141·2.   **21.** $5\frac{1}{2}$.   **22.** $2\frac{2}{7}$.   **23.** $5\frac{3}{4}$.   **24.** 1·44.
25. 1·71.   **26.** 1·91.   **27.** ·76.   **28.** 1·55.   **29.** ·66.
30. 2·63.   **31.** 5·96.   **32.** 2 ft. 6 in.
33. 1 ft. 8 in.; 16 sq. ft. 96 sq. in.   **34.** 1 ft. $11\frac{1}{2}$ in.
35. $759\frac{3}{8}$ sq. ft.   **36.** 60·5 inches.   **37.** 17.60 inches.
38. 17·15 inches.   **39.** 18·6.   **40.** 91·6.
41. (i) ·646 ; (ii) ·965 ; (iii) 2·502.   **43.** 14·6 inches.

XVIII. Page 52.
1. (i) 58·073 m.; (ii) 5807·3 cm.; (iii) ·058073 Km.
2. (i) 3400·902 m.; (ii) 340·0002 Dm.; (iii) 34009·02 dm.
3. (i) 78·01 m.; (ii) 78010 mm.; (iii) 7 Dm. 8 m. 1 cm.
4. (i) ·804602 ; (ii) 804·602 ; (iii) 8 Hm. 4 m. 6 dm. 2 mm.
5. (i) 1 Km.; (ii) 1 Km.
6. (i) 20 Km. 1 m.; (ii) 2 Km. 1 Hm. 8 dm. 4 cm.
7. (i) 5 Km. 8 m. 3 dm.; (ii) 4 Hm. 6 Dm. 9 m. 8 dm. 3 cm.
8. £660. 16s.   **9.** £502.   **10.** £45. 9s. 10d.   **11.** 45·45 grms.
12. (i) ·0000325 ; (ii) ·00005.   **13.** (i) ·000748 ; (ii) 7480.
14. (i) 69042·0489 ; (ii) 6·90420489 ; (iii) 18,709,350.

15. 40 centimes. 16. (i) 5·957 Ha.; (ii) 27·43140625 Ha.
17. 8250 fr. 18. (i) 24 cm.; (ii) 75 cm. 19. 28072.
20. 120 m. 21. (i) 700 litres; (ii) 6482·1 litres.
22. 10298·75. 23. 117·48. 24. 1242 tonneaux.
25. (i) 11·6235 Kgr.; (ii) £48. 8s. $7\frac{1}{2}d.$; (iii.) 2s. 7d. 26. 35·7 Kgr.

XIX. Page 55. 1. (i) 1174·73; (ii.) 2573·27.
2. (i) £33. 0s. 7d.; (ii) £53. 14s. $9\frac{1}{2}d.$ 3. 25·21 francs.
4. £5. 19s. 6d. 5. 1s. 3d. 8. ·13 inch. 9. 51 yds.
10. $161\frac{3}{4}$. 11. 10827 yds.; 12 Km. 69 m. 86 cm.
12. 9 yds. 1 ft. $8\frac{1}{2}$ in. (nearly). 13. 7022. 14. 4·56 ft. per sec.
15. 1s. (nearly). 16. 2·237. 17. 35·937.
18. 953 grammes. 19. 27·56. 20. £84.
21. 836·11. 22. 1·76. 23. 720.
24. 935·41. 25. ·033. 26. 160·9315. 27. 1226.

XX. Page 58. 1. $\frac{7}{200}$. 2. $\frac{2}{45}$. 3. $\frac{7}{125}$.
4. $\frac{1}{16}$. 5. $\frac{9}{160}$. 6. $\frac{9}{175}$. 7. 40%.
8. $33\frac{1}{3}$%. 9. $13\frac{1}{2}$%. 10. $7\frac{1}{2}$%. 11. $37\frac{1}{2}$%.
12. (i) £70; (ii) £69. 6s. 8d.; (iii) £71. 13s. 4d. 13. £240.
14. 6%. 15. £408. 16. £350. 14s. 17. 4s. 2d.
18. 600. 19. £8000. 20. $4\frac{7}{12}$%. 21. 15%.
23. £330. 4s. 6d. 24. $52\frac{1}{12}$%. 25. 47. 26. $83\frac{1}{3}$ acres.

XXI. Page 61. 1. $33\frac{1}{3}$%. 2. $4\frac{2}{3}$%. 3. 12% gain.
4. $26\frac{2}{3}$%. 5. 50%. 6. £21. 17s. 6d. 7. £40. 10s.
8. £16. 9. £52. 10s. 10. £90. 8s. 4d. 11. $24\frac{2}{5}$% loss.
12. £7. 8s. 9d. 13. 30%; £22. 10s. 14. 14s. 15. £428.
16. $62\frac{1}{2}$%. 17. $30\frac{5}{64}$%. 18. 7% loss. 19. 10%.
20. £40. 21. £75. 22. £8. 6s. 8d. 23. £40.
24. £875; $7\frac{11}{27}$%. 25. $18\frac{14}{27}$%. 26. $28\frac{4}{7}$%. 27. £2. 6s. 3d.
28. $40\frac{5}{8}$%. 29. £62. 10s. 30. $4\frac{16}{21}$%.

XXII. Page 63. 1. £2. 11s. $10\frac{1}{4}d.$
2. £1. 5s. $10\frac{1}{2}d.$ 3. £6. 9s. 8d. 4. £16. 16s. $6\frac{1}{4}d.$
5. £20. 13s. $10\frac{1}{2}d.$ 6. £8. 0s. 5d. 7. £27. 8s. 7d.
8. Author, £678. 13s. 4d.; Publisher, £168; Bookseller, £195.
9. £223. 1s. 10d. 10. £1439. 5s. 11. £76. 19s. 9d.

XXIII. Page 66.
3. £40. 10*s*.
6. £126. 11*s*. 3*d*.
9. £27. 9*s*. $2\frac{1}{4}d$.
12. £84. 17*s*. 8*d*.
15. £31. 7*s*. 10*d*.
18. £253. 11*s*. 11*d*.

1. £112. 10*s*.
4. £219. 7*s*. 6*d*.
7. £123. 10*s*. 3*d*.
10. £19. 6*s*. $7\frac{1}{4}d$.
13. £344. 15*s*. 4*d*.
16. £869. 8*s*. 5*d*.
19. £1406. 10*s*. 3*d*.

2. £405.
5. £271. 13*s*. 9*d*.
8. £683. 11*s*. $8\frac{1}{4}d$.
11. £295. 17*s*. 5*d*.
14. £148. 17*s*. 4*d*.
17. £222. 18*s*. 5*d*.

XXIV. Page 68.
4. $6\frac{1}{2}$ years.
8. 24 years.
12. $5\frac{3}{4}\%$.
16. 4 %.
19. £276. 7*s*. 1*d*.
22. £381. 15*s*. 5*d*.

1. 10 years.
5. $2\frac{3}{4}$ years.
9. $5\frac{1}{4}\%$.
13. $4\frac{1}{2}\%$.
17. 8 %.
20. £1525.
23. £252. 1*s*. 8*d*.

2. 9 years.
6. 219 days.
10. $3\frac{1}{3}\%$.
14. $4\frac{2}{3}\%$.
18. £427. 13*s*. $6\frac{1}{2}d$.
21. £1637. 1*s*. 8*d*.
24. £937. 10*s*.

3. $3\frac{3}{4}$ years.
7. 100 days.
11. $3\frac{1}{2}\%$.
15. $3\frac{1}{8}\%$.

XXV. Page 72.
3. £1157. 12*s*. 6*d*.
6. £4104. 18*s*. 9*d*.
9. £48. 9*s*. 6*d*.
12. £161. 5*s*.
15. £54. 14*s*. 3*d*.
18. £48. 4*s*. 11*d*.
21. £666. 13*s*. 4*d*.

1. £243. 7*s*. 2*d*.
4. £494. 5*s*. 5*d*.
7. £89. 9*s*. $10\frac{1}{2}d$.
10. £246. 7*s*. 8*d*.
13. £325. 5*s*. 4*d*.
16. £340. 7*s*. 8*d*.
19. £3125.
22. £3255. 4*s*. 2*d*.

2. £3245. 12*s*. 3*d*.
5. £426. 11*s*. 9*d*.
8. £390. 12*s*. 7*d*.
11. £461. 13*s*.
14. £239. 17*s*. 2*d*.
17. £53. 6*s*. 10*d*.
20. £240.
23. £4166. 13*s*. 4*d*.

XXVI. Page 74.
4. £275.
8. £13. 2*s*. 6*d*.
13. $3\frac{3}{4}\%$.
17. £551. 0*s*. 10*d*.
20. £1250.
22. £23. 6*s*. 9*d*.; £227. 12*s*.

1. £240.
5. £320.
9. £1. 15*s*.
14. £2015.

21. £987. 10*s*.; £987. 13*s*. 1*d*.
23. £3. 15*s*.; £78. 2*s*. 6*d*.

2. £175.
6. £162.
10. £3. 5*s*. 1*d*.
15. 5 %.
18. £12.

3. £1137. 10*s*.
7. £160.
11. £254. 1*s*. 2*d*.
16. £3.
19. £133. 6*s*. 8*d*.

XXVII. Page 77.
3. £2660.
7. £3911. 6*s*. 10*d*.
10. £14. 7*s*.
13. £45.
17. £12,705. 12*s*. 6*d*.; £438. 2*s*. 6*d*.
19. 88 : 91.
22. £4000; £4250; £12,000.

4. £600.
8. £4125. 12*s*. 6*d*.
11. £19. 7*s*.
14. £252.
20. The former; 12*s*. 6*d*.

1. £400.
5. £3136. 10*s*.
15. £1200.
18. The latter.
23. £525; £28.

2. £4000.
6. £2598. 5*s*.
9. £34. 2*s*.
12. £44.
16. £450.
21. £12,480.
24. £3250.

E.C.   G

XXVIII. Page 80. 1. £400. 2. £6000. 3. £1295; £42.
4. $3\frac{4}{7}$ %. 5. £8268. 6s. 8d. 6. $95\frac{3}{8}$. 7. £24 less.
8. £17. 3s. 9d.; gain. 9. £70; £3. 6s. 8d. 10. £132. 18s.; loss.
11. No alteration. 12. £98. 13. £500. 14. £18.
15. £9. 11s. 3d. 16. £2112. 17. £133. 6s. 8d.
18. £5. 19. 155 : 147. 20. £6. 2s.; loss.
21. $2\frac{6}{7}$ years. 22. £3930; £123. 13s. 2d.
23. $3\cdot35$ %; £50. 5s. 3d.; £7680; £211. 4s.
24. £667. 10s. 25. $3\frac{1}{8}$. 26. £11. 18s. 4d.

XXIX. Page 84. 1. 5 days. 2. 60 days. 3. $22\frac{1}{2}$ days.
4. $4\frac{4}{5}$ hours. 5. $\dfrac{15}{38}$ hours. 6. 30 days. 7. 1 hour; 4 min.
8. 32 min. 9. A, $261\frac{3}{17}$ days; B, 120 days.
10. A, 24 shillings; B, 18 shillings. 11. 35 days. 12. £3. 5s.
13. 45 days. 14. £12. 9s. 15. 11 days. 16. 5 : 4.
17. 8 : 9. 18. $28\frac{4}{5}$ miles. 19. $34\frac{1}{6}$ miles. 20. B; $4\frac{23}{48}$ feet.

XXX. Page 87. 1. $32\frac{8}{11}$ min. past 6.
2. (i) $32\frac{8}{11}$ min. past 3; (ii) $17\frac{5}{11}$ min. past 3.
3. At $7\frac{7}{11}$ min. past 4 and at 36 min. past 4. 4. 3 o'clock.
5. At 12 o'clock, 60 days after; 16 min. past 12.
6. 12 min. past 5 P.M. on Thursday.
7. On 24th April at 2 P.M.; 1.44 P.M.; 2.14 P.M.
8. 3 shillings. 9. £12. 10. $1\frac{1}{2}$ cwt. 11. 10 gulden.
12. 2 : 1. 13. 3 : 4. 14. 19 : 26. 15. 19 : 10.
16. 11 : 80. 17. 176 yards. 18. 2 min.; $\frac{1}{2}$ mile.
19. B, by 4 secs.; A, 320 secs.; B, 316 secs. 20. $8\frac{1}{2}$ days.
21. A and C win; B loses; $\dfrac{1}{17}$ min.
22. A, £1. 7s.; B, £1. 2s. 6d.; C, 18s.

ALGEBRA.

CHAPTER I.

DEFINITIONS. SUBSTITUTIONS.

1. ALGEBRA treats of quantities as in Arithmetic, but with greater generality ; for while the quantities used in arithmetical processes are denoted by *figures* which have a single definite value, algebraical quantities are denoted by *symbols* which may have any value we choose to assign to them. The symbols used are letters, and it is understood that throughout the same piece of work a symbol keeps the same value. Thus, when we say "let $a=1$," we do not mean that a must have the value 1 always, but only in the particular example we are considering.

We begin with the definitions of Algebra, premising that the signs $+$, $-$, $\times$, $\div$ will have the same meanings as in Arithmetic.

2. When two or more quantities are multiplied together the result is called the **product**. In Arithmetic the product of 2 and 3 is written 2×3, whereas in Algebra the product of a and b may be written in any of the forms $a \times b$, $a \cdot b$, or ab. The form ab is the most usual. Thus, if $a=2$, $b=3$, the product $ab = a \times b = 2 \times 3 = 6$; but in Arithmetic 23 means "twenty-three," or $2 \times 10 + 3$.

3. Each of the quantities multiplied together to form a product is called a **factor** of the product. Thus 5, a, b, are the factors of the product $5ab$.

4. When one of the factors of a product is a numerical quantity, it is called the **coefficient** of the remaining factors. Thus, in $5ab$, 5 is the coefficient. Sometimes any factor, or factors, of a product may be regarded as the coefficient of the remaining factors. Thus, in the product $6abc$, $6a$ may be called the coefficient of *bc*. A coefficient which involves letters is called a **literal coefficient**.

Note. When the coefficient is unity it is usually omitted. Thus we do not write $1a$, but simply a.

5. An **algebraical expression** is a collection of symbols; it may consist of one or more **terms**. Terms are separated from each other by the signs $+$ and $-$. Thus $7a+5b-3c-x+2y$ is an expression consisting of five terms.

Note. When no sign precedes a term the sign $+$ is understood.

6. Expressions are either **simple** or **compound**. A *simple expression* consists of *one* term, as $5a$. A *compound expression* consists of *two or more* terms. An expression of *two* terms, as $3a-2b$, is called a **binomial** expression; one of *three* terms, as $2a-3b+c$, a **trinomial**; one of *more than three* terms a **multinomial**. Simple expressions are also spoken of as **monomials**.

7. If a quantity be multiplied by itself any number of times, the product is called a **power** of that quantity, and is expressed by writing the number of factors to the right of the quantity and above it. Thus

$a \times a$ is called the *second power* of a, and is written a^2;

$a \times a \times a$ ....*third power* of a, ...a^3;

and so on.

The number which expresses the power of any quantity is called its **index** or **exponent**. Thus 2, 5, 7 are respectively the indices of a^2, a^5, a^7.

Note. a^2 is usually read "a squared"; a^3 is read "a cubed"; a^4 is read "a to the fourth"; and so on.

When the index is unity it is omitted, and we do not write a^1, but simply a. Thus a, $1a$, a^1, $1a^1$ all have the same meaning.

8. It is very important to distinguish between *coefficient* and *index*.

Example 1. If $b=5$, distinguish between $4b^2$ and $2b^4$.

Here $\quad 4b^2 = 4 \times b \times b = 4 \times 5 \times 5 = 100$;

whereas $\quad 2b^4 = 2 \times b \times b \times b \times b = 2 \times 5 \times 5 \times 5 \times 5 = 1250$.

Example 2. If $a=4$, $x=1$, find the value of $5x^a$.

Here $\quad 5x^a = 5 \times x \times x \times x \times x = 5 \times 1 \times 1 \times 1 \times 1 = 5$.

Note. Every power of 1 is 1.

9. In Arithmetic the factors of a product may be written in any order. Thus, for example,

$$3 \times 4 = 4 \times 3,$$

and $\qquad 3 \times 4 \times 5 = 4 \times 3 \times 5 = 4 \times 5 \times 3.$

In like manner in Algebra ab and ba have the same value. Again, the expressions abc, acb, bac, bca, cab, cba have the same value, each denoting the product of the three quantities a, b, c. It is immaterial in what order the factors of a product are written ; it is usual, however, to arrange them in alphabetical order.

Example. If $a=6$, $x=7$, $z=5$, find the value of $\dfrac{13}{10} axz$.

Here $\qquad \dfrac{13}{10} axz = \dfrac{13}{10} \times 6 \times 7 \times 5 = 273.$

EXAMPLES I. a.

If $a=7$, $b=2$, $c=1$, $x=5$, $y=3$, find the value of

1. $14x$.	**2.** x^3.	**3.** $3ax$.	**4.** a^3.	**5.** $5by$.
6. b^5.	**7.** $9b^4$.	**8.** $8bcy$.	**9.** $7y^3$.	**10.** $8x^2$.

If $a=8$, $b=5$, $c=4$, $x=1$, $y=3$, find the value of

11. $9xy$.	**12.** $8b^3$.	**13.** $3x^5$.	**14.** x^8.	**15.** $7y^4$.
16. c^x.	**17.** b^y.	**18.** y^c.	**19.** x^b.	**20.** y^b.

If $a=5$, $b=1$, $c=6$, $x=4$, find the value of

21. $\dfrac{3}{8} x^3$.	**22.** 3^x.	**23.** 8^b.	**24.** $\dfrac{7}{15} acx$.	**25.** $\dfrac{x^5}{64}$.

10. When powers of several different quantities are multiplied together a notation similar to that of Art. 7 is adopted. Thus $aabbbcddd$ is written $a^2 b^3 cd^3$. And conversely $7a^3 cd^2$ has the same meaning as $7 \times a \times a \times a \times c \times d \times d$.

Example 1. If $c=3$, $d=5$, find the value of $16c^4 d^3$.

Here $\quad 16c^4 d^3 = 16 \times 3^4 \times 5^3 = (16 \times 5^3) \times 3^4 = 2000 \times 81 = 162000.$

Example 2. If $p=4$, $q=9$, $r=6$, $s=5$, find the value of $\dfrac{32qr^3}{81p^4}$.

Here $\quad \dfrac{32qr^3}{81p^4} = \dfrac{32 \times 9 \times 6^3}{81 \times 4^5} = \dfrac{32 \times 9 \times 6 \times 6 \times 6}{81 \times 4 \times 4 \times 4 \times 4 \times 4} = \dfrac{3}{4}.$

11. If one factor of a product is equal to 0, the product must be equal to 0, *whatever values the other factors may have.* Thus, if $x=0$, then $ab^3 xy^2 = 0$ whatever be the values of a, b, y. Again if $c=0$, then $c^2 = 0$, $c^3 = 0$, *and every power of* 0 *is* 0.

EXAMPLES I. b.

If $a=7$, $b=2$, $c=0$, $x=5$, $y=3$, find the value of

1. $4ax^2$.　　2. a^3b.　　3. $8b^2y$.　　4. $3xy^2$.　　5. $\frac{3}{4}b^2x$.

6. $\frac{5}{6}b^3y^2$.　　7. $\frac{2}{5}xy^4$.　　8. a^3c.　　9. a^2cy.　　10. $8x^3y$.

If $a=2$, $b=3$, $c=1$, $p=0$, $q=4$, $r=6$, find the value of

11. $\frac{3a^2r}{8b}$.　　12. $\frac{8ab^2}{9q^2}$.　　13. $\frac{6a^3c}{b^2}$.　　14. $\frac{4cr^2}{9a^3}$.　　15. $3a^2b^c$.

16. $\frac{8b^q}{9a^r}$.　　17. $5a^bc^r$.　　18. $\frac{2a^2p}{7r}$.　　19. $\frac{5a^rb^q}{64r^a}$.　　20. $\frac{27a^q}{32}$.

12. In the case of expressions which contain more than one term, each term can be dealt with singly by the rules already given, and by combining the terms the numerical value of the whole expression is obtained. When brackets ( ) are used, they will have the same meaning as in Arithmetic, indicating that the terms enclosed within them are to be considered as one quantity.

Example 1.　When $c=5$, find the value of $c^4-4c+2c^3-3c^2$.

The expression $=5^4-4\times5+2\times5^3-3\times5^2$.

$$=625-20+250-75=780.$$

Example 2.　If $a=2$, $b=0$, $x=5$, $y=3$, find the value of

$$5a^3-ab^2+2x^2y+3bxy.$$

The expression $=(5\times2^3)-0+(2\times5^2\times3)+0$

$$=40+150=190.$$

Example 3.　When $a=5$, $b=3$, $c=1$, find the value of

$$\frac{(a-b)^2}{a+b}+\frac{(b-c)^2}{b+c}+\frac{(a-c)^2}{a+c}.$$

The expression $=\dfrac{(5-3)^2}{5+3}+\dfrac{(3-1)^2}{3+1}+\dfrac{(5-1)^2}{5+1}$

$$=\frac{2^2}{8}+\frac{2^2}{4}+\frac{4^2}{6}=\frac{4}{8}+\frac{4}{4}+\frac{16}{6}=\frac{1}{2}+1+2\frac{2}{3}=4\frac{1}{6}.$$

13. DEFINITION. The **square root** of any proposed expression is that quantity whose square, or second power, is equal to the given expression. Thus the square root of 81 is 9, because $9^2=81$.

The square root of a is denoted by $\sqrt[2]{a}$, or more simply $\sqrt{a}$.

Similarly the **cube, fourth, fifth**, etc., **root** of any expression is that quantity whose third, fourth, fifth, etc., power is equal to the given expression.

The roots are denoted by the signs $\sqrt[3]{}$, $\sqrt[4]{}$, $\sqrt[5]{}$, etc.

Examples. $\sqrt[3]{27}=3$; because $3^3=27$.

$\sqrt[5]{32}=2$; because $2^5=32$.

The sign $\sqrt{\ }$ is sometimes called the **radical sign.**

Example 1. Find the value of $5\sqrt{(6a^3b^4c)}$, when $a=3$, $b=1$, $c=8$.

$5\sqrt{(6a^3b^4c)}=5\times\sqrt{(6\times 3^3\times 1^4\times 8)}=5\times\sqrt{(6\times 27\times 8)}$

$\qquad =5\times\sqrt{1296}=5\times 36=180.$

Note. An expression like $\sqrt{(6a^3b^4c)}$ is sometimes written in the form $\sqrt{6a^3b^4c}$, the line above being used with the same meaning as the brackets to indicate the square root of the expression taken as a whole.

Example 2. Find the value of $\sqrt[3]{\left(\dfrac{ab^4}{8x^3}\right)}$, when $a=9$, $b=3$, $x=5$.

$$\sqrt[3]{\left(\frac{ab^4}{8x^3}\right)}=\sqrt[3]{\frac{9\times 3^4}{8\times 5^3}}=\sqrt[3]{\frac{9\times 81}{8\times 125}}=\sqrt[3]{\frac{9\times 9\times 9}{1000}}=\frac{9}{10}.$$

Example 3. When $p=9$, $r=6$, $k=4$, find the value of

$$\frac{1}{3}\sqrt[3]{\left(\frac{pr}{k^2}\right)}+\sqrt{(5p+3r+1)}-\frac{2r^2}{9k}.$$

$$\frac{1}{3}\sqrt[3]{\left(\frac{pr}{k^2}\right)}+\sqrt{(5p+3r+1)}-\frac{2r^2}{9k}=\frac{1}{3}\sqrt[3]{\frac{54}{16}}+\sqrt{45+18+1}-\frac{2\times 36}{9\times 4}$$

$$=\frac{1}{3}\sqrt[3]{\frac{27}{8}}+\sqrt{64}-2$$

$$=\frac{1}{3}\times\frac{3}{2}+8-2=6\tfrac{1}{2}.$$

EXAMPLES I. c.

If $a=4$, $b=1$, $c=3$, $f=5$, $g=7$, $h=0$, find the value of

1. $3f+5h-7b$.   2. $7c-9h+2a$.   3. $4g-5c-9b$.

4. $3a-9b+c$.   5. $2f-3g+5a$.   6. $3c-4a+7b$.

7. $7c+5b-4a+8h+3g$.   8. $9b+a-3g+4f+7h$.

9. $f^2-3a^2+2c^3$.   10. $b^3-2h^3+3a^2$.   11. $3b^2-2b^3+4h^2-2h^4$.

12. $\dfrac{(a-b)^2}{(c+h)^2}+\dfrac{2(b-h)}{3(g-f)}$.   13. $\dfrac{(a+b+g)^2}{(f+c-h)^2}-\dfrac{7(f-c)^3}{3(b+g)}$.

If $a=8$, $b=6$, $c=1$, $x=9$, $y=4$, find the value of

14. $\tfrac{5}{3}a-\tfrac{1}{9}b^3+\tfrac{7}{8}y^2$.   15. $\tfrac{5}{27}ax-\dfrac{32}{y^2}-\dfrac{6a}{cxy}$.

16. $\sqrt[3]{\left(\dfrac{6cy^4}{x^2}\right)}+2\sqrt{\left(\dfrac{3a^3}{4b^3}\right)}$.   17. $\sqrt[3]{(bxy)}-\tfrac{1}{8}b^2+\dfrac{8x^2}{b^2y}$.

18. $\tfrac{3}{4}ac-\sqrt{\dfrac{b^2}{9y}}-\sqrt{\dfrac{by}{x^2}}$.   19. $\dfrac{5b^2y^3}{12a^2x}-\sqrt[3]{\dfrac{x^4a}{b^2y^2}}+\sqrt{\dfrac{ab^3}{3x}}$.

CHAPTER II.

Negative Quantities. Addition.

14. Definition. Algebraical quantities which are preceded by the sign + are said to be **positive**; those to which the sign − is prefixed are said to be **negative**.

15. In Arithmetic the sum of the positive or **additive** terms is always greater than the sum of the negative or **subtractive** terms; if the reverse were the case the result would have no arithmetical meaning. In Algebra, however, not only may the sum of the negative terms exceed that of the positive, but a negative term may stand alone, and yet have a meaning quite intelligible.

This idea may be made clearer by one or two simple illustrations.

(i.) Suppose a man were to gain £100 and then lose £70, his total *gain* would be £30. But if he first gains £70 and then loses £100 the result of his trading is a *loss* of £30.

The corresponding algebraical statements would be

$$£100 - £70 = +£30,$$
$$£70 - £100 = -£30,$$

and the negative quantity in the second case is interpreted as a *debt*, that is, a sum of money *equal in value but opposite in character* to the positive quantity, or *gain*, in the first case.

(ii.) Again, suppose a man starting from a given point were to row 60 yards up a stream, and then drift down with the current for 40 yards, his distance from the starting point would be 20 yards *up* stream. But if he had rowed 40 yards up stream and then drifted down 60 yards, his distance from the starting point would be 20 yards *down* stream. Thus we see that −20 yards denotes a distance *equal in magnitude but opposite in direction* to that denoted by +20 yards.

(iii.) The freezing point on a Centigrade thermometer is marked zero, and a temperature of 15° C. means 15° *above* the freezing point, while a temperature 15° *below* the freezing point is indicated by −15° C.

16. DEFINITION. When terms do not differ, or when they differ only in their numerical coefficients, they are called **like**, otherwise they are called **unlike**. Thus $3a$, $7a$; $5a^2b$, $2a^2b$; $3a^3b^2$, $-4a^3b^2$ are pairs of like terms ; and $4a$, $3b$; $7a^2$, $9a^2b$ are pairs of unlike terms.

The rules for adding like terms are

Rule I.   *The sum of a number of like terms is a like term.*

Rule II.   *If all the terms are positive, add the coefficients.*

Example.   Find the value of $8a + 5a$.

Here we have to increase 8 things by 5 like things, and the aggregate is 13 of such things ;

for instance,   8 lbs. $+ 5$ lbs. $= 13$ lbs.

Hence also, $8a + 5a = 13a$.

Similarly, $8a + 5a + a + 2a + 6a = 22a$.

Rule III.   *If all the terms are negative, add the coefficients numerically and prefix the minus sign to the sum.*

Example.   To find the sum of $-3x$, $-5x$, $-7x$, $-x$.

Here the word *sum* indicates the aggregate of 4 subtractive quantities of like character.   In other words, we have to *take away* successively 3, 5, 7, 1 like things, and the result is the same as taking away $3 + 5 + 7 + 1$ such things in the aggregate.

Thus the sum of $-3x$, $-5x$, $-7x$, $-x$ is $-16x$.

Rule IV.   *If the terms are not all of the same sign, add together separately the coefficients of all the positive terms and the coefficients of all the negative terms ; the difference of these two results, preceded by the sign of the greater, will give the coefficient of the sum required.*

Example 1.   The sum of $17x$ and $-8x$ is $9x$, for the difference of 17 and 8 is 9, and the greater is positive.

Example 2.   To find the sum of $8a$, $-9a$, $-a$, $3a$, $4a$, $-11a$, a.

The sum of the coefficients of the positive terms is 16.

.... negative ....21.

The difference of these is 5, and the sign of the greater is negative ; hence the required sum is $-5a$.

We need not, however, adhere strictly to this rule, for the terms may be added or subtracted in the order we find most convenient.

This process is called **collecting terms.**

17. When quantities are connected by the signs + and −, the resulting expression is called their **algebraical sum.**

Thus $11a - 27a + 13a = -3a$ states that the algebraical sum of $11a$, $-27a$, $13a$ is equal to $-3a$.

Example. Find the algebraical sum of $\frac{2}{3}a$, $3a$, $-\frac{1}{6}a$, $-2a$,

The sum $= 3\frac{2}{3}a - 2\frac{1}{6}a = 1\frac{1}{2}a = \frac{3}{2}a$.

Note. The sum of two quantities numerically equal but with opposite signs is zero. Thus the sum of $5a$ and $-5a$ is 0.

EXAMPLES II. a.

Find the sum of

1. $2a$, $3a$, $6a$, a, $4a$.   2. $4x$, x, $5x$, $6x$, $8x$.
3. $2p$, p, $4p$, $7p$, $6p$, $12p$.   4. d, $9d$, $3d$, $7d$, $4d$, $6d$, $10d$.
5. $-2x$, $-6x$, $-10x$, $-8x$.   6. $-3b$, $-13b$, $-19b$, $-5b$.
7. $-21y$, $-5y$, $-3y$, $-18y$.   8. $-4m$, $-13m$, $-17m$, $-59m$.
9. $2xy$, $-4xy$, $-3xy$, xy, $7xy$.   10. $5pq$, $-8pq$, $8pq$, $-4pq$.
11. abc, $-3abc$, $2abc$, $-5abc$.   12. $-xyz$, $-2xyz$, $7xyz$, $-xyz$.

Find the value of

13. $-9a^2 + 11a^2 + 3a^2 - 4a^2$.   14. $3b^3 - 2b^3 + 7b^3 - 9b^3$.
15. $a^2b^2 - 7a^2b^2 + 8a^2b^2 + 9a^2b^2$. 16. $a^2x - 11a^2x + 3a^2x - 2a^2x$.
17. $9abcd - 11abcd - 4labcd$.   18. $13pqx - 5xpq - 19qpx$.
19. $\frac{1}{2}x - \frac{1}{3}x + x + \frac{2}{3}x$. 20. $\frac{3}{2}a + \frac{3}{5}a - \frac{1}{2}a$.
21. $-5b + \frac{1}{4}b - \frac{3}{2}b + 2b - \frac{1}{2}b + \frac{7}{4}b$.
22. $-\frac{5}{3}x^2 - 2x^2 - \frac{2}{3}x^2 + x^2 + \frac{1}{2}x^2 + \frac{11}{6}x^2$.

18. When a number of quantities are connected together by the signs + and −, the value of the result is the same in whatever order the terms are taken.

Thus $a - b + c$ is equivalent to $a + c - b$, for in the first of the two expressions b is taken from a, and c added to the result; in the second c is added to a, and b taken from the result. Similarly in the case of any expression we may write the terms in any order we please.

Thus it appears that the expression $a - b$ may be written in the equivalent form $-b + a$.

To illustrate this we may suppose, as in Art. 15, that a represents a gain of a pounds, and $-b$ a loss of b pounds : it is clearly immaterial whether the gain precedes the loss, or the loss precedes the gain.

19. The expression $8+(13+5)$ means that 13 and 5 are to be added and their sum added to 8.   It is clear that 13 and 5 may be added separately or together without altering the result.

Thus $\qquad 8+(13+5)=8+13+5=26.$

Similarly $a+(b+c)$ means that the sum of b and c is to be added to a.

Thus $\qquad a+(b+c)=a+b+c.$

$8+(13-5)$ means that to 8 we are to add the excess of 13 over 5 ; now if we add 13 to 8 we have added too much by 5, and must therefore take 5 from the result.

Thus $\qquad 8+(13-5)=8+13-5=16.$

Similarly $a+(b-c)$ means that to a we are to add b, diminished by c.

Thus $\qquad a+(b-c)=a+b-c.$

In like manner,

$$a+b-c+(d-e-f)=a+b-c+d-e-f.$$

By considering these results we are led to the following rule :

Rule. *When an expression within brackets is preceded by the sign $+$, the brackets can be removed without making any change in the expression.*

20. The expression $a-(b+c)$ means that from a we are to take the sum of b and c.   The result will be the same whether b and c are subtracted separately or in one sum.   Thus

$$a-(b+c)=a-b-c.$$

Again, $a-(b-c)$ means that from a we are to subtract the excess of b over c.   If from a we take b we get $a-b$; but by so doing we shall have taken away too much by c, and must therefore add c to $a-b$.   Thus

$$a-(b-c)=a-b+c.$$

In like manner,

$$a-b-(c-d-e)=a-b-c+d+e.$$

Accordingly the following rule may be enunciated :

Rule. *When an expression within brackets is preceded by the sign $-$, the brackets may be removed if the sign of every term within the brackets be changed.*

21. When two or more *like* terms are to be added together we have seen that they may be collected and the result expressed as a *single* like term. If, however, the terms are *unlike* they cannot be collected ; thus in finding the sum of two unlike quantities a and b, all that can be done is to connect them by the sign of addition and leave the result in the form $a+b$.

Also by the rules for removing brackets, $a+(-b)=a-b$; that is, the algebraic sum of a and $-b$ is written in the form $a-b$.

Example 1.   Find the sum of $3a-5b+2c$; $2a+3b-d$; $-4a+2b$.

The sum $=(3a-5b+2c)+(2a+3b-d)+(-4a+2b)$

$=3a-5b+2c+2a+3b-d-4a+2b$

$=3a+2a-4a-5b+3b+2b+2c-d$

$=a+2c-d,$

by collecting like terms.

The addition is, however, more conveniently effected by the following rule:

Rule. *Arrange the expressions in lines so that the like terms may be in the same vertical columns: then add each column beginning with that on the left.*

$3a-5b+2c$

$2a+3b\quad -d$

$-4a+2b$

$a\quad\quad +2c-d$

The algebraical sum of the terms in the first column is a, that of the terms in the second column is zero. The single terms in the third and fourth columns are brought down without change.

Example 2.   Add together $-5ab+6bc-7ac$; $8ab+3ac-2ad$; $-2ab+4ac+5ad$; $bc-3ab+4ad$.

$-5ab+6bc-7ac$

$8ab\quad\quad +3ac-2ad$

$-2ab\quad\quad +4ac+5ad$

$-3ab+\ bc\quad\quad +4ad$

$-2ab+7bc\quad\quad +7ad$

Here we first rearrange the expressions so that like terms are in the same vertical columns, and then add up each column separately.

EXAMPLES II. b.

Find the sum of

1. $3a+2b-5c$; $-4a+b-7c$; $4a-3b+6c$.

2. $3x+2y+6z$; $x-3y-3z$; $2x+y-3z$.

3. $8l-2m+5n$; $-6l+7m+4n$; $-l-4m-8n$.

4. $5a-7b+3c-4d$; $6b-5c+3d$; $b+2c-d$.

5. $7x-5y-7z$; $4x+y$; $5z$; $5x-3y+2z$.

6. $a-2b+7c+3$; $2b-3c+5$; $3c+2a$; $a-8-7c$.

7. $5-x-y$; $7+2x$; $3y-2z$; $-4+x-2y$.

8. $17ab-13kl-5xy$; $7xy$; $12kl-5ab$; $3xy-4kl-ab$.

9. $3ax+cz-4by$; $7by-8ax-cz$; $-3by+9ax$.

10. $-3ab+7cd-5qr$; $2ry+8qr-cd$; $2cd-3qr+ab-2ry$.

22. Each of the letters composing a term is called a **dimension** of the term, and the number of letters involved is called the **degree** of the term. Thus the product abc is said to be *of three dimensions*, or *of the third degree*; and ax^4 is said to be *of five dimensions*, or *of the fifth degree*. A numerical coefficient is not counted. Thus $8a^2b^5$ and a^2b^6 are each of *seven* dimensions.

The **degree of an expression** is the degree of the term of highest dimensions contained in it; thus $a^4 - 8a^3 + 3a - 5$ is *an expression of the fourth degree*, and 4 is said to be the *highest power* of a in the expression. Similarly $2a^2x + 3ax^3 - 7b^2x^3$ is *an expression of the fifth degree*.

A compound expression is said to be **homogeneous** when all its terms are of the same dimensions. Thus $8a^6 - a^4b^2 + 9ab^5$ is a *homogeneous expression of six dimensions*.

23. Different powers of the same letter are unlike terms; thus the result of adding together $2x^3$ and $3x^2$ cannot be expressed by a single term, but must be left in the form $2x^3 + 3x^2$.

Similarly the algebraical sum of $5a^2b^2$, $-3ab^3$, and $-b^4$ is $5a^2b^2 - 3ab^3 - b^4$. This expression is in its simplest form and cannot be abridged.

In adding together several algebraical expressions containing terms with different powers of the same letter, it will be found convenient to arrange all expressions in *descending* or *ascending* powers of that letter. This will be made clear by the following examples.

Example 1. Add together $3x^3 + 7 + 6x - 5x^2$; $2x^2 - 8 - 9x$; $4x - 2x^3 + 3x^2$; $3x^3 - 9x - x^2$; $x - x^2 - x^3 + 4$.

$3x^3 - 5x^2 + 6x + 7$
$\qquad 2x^2 - 9x - 8$
$-2x^3 + 3x^2 + 4x$
$3x^3 - \quad x^2 - 9x$
$- x^3 - \quad x^2 + \quad x + 4$
——————————————
$3x^3 - 2x^2 - 7x + 3$

In writing down the first expression we put in the first term the highest power of x, in the second term the next highest power, and so on till the last term, in which x does not appear. The other expressions are arranged in the same way, so that in each column we have *like powers of the same letter*.

Example 2. Add together $3ab^2 - 2b^3 + a^3$; $5a^2b - ab^2 - 3a^3$; $8a^3 + 5b^3$; $9a^2b - 2a^3 + ab^2$.

$-2b^3 + 3ab^2 \qquad\qquad + a^3$
$\quad - ab^2 + 5a^2b - 3a^3$
$5b^3 \qquad\qquad\qquad + 8a^3$
$\qquad\quad ab^2 + 9a^2b - 2a^3$
——————————————————
$3b^3 + 3ab^2 + 14a^2b + 4a^3$

Here each expression contains powers of two letters, and is arranged according to *descending* powers of b, and *ascending* powers of a.

EXAMPLES II. c.

Find the sum of the following expressions :

1. $x^2 + 3xy - 3y^2$; $-3x^2 + xy + 2y^2$; $2x^2 - 3xy + y^2$.

2. $2x^2 - 2x + 3$; $-2x^2 + 5x + 4$; $x^2 - 2x - 6$.

3. $5x^3 - x^2 + x - 1$; $2x^2 - 2x + 5$; $-5x^3 + 5x - 4$.

4. $a^3 - a^2b + 5ab^2 + b^3$; $-a^3 - 10ab^2 + b^3$; $2a^2b + 5ab^2 - b^3$.

5. $4m^3 + 2m^2 - 5m + 7$; $3m^3 + 6m^2 - 2$; $-5m^2 + 3m$; $2m - 6$.

6. $ax^3 - 4bx^2 + cx$; $3bx^2 - 2cx - d$; $bx^2 + 2d$; $2ax^3 + d$.

7. $py^2 - 9qy + 7r$; $-2py^2 + 3qy - 6r$: $7qy - 4r$; $3py^2$.

8. $2 - a + 8a^2 - a^3$; $2a^3 - 3a^2 + 2a - 2$; $-3a + 7a^3 - 5a^2$.

9. $1 + 2y - 3y^2 - 5y^3$; $-1 + 2y^2 - y$; $5y^3 + 3y^2 + 4$.

10. $a^2x^3 - 3a^3x^2 + x$; $5x + 7a^3x^2$; $4a^3x^2 - a^2x^3 - 5x$.

11. $3x^3 + 2y^2 - 5x + 2$; $7x^3 - 5y^2 + 7x - 5$; $9x^3 + 11 - 8x + 4y^2$;
     $6x - y^2 - 18x^3 - 7$.

12. $x^2 + 2xy + 3y^2$; $3z^2 + 2yz + y^2$; $x^2 + 3z^2 + 2xz$; $z^2 - 3xy - 3yz$;
     $xy + xz + yz - 6z^2 - 4y^2 - 2x^2$.

13. $-\frac{3}{4}x^3 + 5ax^2 - \frac{5}{8}a^2x$; $x^3 - \frac{37}{8}ax^2 + \frac{1}{2}a^2x$; $-\frac{1}{2}x^3 + \frac{3}{4}a^2x$.

14. $\frac{3}{8}x^2 - \frac{5}{3}xy - 7y^2$; $\frac{2}{3}xy + \frac{18}{5}y^2$; $-\frac{5}{8}x^2 + 4y^2$.

15. $\frac{1}{2}a^3 - 2a^2b - \frac{3}{2}b^3$; $\frac{3}{2}a^2b - \frac{3}{4}ab^2 + 2b^3$; $-\frac{3}{2}a^3 + ab^2 + \frac{1}{2}b^2$.

CHAPTER III.

SUBTRACTION.

24. THE simplest cases of Subtraction have already come under the head of addition of *like* terms, of which some are negative. [Art. 16.]

Thus
$$5a - 3a = 2a,$$
$$3a - 7a = -4a,$$
$$-3a - 6a = -9a,$$

Also by the rule for removing brackets [Art. 20],
$$3a - (-8a) = 3a + 8a = 11a,$$
and
$$-3a - (-8a) = -3a + 8a = 5a,$$

25. In dealing with expressions which contain unlike terms we may proceed as in the following examples.

Example 1.   Subtract $3a - 2b - c$ from $4a - 3b + 5c$.

The difference

$= 4a - 3b + 5c - (3a - 2b - c)$
$= 4a - 3b + 5c - 3a + 2b + c$
$= 4a - 3a - 3b + 2b + 5c + c$
$= a - b + 6c.$

The expression to be subtracted is first enclosed in brackets with a minus sign prefixed, then on removal of the brackets the like terms are combined by the rules already explained in Art. 16.

It is, however, more convenient to arrange the work as follows, the signs of all the terms in the lower line being changed.

$$4a - 3b + 5c$$
$$-3a + 2b + c$$

by *addition*, $\quad a - b + 6c$

The like terms are written in the same vertical column, and each column is treated separately.

Rule. *Change the sign of every term in the expression to be subtracted, and add to the other expression.*

Note. It is not necessary that in the expression to be subtracted the signs should be *actually* changed; the operation of changing signs ought to be performed mentally.

Example 2.   From $5x^2 + xy$ take $2x^2 + 8xy - 7y^2$.

$$5x^2 + xy$$
$$2x^2 + 8xy - 7y^2$$
$$3x^2 - 7xy + 7y^2$$

In the first column we combine mentally $5x^2$ and $-2x^2$, the algebraic sum of which is $3x^2$. In the last column the sign of the term $-7y^2$ has to be changed before it is put down in the result.

Terms containing different powers of the same letter being *unlike* must stand in different columns.

Example 3.   Subtract $3x^2 - 2x$ from $1 - x^3$.

$$\begin{array}{l} -x^3 \qquad\qquad +1 \\ \underline{\quad 3x^2 - 2x \quad} \\ -x^3 - 3x^2 + 2x + 1 \end{array}$$ In the first and last columns, as there is nothing to be subtracted, the terms are put down without change of sign. In the second and third columns each sign has to be changed.

The re-arrangement of terms in the first line is not *necessary*, but it is convenient, because it gives the result of subtraction in descending powers of x.

EXAMPLES III.

Subtract

1. $a + 2b - c$ from $2a + 3b + c$.   2. $2a - b + c$ from $3a - 5b - c$.

3. $x + 8y + 8z$ from $10x - 7y - 6z$.

4. $-m - 3n + p$ from $-2m + n - 3p$.

5. $3p - 2q + r$ from $4p - 7q + 3r$.   6. $3x - 5y - 7z$ from $2x + 3y - 4z$.

7. $-2x - 5y$ from $x + 3y - 2z$.   8. $m - 2n - p$ from $m + 2n$.

9. $3ab + 6cd - 3ac - 5bd$ from $3ab + 5cd - 4ac - 6bd$.

10. $-2pq - 3qr + 4rs$ from $qr - 4rs$.

11. $-mn + 11np - 3pm$ from $-11np$.

From

12. $x^3 - 3x^2 + x$ take $-x^3 + 3x^2 - x$.

13. $-2x^3 - x^2$ take $x^3 - x^2 - x$.   14. $a^3 + b^3 - 3abc$ take $b^3 - 2abc$.

15. $-8 + 6bc + b^2c^2$ take $4 - 3bc - 5b^2c^2$.

16. $p^3 + r^3 - 3pqr$ take $r^3 + q^3 + 3pqr$.

17. $1 - 3x^2$ take $x^3 - 3x^2 + 1$.   18. $x^3 + 11x^2 + 4$ take $8x^2 - 5x - 3$.

19. $a^3 + 5 - 2a^2$ take $8a^3 + 3a^2 - 7$.

20. $1 - 2x + 3x^2$ take $7x^3 - 4x^2 + 3x + 1$.

21. $1 - x + x^5 - x^4 - x^3$ take $x^4 - 1 + x - x^2$.

22. $-8mn^2 + 15m^2n + n^3$ take $m^3 - n^3 + 8mn^2 - 7m^2n$.

23. $\frac{3}{8}x^2 - \frac{2}{3}ax$ take $\frac{1}{3} - \frac{1}{4}x^2 - \frac{5}{6}ax$.

24. $\frac{3}{4}x^3 - \frac{1}{3}xy^2 - y^2$ take $\frac{1}{2}x^2y - \frac{5}{6}y^2 - \frac{1}{3}xy^2$.

25. $\frac{1}{8}a^3 - 2ax^2 - \frac{1}{3}a^2x$ take $\frac{1}{3}a^2x + \frac{1}{4}a^3 - \frac{3}{2}ax^2$.

CHAPTER IV.

MULTIPLICATION.

26. MULTIPLICATION in its primary sense signifies repeated addition.

Thus $\qquad 3 \times 4 = 3$ taken 4 times

$$= 3 + 3 + 3 + 3.$$

Here the multiplier contains 4 units, and the number of times we take 3 is the same as the number of units in 4.

Again $a \times b = a$ taken b times

$$= a + a + a + a + \dots, \text{ the number of terms being } b.$$

Also $3 \times 4 = 4 \times 3$; and so long as a and b denote positive whole numbers it is easy to shew that

$$a \times b = b \times a.$$

27. When the quantities to be multiplied together are not positive whole numbers, we may define multiplication as *an operation performed on one quantity which when performed on unity produces the other.* For example, to multiply $\frac{4}{5}$ by $\frac{3}{7}$, we perform on $\frac{4}{5}$ that operation which when performed on unity gives $\frac{3}{7}$; that is, we must divide $\frac{4}{5}$ into seven equal parts and take three of them. Now each part will be equal to $\dfrac{4}{5 \times 7}$, and the result of taking three of such parts is expressed by $\dfrac{4 \times 3}{5 \times 7}$.

Hence $\qquad \dfrac{4}{5} \times \dfrac{3}{7} = \dfrac{4 \times 3}{5 \times 7}.$

Also, by Art. 26, $\qquad \dfrac{4 \times 3}{5 \times 7} = \dfrac{3 \times 4}{7 \times 5} = \dfrac{3}{7} \times \dfrac{4}{5}.$

$$\therefore \quad \dfrac{4}{5} \times \dfrac{3}{7} = \dfrac{3}{7} \times \dfrac{4}{5}.$$

The reasoning is clearly general, and we may now say that $a \times b = b \times a$, where a and b are any *positive* quantities, *integral or fractional*.

In the same way it easily follows that

$$abc = a \times b \times c$$
$$= (a \times b) \times c = (b \times a) \times c = bac$$
$$= b \times (a \times c) = b \times c \times a = bca;$$

that is, *the factors of a product may be taken in any order.*

28.   Since, by definition, $a^3 = aaa$, and $a^5 = aaaaa$;

$\therefore$ $a^3 \times a^5 = aaa \times aaaaa = aaaaaaaa = a^8 = a^{3+5}$;

that is, the index of a in the product is the sum of the indices of a in the factors of the product.

Again, $5a^2 = 5aa$, and $7a^3 = 7aaa$;

$\therefore$ $5a^2 \times 7a^3 = 5 \times 7 \times aaaaa = 35a^5$.

When the expressions to be multiplied together contain powers of different letters, a similar method is used.

Example. $5a^3b^2 \times 8a^2bx^3 = 5aaabb \times 8aabxxx = 40a^5b^3x^3$.

Note.   The beginner must be careful to observe that in this process of multiplication *the indices of one letter cannot combine in any way with those of another.*   Thus the expression $40a^5b^3x^3$ admits of no further simplification.

29.   Rule.   *To multiply two simple expressions together, multiply the coefficients together and prefix their product to the product of the different letters, giving to each letter an index equal to the sum of the indices that letter has in the separate factors.*

The rule may be extended to cases where more than two expressions are to be multiplied together.

Example 1.   Find the product of x^2, x^3, and x^8.

The product $= x^2 \times x^3 \times x^8 = x^{2+3} \times x^8 = x^{2+3+8} = x^{13}$.

The product of three or more expressions is called **the continued product.**

Example 2.   Find the continued product of $5x^2y^3$, $8y^2z^5$, and $3xz^4$.

The product $= 5x^2y^3 \times 8y^2z^5 \times 3xz^4 = 120x^3y^5z^9$.

30.   By definition, $(a+b)m = (a+b)$ taken m times

$= (a+a+a+\ldots\text{taken } m \text{ times})$,

together with $(b+b+b+\ldots\text{taken } m \text{ times})$,

$= am + bm$.

Also $(a-b)m = (a-b)$ taken m times

$= (a+a+a+\ldots\text{taken } m \text{ times})$

diminished by $(b+b+b+\ldots\text{taken } m \text{ times})$,

$= am - bm$.

Similarly $(a-b+c)m = am - bm + cm$.

Thus it appears that *the product of a compound expression by a single factor is the algebraic sum of the partial products of each term of the compound expression by that factor.*

Examples. $3(2a+3b-4c) = 6a + 9b - 12c$.

$(4x^2 - 7y - 8z^3) \times 3xy^2 = 12x^3y^2 - 21xy^3 - 24xy^2z^3$.

EXAMPLES IV. a.

Find the value of

1. $5x \times 7$. 2. $3 \times 2b$. 3. $x^2 \times x^3$. 4. $5x \times 6x^2$.
5. $6c^3 \times 7c^4$. 6. $9y^2 \times 5y^5$. 7. $3m^3 \times 5m^5$. 8. $4a^6 \times 6a^4$.
9. $6ax \times 5ax$. 10. $3qr \times 4qr$. 11. $a^3x \times a^4x^3$. 12. $3x^3y^2 \times 4y^5$.
13. $a^2 \times a^3b \times 5ab^4$. 14. $6x^3y \times xy \times 9x^4y^2$.
15. $7a^2 \times 3b^3 \times 5c^4$. 16. $3abcd \times 5bca^2 \times 4cabd$.

Multiply

17. $ab - ac$ by a^2c. 18. $x^2y - x^3z + 4yz^5$ by x^3yz^2.
19. $5a^2 - 3b^2$ by $3ab^2c^4$. 20. $a^2b - 5ab + 6a$ by $3a^3b$.
21. $xy^2 - 3x^2z - 2$ by $3yz$. 22. $a^3 - 3a^2x$ by $2a^2bx$.

31. Since $\qquad (a-b)m = am - bm$, $\qquad\qquad$ [Art. 30.]
by putting $c - d$ in the place of m, we have
$$(a-b)(c-d) = a(c-d) - b(c-d)$$
$$= (c-d)a - (c-d)b$$
$$= (ac - ad) - (bc - bd)$$
$$= ac - ad - bc + bd.$$

If we consider each term on the right-hand side of this result, and the way in which it arises, we find that
$$(+a) \times (+c) = +ac.$$
$$(-b) \times (-d) = +bd.$$
$$(-b) \times (+c) = -bc.$$
$$(+a) \times (-d) = -ad.$$

These results enable us to state what is known as the **Rule of Signs** in multiplication.

Rule of Signs. *The product of two terms with like signs is positive; the product of two terms with unlike signs is negative.*

Example 1. Multiply $4a$ by $-3b$.
By the rule of signs the product is negative; also $4a \times 3b = 12ab$;
$$\therefore \quad 4a \times (-3b) = -12ab.$$

Example 2. Multiply $-5ab^3x$ by $-ab^3x$.
Here the absolute value of the product is $5a^2b^6x^2$, and by the rule of signs the product is positive ;
$$\therefore \quad (-5ab^3x) \times (-ab^3x) = 5a^2b^6x^2.$$

Example 3.   Find the continued product of $3a^2b$, $-2a^3b^2$, $-ab^4$.

$3a^2b \times (-2a^3b^2) = -6a^5b^3$;
$(-6a^5b^3) \times (-ab^4) = +6a^6b^7$.

This result, however, may be written down at once : for

$3a^2b \times 2a^3b^2 \times ab^4 = 6a^6b^7$,

Thus the complete product is $6a^6b^7$.

and by the rule of signs the required product is positive.

Example 4.   Multiply $6a^3 - \frac{5}{3}a^2b - \frac{4}{5}ab^2$ by $-\frac{3}{4}ab^2$.

The product is the algebraical sum of the partial products formed according to the rule enunciated in Art. 30 ; thus

$$(6a^3 - \tfrac{5}{3}a^2b - \tfrac{4}{5}ab^2) \times (-\tfrac{3}{4}ab^2) = -\tfrac{9}{2}a^4b^2 + \tfrac{5}{4}a^3b^3 + \tfrac{3}{5}a^2b^4.$$

EXAMPLES IV. b.

Multiply together

1. a, -2.   2. -3, $4x$.   3. $-x^2$, $-x^3$.   4. $-5m$, $3m^3$.
5. $-4q$, $3q^2$.   6. $-4y^3$, $-4y^3$.   7. $-3m^3$, $3m^3$.   8. $4x^4$, $-4x^4$.
9. $-3ab$, $-4ac$, $3bc$.   10. $-2a^3$, $-3a^2b$, -6.   11. $-2p$, $-3q$, $4s$, $-t$.

Find the product of

12. $2a - 3b + 4c$ and $-\frac{3}{2}a$.   13. $3x - 2y - 4$ and $-\frac{5}{6}x$.
14. $\frac{2}{3}a - \frac{1}{6}b - c$ and $-\frac{3}{8}ax$.   15. $\frac{6}{7}a^2x^2 - \frac{3}{2}ax^3$ and $-\frac{7}{3}a^3x$.
16. $-\frac{5}{3}a^2x^2$ and $-\frac{3}{5}a^2 + ax - \frac{2}{5}x^2$.   17. $-\frac{7}{2}xy$ and $3x^2 + \frac{2}{7}xy$.

32.   *To find the product of* a+b *and* c+d.

From Art. 30, $(a+b)m = am + bm$;

replacing m by $c+d$, we have

$$(a+b)(c+d) = a(c+d) + b(c+d)$$
$$= (c+d)a + (c+d)b$$
$$= ac + ad + bc + bd.$$

Similarly it may be shewn that

$$(a-b)(c+d) = ac + ad - bc - bd ;$$
$$(a+b)(c-d) = ac - ad + bc - bd ;$$
$$(a-b)(c-d) = ac - ad - bc + bd.$$

33.   When one or both of the expressions to be multiplied together contain more than two terms a similar method may be used.   For instance

$$(a-b+c)m = am - bm + cm ;$$

replacing m by $x - y$, we have

$$(a - b + c)(x - y) = a(x - y) - b(x - y) + c(x - y)$$
$$= (ax - ay) - (bx - by) + (cx - cy)$$
$$= ax - ay - bx + by + cx - cy.$$

34. The preceding results enable us to state the general rule for multiplying together any two compound expressions.

Rule. *Multiply each term of the first expression by each term of the second. When the terms multiplied together have like signs, prefix to the product the sign +, when unlike prefix − ; the algebraical sum of the partial products so formed gives the complete product.*

Example 1. Multiply $x + 8$ by $x + 7$.

The product
$$= (x + 8)(x + 7)$$
$$= x^2 + 8x + 7x + 56$$
$$= x^2 + 15x + 56.$$

The operation is more conveniently arranged as follows :

$$
\begin{array}{l}
x + 8 \\
x + 7 \\
\hline
x^2 + 8x \\
\quad + 7x + 56 \\
\hline
\end{array}
$$

by addition, $x^2 + 15x + 56.$

We begin on the left and work to the right, placing the second result one place to the right, so that like terms may stand in the same vertical column.

Example 2. Multiply $2x - 3y$ by $4x - 7y$.

$$
\begin{array}{l}
2x - 3y \\
4x - 7y \\
\hline
8x^2 - 12xy \\
\quad - 14xy + 21y^2 \\
\hline
\end{array}
$$

by addition, $8x^2 - 26xy + 21y^2.$

EXAMPLES IV. c.

Find the product of

1. $a + 7$, $a + 5$.	2. $x - 3$, $x + 4$.	3. $a - 6$, $a - 7$.
4. $y - 4$, $y + 4$.	5. $x + 9$, $x - 8$.	6. $c - 8$, $c + 8$.
7. $p - 10$, $p + 10$.	8. $d + 7$, $d + 7$.	9. $x - 4$, $-x + 4$.
10. $-y + 3$, $-y - 3$.	11. $-a + 4$, $-a + 5$.	12. $-y - 7$, $-y - 7$.
13. $2a - 5$, $3a + 2$.	14. $x - 7$, $2x + 5$.	15. $3x - 4$, $2x + 3$.
16. $x - 3a$, $2x + 3a$.	17. $3a - 2b$, $2a + 3b$.	18. $5c + 4d$, $5c - 4d$.
19. $3x - 5y$, $4x + y$.	20. $2y - 3z$, $2y + 3z$.	21. $xy + 2b$, $xy - 2b$.

35. We shall now give a few examples of greater difficulty.

Example 1.   Find the product of $3x^2 - 2x - 5$ and $2x - 5$.

$$3x^2 - \ 2x \ -5$$
$$2x \ - \ 5$$
$$\overline{6x^3 - \ 4x^2 - 10x}$$
$$\ \ -15x^2 + 10x + 25$$
$$\overline{6x^3 - 19x^2 \ \ \ \ \ \ + 25.}$$

Each term of the first expression is multiplied by $2x$, the first term of the second expression; then each term of the first expression is multiplied by -5; like terms are placed in the same columns and the results added.

Example 2.   Multiply $a - b + 3c$ by $a + 2b$.

$$a \ - \ \ b + 3c$$
$$a \ + \ 2b$$
$$\overline{a^2 - \ ab + 3ac}$$
$$\ \ \ \ \ 2ab \ \ \ \ \ \ \ - 2b^2 + 6bc$$
$$\overline{a^2 + \ ab + 3ac - 2b^2 + 6bc.}$$

36.   If the expressions are not arranged according to powers, ascending or descending, of some common letter, a re-arrangement will be found convenient.

Example.   Find the product of $2a^2 + 4b^2 - 3ab$ and $3ab - 5a^2 + 4b^2$.

$$2a^2 - \ 3ab \ + 4b^2$$
$$- \ 5a^2 + \ 3ab \ + 4b^2$$
$$\overline{-10a^4 + 15a^3b - 20a^2b^2}$$
$$\ \ \ \ + \ 6a^3b - \ 9a^2b^2 + 12ab^3$$
$$\ \ \ \ \ \ \ \ \ \ \ 8a^2b^2 - 12ab^3 + 16b^4$$
$$\overline{-10a^4 + 21a^3b - 21a^2b^2 \ \ \ \ \ \ + 16b^4.}$$

The re-arrangement is not *necessary*, but convenient, because it makes the collection of like terms more easy.

37.   When the coefficients are fractional we use the ordinary process of Multiplication, combining the fractional coefficients by the rules of Arithmetic.

Example.   Multiply $\frac{1}{3}a^2 - \frac{1}{2}ab + \frac{2}{3}b^2$ by $\frac{1}{2}a + \frac{1}{3}b$.

$$\tfrac{1}{3}a^2 - \ \tfrac{1}{2}ab \ + \tfrac{2}{3}b^2$$
$$\tfrac{1}{2}a \ + \ \tfrac{1}{3}b$$
$$\overline{\tfrac{1}{6}a^3 - \ \tfrac{1}{4}a^2b + \tfrac{1}{3}ab^2}$$
$$\ \ \ \ + \ \tfrac{1}{9}a^2b - \tfrac{1}{6}ab^2 + \tfrac{2}{9}b^3$$
$$\overline{\tfrac{1}{6}a^3 - \tfrac{5}{36}a^2b + \tfrac{1}{6}ab^2 + \tfrac{2}{9}b^3}$$

EXAMPLES IV. d.

Multiply

1. $x^2 - 3x - 2$ by $2x - 1$.   2. $4a^2 - a - 2$ by $2a + 3$.

3. $2y^2 - 3y + 1$ by $3y - 1$.   4. $3x^2 + 4x + 5$ by $4x - 5$.

5. $3x^2 - 2x + 7$ by $2x - 7$.   6. $5c^2 - 4c + 3$ by $-2c + 1$.

7. $2a^2 - 3a - 6$ by $a^2 - a + 2$.   8. $2k^2 - 3k - 1$ by $3k^2 - k - 1$.

9. $x^2 - xy + y^2$ by $x^2 + xy + y^2$.

10. $a^2 - 2ax + 2x^2$ by $a^2 + 2ax + 2x^2$.

11. $x^3 - 3x^2 - x$ by $x^2 - 3x + 1$.

12. $a^3 - 6a + 5$ by $a^3 + 6a - 5$.

13. $2y^4 - 4y^2 + 1$ by $2y^4 - 4y^2 - 1$.

14. $2x + 2x^3 - 3x^2$ by $3x + 2 + 2x^2$.

15. $a^3 + b^3 - a^2b^2$ by $a^2b^2 - a^3 + b^3$.

16. $a^3 + x^3 + 3ax^2 + 3a^2x$ by $a^3 + 3ax^2 - x^3 - 3a^2x$.

17. $a^4 + 1 + 6a^2 - 4a^3 - 4a$ by $a^3 - 1 + 3a - 3a^2$.

18. $x^4 + 6x^2y^2 + y^4 - 4x^3y - 4xy^3$ by $-x^4 - y^4 - 6x^2y^2 - 4xy^3 - 4x^3y$.

19. $\frac{2}{3}x^2 + xy + \frac{3}{2}y^2$ by $\frac{1}{3}x - \frac{1}{2}y$.

20. $\frac{3}{2}x^2 - ax - \frac{2}{3}a^2$ by $\frac{3}{4}x^2 - \frac{1}{2}ax + \frac{1}{3}a^2$.

21. $\frac{1}{2}x^2 - \frac{2}{3}x - \frac{3}{4}$ by $\frac{1}{2}x^2 + \frac{2}{3}x - \frac{3}{4}$.

22. $\frac{2}{3}ax + \frac{2}{3}x^2 + \frac{1}{3}a^2$ by $\frac{3}{4}a^2 + \frac{3}{2}x^2 - \frac{3}{2}ax$.

CHAPTER V.

DIVISION.

38. WHEN a quantity a is divided by the quantity b, the **quotient** is defined to be that which when multiplied by b produces a. This operation of division is denoted by $a \div b$, $\dfrac{a}{b}$, or a/b; in each of these modes of expression a is called the **dividend,** and b the **divisor.**

Division is thus the inverse of multiplication, and
$$(a \div b) \times b = a.$$

This statement may also be expressed verbally as follows :
$$quotient \times divisor = dividend,$$
or
$$\frac{dividend}{divisor} = quotient.$$

Example 1. Since the product of 4 and x is $4x$, it follows that when $4x$ is divided by x the quotient is 4,
or otherwise, $\qquad\qquad 4x \div x = 4.$

Example 2. Divide $27a^5$ by $9a^3$.

The quotient $= \dfrac{27a^5}{9a^3} = \dfrac{27aaaaa}{9aaa}$

$\qquad\qquad\quad = 3aa = 3a^2$

We remove from the divisor and dividend the factors common to both, just as in arithmetic.

Therefore $\qquad\qquad 27a^5 \div 9a^3 = 3a^2.$

Example 3. Divide $35a^3b^2c^3$ by $7ab^2c^2$.

The quotient $= \dfrac{35aaa \;.\; bb \;.\; ccc}{7a \;.\; bb \;.\; cc} = 5aa \;.\; c = 5a^2c.$

In each of these cases it should be noticed that *the index of any letter in the quotient is the difference of the indices of that letter in the dividend and divisor.*

39. It is easy to prove that *the rule of signs holds for division.*

Thus $\qquad ab \div a = \dfrac{ab}{a} = \dfrac{a \times b}{a} = b.$

$$-ab \div a = \frac{-ab}{a} = \frac{a \times (-b)}{a} = -b.$$

$$ab \div (-a) = \frac{ab}{-a} = \frac{(-a) \times (-b)}{-a} = -b.$$

$$-ab \div (-a) = \frac{-ab}{-a} = \frac{(-a) \times b}{-a} = b.$$

Hence in division as well as multiplication
like signs produce +, *unlike signs produce* −.

Rule. To divide one simple expression by another :

The index of each letter in the quotient is obtained by subtracting its index in the divisor from its index in the dividend.

To the result so obtained prefix with its proper sign the quotient of the coefficient of the dividend by that of the divisor.

Example 1.   Divide $84a^5x^3$ by $-12a^4x$.

The quotient $= (-7) \times a^{5-4}x^{3-1}$	Or at once mentally,
$\qquad = -7ax^2$	$84a^5x^3 \div (-12a^4x) = -7ax^2.$

Example 2. $-45a^6b^2x^4 \div (-9a^3bx^2) = 5a^3bx^2.$

Rule. *To divide a compound expression by a single factor, divide each term separately by that factor, and take the algebraic sum of the partial quotients so obtained.*

This follows at once from Art. 30.

Examples. $(9x - 12y + 3z) \div (-3) = -3x + 4y - z.$

$\qquad\qquad (36a^3b^2 - 24a^2b^5 - 20a^4b^2) \div 4a^2b = 9ab - 6b^4 - 5a^2b.$

$\qquad\qquad (2x^2 - 5xy + \tfrac{3}{2}x^2y^3) \div -\tfrac{1}{2}x = -4x + 10y - 3xy^3.$

EXAMPLES V. a.

Divide

1. $2x^3$ by x^2.
2. $6a^5$ by $3a$.
3. $5a^7$ by a^4.
4. $21b^7$ by $7b^3$.
5. $4p^2q^3$ by $-2pq$.
6. $-l^3m^2$ by $-lm$.
7. $-48x^9$ by $-6x^3$.
8. $35z^{11}$ by $-7z^5$.
9. $-7a^3b$ by $-7b$.
10. $-45a^4b^3c^{15}$ by $9a^2b^3c^{10}$.
11. $-x^3y^4z^5$ by $-x^3yz^5$.
12. $3x^2 - 2x$ by x.
13. $5a^3b - 7ab^3$ by ab.
14. $x^2 - xy - xz$ by $-x$.
15. $10a^3 - 5a^2b + a$ by $-a$.
16. $4x^3 + 36ax^2 - 16x$ by $-4x$.
17. $3a^3 - 9a^2b - 6ab^2$ by $-3a$.
18. $-3a^2 + \tfrac{9}{2}ab - 6ac$ by $-\tfrac{3}{2}a$.
19. $\tfrac{1}{2}x^5y^2 - 3x^3y^4$ by $-\tfrac{3}{2}x^3y^2$.
20. $-\tfrac{5}{2}x^2 + \tfrac{5}{3}xy + \tfrac{10}{3}x$ by $-\tfrac{5}{6}x$.
21. $-2a^5x^3 + \tfrac{7}{2}a^4x^4$ by $\tfrac{7}{3}a^3x$.

40. To divide one compound expression by another.

Rule. 1. *Arrange divisor and dividend in ascending or descending powers of some common letter.*

2. *Divide the term on the left of the dividend by the term on the left of the divisor, and put the result in the quotient.*

3. *Multiply the* WHOLE *divisor by this quotient, and put the product under the dividend.*

4. *Subtract and bring down from the dividend as many terms as may be necessary.*

Repeat these operations till all the terms from the dividend are brought down.

Example 1.   Divide $x^2 + 11x + 30$ by $x + 6$.

Arrange the work thus : $x + 6\,)\,x^2 + 11x + 30\,($

divide x^2, the first term of the dividend, by x, the first term of the divisor ; the quotient is x.   Multiply the *whole* divisor by x, and put the product $x^2 + 6x$ under the dividend.   We then have

$$x + 6\,)\,x^2 + 11x + 30\,(\,x$$
$$\underline{x^2 +\ 6x}$$

by subtraction, $5x + 30$

On repeating the process above explained, we find that the **next** term in the quotient is $+5$.

The entire operation is more compactly written as follows :

$$x + 6\,)\,x^2 + 11x + 30\,(\,x + 5$$
$$\underline{x^2 +\ 6x}$$
$$5x + 30$$
$$\underline{5x + 30}$$

Example 2.   Divide $16a^3 - 46a^2 + 39a - 9$ by $8a - 3$.

$$8a - 3\,)\,16a^3 - 46a^2 + 39a - 9\,(\,2a^2 - 5a + 3$$
$$\underline{16a^3 -\ 6a^2}$$
$$-40a^2 + 39a$$
$$\underline{-40a^2 + 15a}$$
$$24a - 9$$
$$\underline{24a - 9}$$

Thus the quotient is $2a^2 - 5a + 3$.

EXAMPLES V. b.

Divide

1. $a^2 + 2a + 1$ by $a + 1$.
2. $b^2 + 3b + 2$ by $b + 2$.
3. $x^2 + 5x - 6$ by $x - 1$.
4. $x^2 + 2x - 8$ by $x - 2$.
5. $m^2 + 7m - 78$ by $m - 6$.
6. $x^2 + ax - 30a^2$ by $x + 6a$.
7. $a^2 + 9ab - 36b^2$ by $a + 12b$.
8. $-x^2 + 18x - 45$ by $x - 15$.

9. $2x^2 - 13x - 24$ by $2x + 3$.   10. $12a^2 + ax - 6x^2$ by $3a - 2x$.

11. $-5x^2 + xy + 6y^2$ by $-x - y$.   12. $6a^2 - ac - 35c^2$ by $2a - 5c$.

13. $4m^2 - 49n^2$ by $2m + 7n$.   14. $-25x^2 + 49y^2$ by $-5x + 7y$.

15. $-2x^3 + 13x^2 - 17x + 10$ by $-x + 5$.

16. $6x^3y - x^2y^2 - 7xy^3 + 12y^4$ by $2x + 3y$.

41. The process of Art. 40 is applicable to cases in which the divisor consists of more than two terms.

Example 1.   Divide $a^4 - 2a^3 - 7a^2 + 8a + 12$ by $a^2 - a - 6$.

$$a^2 - a - 6 \,)\, a^4 - 2a^3 - 7a^2 + 8a + 12 \,(\, a^2 - a - 2$$
$$\underline{a^4 - \ a^3 - 6a^2}$$
$$-\ a^3 - \ a^2 + 8a$$
$$\underline{-\ a^3 + \ a^2 + 6a}$$
$$-2a^2 + 2a + 12$$
$$\underline{-2a^2 + 2a + 12}$$

Example 2.   Divide $4x^3 - 5x^2 + 6x^5 - 15 - x^4 - x$ by $3 + 2x^2 - x$.

First arrange each of the expressions in descending powers of x.

$$2x^2 - x + 3 \,)\, 6x^5 - \ x^4 + 4x^3 - \ 5x^2 - x - 15 \,(\, 3x^3 + x^2 - 2x - 5$$
$$\underline{6x^5 - 3x^4 + 9x^3}$$
$$2x^4 - 5x^3 - \ 5x^2$$
$$\underline{2x^4 - \ x^3 + \ 3x^2}$$
$$-4x^3 - \ 8x^2 - \ x$$
$$\underline{-4x^3 + \ 2x^2 - 6x}$$
$$-10x^2 + 5x - 15$$
$$\underline{-10x^2 + 5x - 15}$$

Example 3.   Divide $a^3 + b^3 + c^3 - 3abc$ by $a + b + c$.

$$a + b + c \,)\, a^3 - 3abc + \ b^3 + c^3 \,(\, a^2 - ab - ac + b^2 - bc + c^2$$
$$\underline{a^3 + \ a^2b + a^2c}$$
$$-\ a^2b - a^2c - 3abc$$
$$\underline{-\ a^2b - ab^2 - \ abc}$$
$$-\ a^2c + \ ab^2 - 2abc$$
$$\underline{-\ a^2c \qquad\quad - abc - ac^2}$$
$$ab^2 - \ abc + ac^2 + b^3$$
$$\underline{ab^2 \qquad\qquad + b^3 + b^2c}$$
$$-\ abc + ac^2 - b^2c$$
$$\underline{-\ abc \qquad - b^2c - bc^2}$$
$$ac^2 + bc^2 + \ c^3$$
$$\underline{ac^2 + bc^2 + \ c^3}$$

Note. Sometimes it will be found more convenient to arrange the expressions in *ascending* powers of a common letter.

42. When the coefficients are fractional the ordinary process may still be employed.

Example. Divide $\frac{1}{4}x^3 + \frac{1}{12}xy^2 + \frac{1}{12}y^3$ by $\frac{1}{2}x + \frac{1}{3}y$.

$$\frac{1}{2}x + \frac{1}{3}y \,\big)\, \frac{1}{4}x^3 + \frac{1}{12}xy^2 + \frac{1}{12}y^3 \,\big(\, \frac{1}{2}x^2 - \frac{1}{3}xy + \frac{1}{4}y^2$$

$$\underline{\frac{1}{4}x^3 + \frac{1}{6}\,x^2y}$$

$$-\frac{1}{6}\,x^2y + \frac{1}{12}xy^2$$

$$\underline{-\frac{1}{6}\,x^2y - \frac{1}{9}\,xy^2}$$

$$\frac{1}{8}\,xy^2 + \frac{1}{12}y^3$$

$$\underline{\frac{1}{8}\,xy^2 + \frac{1}{12}y^3}$$

In the examples given hitherto the divisor has been exactly contained in the dividend. When the division is not exact the work should be carried on until the remainder is of lower dimensions [Art. 22] than the divisor.

EXAMPLES V. c.

Divide

1. $a^3 - 6a^2 + 11a - 6$ by $a^2 - 4a + 3$.
2. $y^3 + y^2 - 9y + 12$ by $y^2 - 3y + 3$.
3. $6a^3 - 5a^2 - 9a - 2$ by $2a^2 - 3a - 1$.
4. $12x^3 - 8ax^2 - 27a^2x + 18a^3$ by $6x^2 - 13ax + 6a^2$.
5. $16x^3 + 14x^2y - 129xy^2 - 15y^3$ by $8x^2 + 27xy + 3y^2$.
6. $3x^4 - 10x^3 + 12x^2 - 11x + 6$ by $3x^2 - x + 3$.
7. $x^3 - x^2 - 8x - 13$ by $x^2 + 3x + 3$.
8. $21m^3 - 27m - 26m^2 + 20$ by $3m + 7m^2 - 4$.
9. $3y^4 - 4y^3 + 10y^2 + 3y - 2$ by $y^3 - y^2 + 3y + 2$.
10. $28x^4 + 69x + 2 - 71x^3 - 35x^2$ by $4x^2 + 6 - 13x$.
11. $x^3 - 8a^3$ by $x^2 + 2ax + 4a^2$.   12. $y^4 + 9y^2 + 81$ by $y^2 - 3y + 9$.
13. $x^4 + 4y^4$ by $x^2 + 2xy + 2y^2$.   14. $9a^4 - 4a^2 + 4$ by $3a^2 - 4a + 2$.
15. $a^8 + 64$ by $a^4 - 4a^2 + 8$.
16. $4m^5 - 29m - 36 + 8m^2 - 7m^3 + 6m^4$ by $m^3 - 2m^2 + 3m - 4$.
17. $15x^4 + 22 - 32x^3 - 30x + 50x^2$ by $3 - 4x + 5x^2$.
18. $a^3b^3 + ab - 9 - b^4 + 3b^3 + 3b - a^4 - 3a^3 - 3a$ by $3 - b + a^3$.
19. $x^8 + 1$ by $x^3 + x^2 + x + 1$.   20. $2a^6 + 2$ by $a^3 + 2a^2 + 2a + 1$.
21. $a^3 + 3a^2b + b^3 - 1 + 3ab^2$ by $a + b - 1$.
22. $1 - a^3 - 8x^3 - 6ax$ by $1 - a - 2x$.
23. $\frac{1}{8}a^3 - \frac{9}{4}a^2x + \frac{27}{2}ax^2 - 27x^3$ by $\frac{1}{2}a - 3x$.
24. $36x^2 + \frac{1}{9}y^2 + \frac{1}{4} - 4xy - 6x + \frac{1}{3}y$ by $6x - \frac{1}{3}y - \frac{1}{2}$.
25. $\frac{8}{27}a^5 - \frac{243}{512}ax^4$ by $\frac{2}{3}a - \frac{3}{4}x$.

43. The following examples in division may be easily verified; they are of great importance and should be carefully noticed.

$$\text{I.}\begin{cases}\dfrac{x^2-y^2}{x-y}=x+y,\\[2ex]\dfrac{x^3-y^3}{x-y}=x^2+xy+y^2,\\[2ex]\dfrac{x^4-y^4}{x-y}=x^3+x^2y+xy^2+y^3.\end{cases}$$

and so on; the divisor being $x-y$, the terms in the quotient *all positive*, and the index in the dividend *either odd or even.*

$$\text{II.}\begin{cases}\dfrac{x^3+y^3}{x+y}=x^2-xy+y^2,\\[2ex]\dfrac{x^5+y^5}{x+y}=x^4-x^3y+x^2y^2-xy^3+y^4,\\[2ex]\dfrac{x^7+y^7}{x+y}=x^6-x^5y+x^4y^2-x^3y^3+x^2y^4-xy^5+y^6,\end{cases}$$

and so on; the divisor being $x+y$, the terms in the quotient *alternately positive and negative,* and the index in the dividend *always odd.*

$$\text{III.}\begin{cases}\dfrac{x^2-y^2}{x+y}=x-y,\\[2ex]\dfrac{x^4-y^4}{x+y}=x^3-x^2y+xy^2-y^3,\\[2ex]\dfrac{x^6-y^6}{x+y}=x^5-x^4y+x^3y^2-x^2y^3+xy^4-y^5,\end{cases}$$

and so on; the divisor being $x+y$, the terms in the quotient *alternately positive and negative,* and the index in the dividend *always even.*

IV. The expressions x^2+y^2, x^4+y^4, x^6+y^6, ... (where the index is *even*, and the terms *both positive*) are *never* exactly divisible by $x+y$ or by $x-y$.

All these different cases may be more concisely stated as follows:

(1) x^n-y^n is divisible by $x-y$ if n be *any* whole number.
(2) x^n+y^n is divisible by $x+y$ if n be any *odd* whole number.
(3) x^n-y^n is divisible by $x+y$ if n be any *even* whole number.
(4) x^n+y^n is never divisible by $x+y$ or by $x-y$, when n is an *even* whole number.

CHAPTER VI.

Removal and Insertion of Brackets.

44. Quantities are sometimes enclosed within brackets to indicate that they must all be operated upon in the same way. Thus in the expression $2a - 3b - (4a - 2b)$ the brackets indicate that the expression $4a - 2b$ *treated as a whole* has to be subtracted from $2a - 3b$.

In removing brackets we apply the rules given in Art. 20.

Example. Simplify, by removing brackets, the expression
$$(2a - 3b) - (3a + 4b) - (b - 2a).$$
The expression $= 2a - 3b - 3a - 4b - b + 2a$
$$= a - 8b, \quad \text{by collecting like terms.}$$

Sometimes it is convenient to enclose within brackets part of an expression already enclosed within brackets. For this purpose it is usual to employ brackets of different forms. The brackets in common use are $(\ )$, $\{\ \}$, $[\ ]$. When there are two or more pairs of brackets to be removed, it is generally best to begin with the innermost pair. In dealing with each pair in succession we apply the rules quoted above.

Example. Simplify, by removing brackets, the expression
$$a - 2b - [4a - 6b - \{3a - c + (2a - 4b + c)\}].$$

Removing the brackets one by one,

the expression $= a - 2b - [4a - 6b - \{3a - c + 2a - 4b + c\}]$
$$= a - 2b - [4a - 6b - 3a + c - 2a + 4b - c]$$
$$= a - 2b - 4a + 6b + 3a - c + 2a - 4b + c$$
$$= 2a, \quad \text{by collecting like terms.}$$

EXAMPLES VI. a.

Simplify by removing brackets and collecting like terms :

1. $(x - 3y) + (2x - 4y) - (x - 8y).$ 2. $(x - 3y + 2z) - (z - 4y + 2x).$
3. $2a + (b - 3a) - (4a - 8b) - (6b - 5a).$
4. $m - (n - p) - (2m - 2p + 3n) - (n \quad m + 2p).$
5. $5x - (7y + 3x) - (2y + 7x) - (3x + 8y).$

6. $(m^2 - 2n^2) - (2n^2 - 3m^2) - (3m^2 - 4n^2)$.

7. $(a + 3b) - (b - 3a) - \{a + 2b - (2a - b)\}$.

8. $p^2 - 2q^2 - (q^2 + 2p^2) - \{p^2 + 3q^2 - (2p^2 - q^2)\}$.

9. $x - [y + \{x - (y - x)\}]$.   **10.** $(a - b) - \{a - b - (a + b) - (a - b)\}$.

11. $3x - y - [x - (2y - z) - \{2x - (y - z)\}]$.

12. $[3a - \{2a - (a - b)\}] - [4a - \{3a - (2a - b)\}]$.

45. A coefficient placed before any bracket indicates that every term of the expression within the bracket is to be multiplied by that coefficient; but when there are two or more brackets to be considered, a prefixed coefficient must be used as a multiplier only when its own bracket is being removed.

Example 1. $2x + 3(x - 4) = 2x + 3x - 12 = 5x - 12$.

Example 2. $7x - 2(x - 4) = 7x - 2x + 8 = 5x + 8$.

Example 3.   Simplify $5a - 4[10a + 3\{x - a - 2(a + x)\}]$.

The expression

$= 5a - 4[10a + 3\{x - a - 2a - 2x\}]$

$= 5a - 4[10a + 3\{ - x - 3a\}]$

$= 5a - 4[10a - 3x - 9a]$

$= 5a - 4[a - 3x]$

$= 5a - 4a + 12x$

$= a + 12x$.

On removing the innermost bracket each term is multiplied by -2. Then before multiplying by 3, the expression within its bracket is simplified. The other steps will be easily seen.

46. Sometimes a line called a **vinculum** is drawn over the symbols to be connected; thus $a - \overline{b + c}$ is used with the same meaning as $a - (b + c)$, and hence $a - \overline{b + c} = a - b - c$.

Example 4.   Find the value of

$$84 - 7[- 11x - 4\{ - 17x + 3(8 - \overline{9 - 5x})\}].$$

The expression $= 84 - 7[- 11x - 4\{ - 17x + 3(8 - 9 + 5x)\}]$

$= 84 - 7[- 11x - 4\{ - 17x + 3(5x - 1)\}]$

$= 84 - 7[- 11x - 4\{ - 17x + 15x - 3\}]$

$= 84 - 7[- 11x - 4\{ - 2x - 3\}]$

$= 84 - 7[- 11x + 8x + 12]$

$= 84 - 7[- 3x + 12]$

$= 84 + 21x - 84$.

$= 21x$.

After a little practice the number of steps may be considerably diminished.

Insertion of Brackets.

47. The rules for insertion of brackets are the converse of those given on page 9, and may be easily deduced from them.

For the following equivalents have been established in Arts. 19 and 20 :

$$a+b-c=a+(b-c),$$
$$a-b-c=a-(b+c),$$
$$a-b+c=a-(b-c).$$

From these results the rules follow.

Rule. 1. *Any part of an expression may be enclosed within brackets and the sign + prefixed, the sign of every term within the brackets remaining unaltered.*

Examples. $a-b+c-d-e=a-b+(c-d-e).$
$x^2-ax+bx-ab=(x^2-ax)+(bx-ab).$

Rule. 2. *Any part of an expression may be enclosed within brackets and the sign − prefixed, provided the sign of every term within the brackets be changed.*

Examples. $a-b+c-d-e=a-(b-c)-(d+e).$
$xy-ax-by+ab=(xy-by)-(ax-ab).$

48. The terms of an expression can be bracketed in various ways.

Example. The expression $ax-bx+cx-ay+by-cy$
may be written $(ax-bx)+(cx-ay)+(by-cy),$
or $(ax-bx+cx)-(ay-by+cy),$
or $(ax-ay)-(bx-by)+(cx-cy).$

49. When every term of an expression is divisible by a common factor, the expression may be simplified by dividing each term by this factor, and enclosing the quotient within brackets, the common factor being placed outside as a coefficient.

Thus $3x-21=3(x-7)$;
and $x^2-2ax+4a^2=x^2-2a(x-2a).$

EXAMPLES VI. b.

Simplify by removing brackets :

1. $16-3(2x-3)-(2x+3).$ **2.** $4(x+3)-2(7+x)+2.$
3. $8(x-3)-(6-2x)-2(x+2)+5(5-x).$
4. $2x-5\{3x-7(4x-9)\}.$ **5.** $4x-3\{x-(1-y)+2(1-x)\}.$

6. $x - (y - z) - [x - y - z - 2(y + z)]$.

7. $5x + 4(y - 2z) - 4\{x + 2(y - z)\}$. **8.** $a + \{-2b + 3(c - \overline{d - e})\}$.

9. $3p - \{5q - [6q + 2(10q - p)]\}$. **10.** $3x - 2[2x - \{2(x - y) - y\} - y]$.

11. $2[4x - \{2y + (2x - y) - (x + y)\}] - 2(-x - \overline{y - x})$.

12. $20(2 - x) + 3(x - 7) - 2[x + 9 - 3\{9 - 4(2 - x)\}]$.

13. $-4(a + y) + 24(b - x) - 2[x + y + a - 3\{y + a - 4(b + x)\}]$.

14. Multiply $2x - 3y - 4(x - 2y) + 5\{3x - 2(x - y)\}$

by $4x - (y - x) - 3\{2y - 3(x + y)\}$.

In each of the following expressions bracket the powers of x so that the signs before the brackets may be (1) positive, (2) negative.

15. $ax^4 + 2x^3 - cx^2 + 2x^2 - bx^3 - x^4$. **16.** $ax^2 + a^2x^3 - bx^2 - 5x^2 - cx^3$.

Substitution of Negative Quantities.

50. To further illustrate the use of the rule of signs, we add some examples in substitution where some of the symbols denote negative quantities.

Example 1. If $a = -4$, find the value of a^3.

Here $a^3 = (-4)^3 = (-4) \times (-4) \times (-4) = -64$.

By repeated applications of the rule of signs it may easily be shewn that any *odd* power of a negative quantity is *negative*, and any *even* power of a negative quantity is positive.

Example 2. If $a = -1$, $b = 3$, $c = -2$, find the value of $-3a^4bc^3$.

Here $-3a^4bc^3 = -3 \times (-1)^4 \times 3 \times (-2)^3$ | We write down at
$= -3 \times (+1) \times 3 \times (-8)$ | once $(-1)^4 = +1$, and
$= 72$. | $(-2)^3 = -8$.

Example 3. If $2x = 3$, $4y = 3$, $z = -2$ find the value of $\sqrt{(8y + 2z + 7)} + \sqrt{(6x - 8y + z)}$.

Here $x = \dfrac{3}{2}$, $y = \dfrac{3}{4}$, $z = -2$;

$\therefore$ $8y + 2z + 7 = 8 \times \dfrac{3}{4} + 2(-2) + 7 = 6 - 4 + 7 = 9$,

and $6x - 8y + z = 6 \times \dfrac{3}{2} - 8 \times \dfrac{3}{4} + (-2) = 9 - 6 - 2 = 1$;

$\therefore$ the whole expression $= \sqrt{9} + \sqrt{1} = 4$.

EXAMPLES VI. c.

If $a = -2$, $b = 3$, $c = -1$, $x = -5$, $y = 4$, find the value of

1. $3a^2b$.   2. $8abc^2$.   3. $-5c^3$.   4. $6a^2c^2$.
5. $4c^3y$.   6. $3a^2c$.   7. $-4a^2c^4$.   8. $3c^3x^3$.
9. $5a^2x^2$.   10. $-7c^4xy$.   11. $-8ax^3$.   12. $7a^5c^4$.

If $a = -4$, $b = -3$, $c = -1$, $f = 0$, $x = 4$, $y = 1$, find the value of

13. $3a^2 + bx - 4cy$.   14. $3b^2y^4 - 4b^2f - 6c^4x$.
15. $2\sqrt{(ac)} - 3\sqrt{(xy)} + \sqrt{(b^2c^4)}$.   16. $3\sqrt{(acx)} - 2\sqrt{(b^2y)} - 6\sqrt{(c^2y)}$.
17. $7\sqrt{a^2x} - 3\sqrt{b^4c^2} + 5\sqrt{f^2}$.   18. $3c\sqrt{3bc} - 5\sqrt{4c^2} - 2c\sqrt{3bc}$.

If $2x = 1$, $3y = -4$, $2z = 7$, find the value of

19. $x^2y^3 + 2y^2z - 2yz^2$.   20. $\dfrac{1}{x+y} + \dfrac{2}{y+z}$.

If $x = -2$, $y = 3$, $4a = -1$, find the value of

21. $\dfrac{x}{a-y} - \dfrac{y}{a+2x}$.   22. $3xy + 4a^2 + \sqrt{10 - xy}$.

23. If $a = 10$, $b = -\dfrac{1}{4}$, $c = -\dfrac{1}{5}$, find the value of $a^4b^2c^3\sqrt{b^2 - c^2}$.

24. If $x = 1$, $y = -3$, $z = 1$, find the value of
$$\sqrt{(x^2 + y^3 + z)(x - y - 3z)} \div \sqrt[3]{xy^3z^2}.$$

25. If $a = 1$, $b = 2$, $c = -\dfrac{1}{2}$, $d = 0$, find the value of
$$\dfrac{a - b + c}{a - b - c} - \dfrac{ad - bc}{bd - ac} - \sqrt{\left(\dfrac{b^3}{a^3} - \dfrac{a^3}{c^3}\right)}.$$

26. When $a = 1$, $b = -1$, $c = 2$, evaluate the expression
$$\sqrt{3a^3(b - c)} + 3b^3(c - a) + 3c^3(a - b).$$

27. Find the value of
$$\sqrt{(x^2 + y^3 + z)(x - y - 3z)} \div \sqrt[3]{xy^3z^2}$$
when $x = -1$, $y = -3$, $z = 1$.

28. When $a = 4$, $b = -2$, $c = \dfrac{3}{2}$, $d = -1$, find the value of
(1) $a^3 - b^3 - (a - b)^3 - 11(3b + 2c)\left(2c^2 - \dfrac{d^2}{2}\right)$;
(2) $\sqrt[3]{4c^2 - a(a - 2b - d)} - \sqrt[3]{b^4c + 11b^3d^2}$.

CHAPTER VII.

SIMPLE EQUATIONS.

51. An equation is a statement that two algebraical expressions are equal.

The parts of an equation separated by the sign of equality are called **members** or sides of the equation and are distinguished as the *right side* and the *left side*.

52. If the expressions are *always* equal, for *any* values of the symbols involved, the equation is called an **identity.** Thus $x+3+x+4=2x+7$ is an identity.

If the two expressions are only equal for particular values of the symbols, the equation is called an **equation of condition**; it is in this sense that the word equation is generally used. Thus the equation $2+x=2x-1$, will be found to be only true when $x=3$, and the value 3 is said to **satisfy** the equation. The object of the present chapter is to shew how to find the value which satisfies an equation of the simplest kind.

53. The letter whose value it is required to find is called the **unknown quantity.** The process of finding its value is called **solving the equation.** The value so found is called the **root** or the **solution** of the equation.

54. An equation which involves the unknown quantity in the first degree only is called a **simple equation.** It is usual to denote the unknown quantity by the letter x.

The process of solving a simple equation depends only upon the following **axioms**:

1. If to equals we add equals the sums are equal.
2. If from equals we take equals the remainders are equal.
3. If equals are multiplied by equals the products are equal.
4. If equals are divided by equals the quotients are equal.

55. Consider the equation $7x = 14$.

It is required to find what numerical value x must have consistent with this statement.

Dividing both sides by 7 we get

$$x = 2, \hspace{3cm} \text{[Axiom 4.]}$$

Similarly, if $\qquad\qquad \dfrac{x}{2} = -6,$

multiplying both sides by 2, we get

$$x = -12, \hspace{3cm} \text{[Axiom 3.]}$$

Again, in the equation $7x - 2x - x = 23 + 15 - 10$, by collecting terms, we have $\qquad\qquad 4x = 28.$

$$\therefore \quad x = 7.$$

56. To solve $\qquad 3x - 8 = x + 12.$

Subtract x from both sides of the equation, and we get

$$3x - x - 8 = 12, \hspace{2.5cm} \text{[Axiom 2.]}$$

Adding 8 to both sides, we have

$$3x - x = 12 + 8, \hspace{2.5cm} \text{[Axiom 1.]}$$

Thus we see that $+x$ has been removed from one side, and appears as $-x$ on the other; and -8 has been removed from one side and appears as $+8$ on the other.

It is evident that similar steps may be employed in all cases. Hence we may enunciate the following rule :

Rule. *Any term may be transposed from one side of the equation to the other by changing its sign.*

It appears from this that *we may change the sign of every term in an equation;* for this is equivalent to transposing all the terms, and then making the right and left-hand members change places.

57. To solve $\dfrac{x}{2} - 3 = \dfrac{x}{4} + \dfrac{x}{5}.$

Here it will be convenient to begin by clearing the equation of *fractional* coefficients. This can always be done by multiplying both sides of the equation by the least common multiple of the denominators. [Axiom 3.]

Thus, multiplying by 20,

$$10x - 60 = 5x + 4x\,;$$

transposing, $\qquad\qquad 10x - 5x - 4x = 60\,;$

$$\therefore \quad x = 60.$$

58. We can now give a general rule for solving any simple equation with one unknown quantity.

Rule. *First, if necessary, clear of fractions; then transpose all the terms containing the unknown quantity to one side of the equation, and the known quantities to the other. Collect the terms on each side; divide both sides by the coefficient of the unknown quantity and the value required is obtained.*

Example 1. Solve $5(x-3) - 7(6-x) + 3 = 24 - 3(8-x)$.

Removing brackets, $5x - 15 - 42 + 7x + 3 = 24 - 24 + 3x$;

transposing $\qquad 5x + 7x - 3x = 24 - 24 + 15 + 42 - 3$;

$$\therefore \quad 9x = 54 ;$$

$$\therefore \quad x = 6.$$

Example 2. Solve $(x+1)(2x-1) - 5x = (2x-3)(x-5) + 47$.

Forming the products, we have

$$2x^2 + x - 1 - 5x = 2x^2 - 13x + 15 + 47.$$

Erasing the term $2x^2$ on each side, and transposing,

$$x - 5x + 13x = 15 + 47 + 1 ;$$

$$\therefore \quad 9x = 63 ;$$

$$\therefore \quad x = 7.$$

EXAMPLES VII. a.

Solve the following equations :

1. $6x + 3 = 15$. **2.** $5x - 7 = 28$. **3.** $13 = 7 + 2x$.

4. $15 = 37 - 11x$. **5.** $4x - 7 = 11$. **6.** $7x = 18 - 2x$.

7. $3x - 18 = 7 - 2x$. **8.** $0 = 11 - 2x + 7 - 10x$.

9. $5x - 17 + 3x - 5 = 6x - 7 - 8x + 115$.

10. $15 - 7x - 9x - 28 + 14x - 17 = 21 - 3x + 13 - 9x + 8x$.

11. $5(x-3) = 4(x-2)$. **12.** $11(5 - 4x) = 7(5 - 6x)$.

13. $3 - 7(x-1) = 5 - 4x$. **14.** $5 - 4(x-3) = x - 2(x-1)$.

15. $8(x-3) - 2(3-x) = 2(x+2) - 5(5-x)$.

16. $\dfrac{1}{2}x + \dfrac{1}{3}x = x - 3$. **17.** $\dfrac{1}{2}x - \dfrac{1}{3}x = \dfrac{1}{4}x + \dfrac{1}{2}$.

18. $x - \dfrac{x}{4} - \dfrac{1}{2} = 3 + \dfrac{x}{4}$. **19.** $\dfrac{1}{2}x - \dfrac{3}{4}x - 1\dfrac{1}{3} = \dfrac{1}{6}x + 2$.

20. $(x+3)(2x-3) - 6x = (x-4)(2x+4) + 12$.

21. $(2x+1)(2x+6) - 7(x-2) = 4(x+1)(x-1) - 9x$.

22. $(3x+1)^2 + 6 + 18(x+1)^2 = 9x(3x-2) + 65$.

59. We shall now give some equations of greater difficulty.

Example 1.   Solve $5x - (4x - 7)(3x - 5) = 6 - 3(4x - 9)(x - 1)$.

Simplifying, we have
$$5x - (12x^2 - 41x + 35) = 6 - 3(4x^2 - 13x + 9) ;$$
and by removing brackets
$$5x - 12x^2 + 41x - 35 = 6 - 12x^2 + 39x - 27.$$
Erase the term $-12x^2$ on each side and transpose ;

thus $5x + 41x - 39x = 6 - 27 + 35 ;$
$$\therefore \quad 7x = 14 ;$$
$$\therefore \quad x = 2.$$

Note. Since the $-$ sign before a bracket affects every term within it, in the first line of work we do not remove the brackets until we have formed the products.

Example 2.   Solve $4 - \dfrac{x - 9}{8} = \dfrac{x}{22} - \dfrac{1}{2}$.

Multiply by 88, the least common multiple of the denominators ;
$$352 - 11(x - 9) = 4x - 44 ;$$
removing brackets, $352 - 11x + 99 = 4x - 44 ;$
transposing, $-11x - 4x = -44 - 352 - 99 ;$
collecting terms and changing signs, $15x = 495 ;$
$$\therefore \quad x = 33.$$

Note. In this equation $-\dfrac{x - 9}{8}$ is regarded as a single term with the minus sign before it. In fact it is equivalent to $-\dfrac{1}{8}(x - 9)$, the *vinculum* or line between the numerator and denominator having the same effect as a bracket.

In certain cases it will be found more convenient not to multiply throughout by the L.C.M. of the denominator, but to clear of fractions in two or more steps.

Example 3.   Solve $\dfrac{x - 4}{3} + \dfrac{2x - 3}{35} = \dfrac{5x - 32}{9} - \dfrac{x + 9}{28}$.

Multiplying throughout by 9, we have
$$3x - 12 + \frac{18x - 27}{35} = 5x - 32 - \frac{9x + 81}{28} ;$$
transposing, $\dfrac{18x - 27}{35} + \dfrac{9x + 81}{28} = 2x - 20.$

Now clear of fractions by multiplying by $5 \times 7 \times 4$ or 140 ;
$$\therefore \quad 72x - 108 + 45x + 405 = 280x - 2800 ;$$
$$\therefore \quad 2800 - 108 + 405 = 280x - 72x - 45x ;$$
$$\therefore \quad 3097 = 163x ;$$
$$\therefore \quad x = 19.$$

60. To solve equations whose coefficients are decimals, we may express the decimals as vulgar fractions, and proceed as before; but it is often found more simple to work entirely in decimals.

Example. Solve $\cdot375x - 1\cdot875 = \cdot12x + 1\cdot185$.

Transposing, $\quad\cdot375x - \cdot12x = 1\cdot185 + 1\cdot875$;

collecting terms, $\quad(\cdot375 - \cdot12)x = 3\cdot06$;

that is, $\qquad\qquad \cdot255x = 3\cdot06$;

$$\therefore\quad x = \frac{3\cdot06}{\cdot255}$$

$$= 12.$$

EXAMPLES VII. b.

Solve the equations:

1. $(x+15)(x-3) - (x-3)^2 = 30 - 15(x-1)$.

2. $21 - x(2x+1) + 2(x-4)(x+2) = 0$.

3. $3(x+5) - 3(2x-1) = 32 - 4(x-5)^2 + 4x^2$.

4. $(x-6)(2x-9) - (11-2x)(7-x) = 5x - 4 - 7(x-2)$.

5. $\dfrac{x-1}{5} + \dfrac{x-9}{2} = 3$. **6.** $\dfrac{6x-2}{9} + \dfrac{3x+5}{18} = \dfrac{1}{3}$.

7. $\dfrac{10x+1}{5} - 1 = 5x - 2$. **8.** $\dfrac{x-6}{4} - \dfrac{x-4}{6} = 1 - \dfrac{x}{10}$.

9. $\dfrac{x+12}{6} - x = 6\frac{1}{2} - \dfrac{x}{12}$. **10.** $\dfrac{11-6x}{5} - \dfrac{9-7x}{2} = \dfrac{5(x-1)}{6}$.

11. $\dfrac{47-6x}{5} - (x-6) = \dfrac{4(x-7)}{15}$. **12.** $\dfrac{1-2x}{7} - \dfrac{2-3x}{8} = 1\frac{1}{2} + \dfrac{x}{4}$.

13. $\dfrac{1}{6}(x+4) - \dfrac{1}{2}(x-3) = \dfrac{1}{2}(3x-5) - \dfrac{1}{4}(x-6) - \dfrac{1}{5}(x-2)$.

14. $\dfrac{1}{3}(x+4) - \dfrac{1}{9}(20-x) = \dfrac{1}{18}(5x-1) - \dfrac{1}{6}(5x-13) + 8$.

15. $5 - \dfrac{10x+1}{27} - \dfrac{x}{8} = \dfrac{13x+4}{18} - \dfrac{5(x-4)}{4}$.

16. $\dfrac{x+4}{39} - \dfrac{1}{5}(1-x) = 2 - \dfrac{3}{26}(6-5x) - \dfrac{1}{5}(x+4)$.

17. $\dfrac{3}{11} + \dfrac{1}{44}x = \dfrac{1}{2}\left(\dfrac{4}{11} - \dfrac{x}{33}\right) - \dfrac{5}{66} + \dfrac{1}{3}\left(1 - \dfrac{x}{22}\right)$.

18. $\cdot7x - 3\cdot35 = 6\cdot4 - 3\cdot2x$. **19.** $\cdot5x + \cdot25 + \cdot1 + 1\cdot25 = \cdot4x$.

20. $3\cdot25x - \cdot75x = 9 + 1\cdot5x$. **21.** $\cdot2x - \cdot01x + \cdot005x = 11\cdot7$.

22. $\cdot5x - \cdot6x = \cdot75x - 11$. **23.** $\cdot4x - \cdot83x = \cdot7 - \cdot3$.

[*Chapter XVI. will furnish further practice in Simple Equations.*]

CHAPTER VIII.

SYMBOLICAL EXPRESSION.

61. IN solving algebraical problems the chief difficulty of the student is to express the conditions of the question by means of symbols so as to form an equation which may be solved by the methods already explained. Practice alone can give the requisite facility, but the easy examples in this chapter will furnish the student with a useful introduction to the problems in Chapter IX.

Example 1. By how much does x exceed 17?
Take a numerical instance; "by how much does 27 exceed 17?"
The answer obviously is 10, which is equal to $27 - 17$.
Hence the excess of x over 17 is $x - 17$.
Similarly the defect of x from 17 is $17 - x$.

Example 2. If x is one *part* of 45 the other part is $45 - x$.

Example 3. If x is one *factor* of 45 the other factor is $\dfrac{45}{x}$.

Example 4. How far can a man walk in a hours at the rate of 4 miles an hour?
In 1 hour he walks 4 miles.
In a hours he walks a times as far, that is, $4a$ miles.

Example 5. If £20 is divided equally among y persons, the share of each is the total sum divided by the number of persons, or £$\dfrac{20}{y}$.

Example 6. Out of a purse containing £x and y florins a man spends z shillings; express in pence the sum left.
$$£x = 20x \text{ shillings,}$$
and
$$y \text{ florins} = 2y \text{ shillings;}$$
$$\therefore \text{ the sum left} = (20x + 2y - z) \text{ shillings}$$
$$= 12(20x + 2y - z) \text{ pence.}$$

EXAMPLES VIII. a.

1. By how much does x exceed 5?

2. By how much is y less than 15?

3. What must be added to 6 to make b?

4. What is the quotient when 3 is divided by a?

5. By how much does $6x$ exceed $2x$?

6. The sum of two numbers is x and one of the numbers is 10; what is the other?

7. The sum of three numbers is 100; if one of them is 25 and another is x, what is the third?

8. The product of two factors is $4x$; if one of the factors is 4, what is the other?

9. How many times is x contained in $2y$?

10. The difference of two numbers is 8, and the greater of them is a; what is the other?

11. The sum of 12 equal numbers is $48x$; what is the value of each number?

12. If there are x numbers each equal to $2a$, what is their sum?

13. If there are x numbers each equal to p, what is their product?

14. If there are n books each worth y shillings, what is the total cost?

15. How many books each worth 2 shillings can be bought for y shillings?

16. What is the price in pence of n oranges at sixpence a score?

17. If I spend n shillings out of a sum of £5, how many shillings have I left?

18. What is the daily wage in shillings of a man who earns £6 in p weeks, working 6 days a week?

19. If x persons combine to pay a bill of £y, what is the share of each in shillings?

20. How many hours will it take to travel x miles at 10 miles an hour?

21. How far can I walk in p hours at the rate of q miles an hour?

22. If I can walk m miles in n days, what is my rate per day?

23. How many days will it take to travel y miles at x miles a day?

62. We subjoin a few harder examples worked out in full.

Example 1.   What is (1) the sum, (2) the product of three consecutive numbers of which the least is n?

The two numbers consecutive to n are $n+1$ and $n+2$;

$$\therefore \text{ the sum} = n + (n+1) + (n+2)$$
$$= 3n + 3.$$

And the product　　　　$= n(n+1)(n+2).$

Example 2.   A boy is x years old, and five years hence his age will be half that of his father: how old is the father now?

In five years the boy will be $x+5$ years old; therefore his father will then be $2(x+5)$, or $2x+10$ years old; his present age must therefore be $2x+10-5$ or $2x+5$ years.

Example 3. A and B are playing for money; A begins with £p and B with q shillings. B wins £x; express by an equation the fact that A has now three times as much as B.

What B has won A has lost;

$\therefore$ A has $p-x$ pounds, that is $20(p-x)$ shillings,

B has q shillings $+x$ pounds, that is $q+20x$ shillings.

Thus the required equation is $20(p-x)=3(q+20x)$.

Example 4.   How many men will be required to do in p hours what q men do in np hours?

np hours is the time occupied by q men;

$\therefore$ 1 hour .... $q \times np$ men;

that is, p hours.................. ..... $\dfrac{q \times np}{p}$ men.

Therefore the required number of men is qn.

EXAMPLES VIII. b.

1.   Write down three consecutive numbers of which a is the least.

2.   Write down four consecutive numbers of which b is the greatest.

3.   What is the next odd number after $2n-1$?

4.   What is the even number next before $2n$?

5.   How old is a man who will be x years old in 15 years?

6.   How old was a man x years ago if his present age is n years?

7. In $2x$ years a man will be y years old, what is his present age?

8. How old is a man who in x years will be twice as old as his son now aged 20 years?

9. In 5 years a boy will be x years old; what is the present age of his father if he is twice as old as his son?

10. A has £m and B has n shillings; after A has won 3 shillings from B, each has the same amount. Express this in algebraical symbols.

11. A has 25 shillings and B has 13 shillings; after B has won x shillings he then has four times as much as A. Express this in algebraical symbols.

12. How many miles can a man walk in 30 minutes if he walks 1 mile in x minutes?

13. How long will it take a man to walk p miles if he walks 15 miles in q hours?

14. How far can a pigeon fly in x hours at the rate of 2 miles in 7 minutes?

15. If x men do a work in $5x$ hours, how many men will be required to do the same work in y hours?

16. If a bill is shared equally among n persons, and each pays 6s. 8d., how many pounds does the bill amount to?

17. A man has £x in his purse, he pays away 25 shillings, and receives y pence; express in shillings the sum he has left.

18. How many pounds does a man save in a year, if he earns £x a week and spends y shillings a calendar month?

19. What is the total cost of $6x$ nuts and $4x$ plums, when x plums cost a shilling and plums are three times as expensive as nuts?

20. If x horses eat m lbs. of corn in ab days, how long will am lbs. last bx horses?

21. If p is the cost in pence of k lbs. of tea, how many shillings will be required to buy m oz.?

22. A person buys goods for £a and sells them for £$(a + b)$: what is his gain per cent.?

23. A person buys goods for £$(a + b)$ and sells them for £a: what is his loss per cent.?

24. If a person buys an article for £a and sells it at a gain of b per cent., how much does he obtain for it?

CHAPTER IX.

63. THE principles of the last chapter may now be employed to solve various problems.

The method of procedure is as follows :

Represent the unknown quantity by a symbol x, and express in symbolical language the conditions of the question ; we thus obtain a simple equation which can be solved by the methods already given in Chapter VII.

Example I. Find two numbers whose sum is 28, and whose difference is 4.

Let x be the smaller number, then $x+4$ is the greater.

Their sum is $x+(x+4)$, which is to be equal to 28.

Hence
$$x+x+4=28 ;$$
$$2x=24 ;$$
$$\therefore \ x=12.$$
and
$$x+4=16,$$
so that the numbers are 12 and 16.

The student is advised to test his solution by finding whether it satisfies the data of the question or not.

Example II. Divide £47 between A, B, C, so that A may have £10 more than B, and B £8 more than C.

Let x represent the *number* of pounds that C has; then B has $x+8$ pounds, and A has $x+8+10$ pounds.

Hence
$$x+(x+8)+(x+8+10)=47 ;$$
$$x+x+8+x+8+10=47,$$
$$3x=21 ;$$
$$\therefore \ x=7 ;$$
so that C has £7, B £15, A £25.

EXAMPLES IX. a.

1. Six times a number increased by 11 is equal to 65; find it.

2. Find a number which when multiplied by 11 and then diminished by 18 is equal to 15.

3. If 3 be added to a number, and the sum multiplied by 12, the result is 84; find the number.

4. One number exceeds another by 3, and their sum is 27; find them.

5. Find two numbers whose sum is 30, such that one of them is greater than the other by 8.

6. Find two numbers which differ by 10, so that one is three times the other.

7. Find two numbers whose sum is 19, such that one shall exceed twice the other by 1.

8. Find two numbers whose sum shall be 26 and their difference 8.

9. Divide £100 between A and B so that B may have £30 more than A.

10. Divide £66 between A, B, and C so that B may have £8 more than A, and C £14 more than B.

11. A, B, and C have £72 between them; C has twice as much as B, and B has £4 less than A; find the share of each.

12. How must a sum of 73 rupees be divided between A, B, and C, so that B may have 8 rupees less than A and 4 rupees more than C?

Example III. Divide 60 into two parts, so that three times the greater may exceed 100 by as much as eight times the less falls short of 200.

Let x be the greater part, then $60 - x$ is the less.

Three times the greater part is $3x$, and its excess over 100 is

$$3x - 100.$$

Eight times the less is $8(60 - x)$, and its defect from 200 is

$$200 - 8(60 - x).$$

Whence the symbolical statement of the question is

$$3x - 100 = 200 - 8(60 - x);$$
$$3x - 100 = 200 - 480 + 8x,$$
$$480 - 100 - 200 = 8x - 3x,$$
$$5x = 180;$$
$$\therefore \quad x = 36, \text{ the greater part,}$$

and
$$60 - x = 24, \text{ the less.}$$

Example IV. A is 4 years older than B, and half A's age exceeds one-sixth of B's age by 8 years; find their ages.

Let x be the *number* of years in B's age, then A's age is $x+4$ years.

One-half of A's age is represented by $\frac{1}{2}(x+4)$ years, and one-sixth of B's age by $\frac{1}{6}x$ years.

Hence $\qquad\qquad \frac{1}{2}(x+4) - \frac{1}{6}x = 8$;

multiplying by 6, $\qquad 3x + 12 - x = 48$;

$$\therefore\ \ 2x = 36\ ;$$
$$\therefore\ \ \ x = 18.$$

Thus B's age is 18 years, and A's age is 22 years.

13. Divide 75 into two parts, so that three times one part may be double of the other.

14. Divide 122 into two parts, such that one may be as much above 72 as twice the other is below 60.

15. A certain number is doubled and then increased by 5, and the result is less by 1 than three times the number; find it.

16. How much must be added to 28 so that the resulting number may be eight times the added part?

17. Find the number whose double exceeds its half by 9.

18. What is the number whose seventh part exceeds its eighth part by 1?

19. Divide 48 into two parts, so that one part may be three-fifths of the other.

20. If A, B, and C have £76 between them, and A's money is double of B's, and C's one-sixth of B's, what is the share of each?

21. Divide £511 between A, B, and C, so that B's share shall be one-third of A's, and C's share three-fourths of A's and B's together.

22. B is 16 years younger than A, and one-half B's age is equal to one-third of A's; how old are they?

23. A is 8 years younger than B, and 24 years older than C; one-sixth of A's age, one-half of B's, and one-third of C's together amount to 38 years; find their ages.

24. Find two consecutive numbers whose product exceeds the square of the smaller by 7. [See Art. 62, Ex. 1.]

25. The difference between the squares of two consecutive numbers is 31; find the numbers.

64. We shall now give examples of somewhat greater difficulty.

Example I. A has £9, and B has 4 guineas; after B has won from A a certain sum, A has then five-sixths of what B has; how much did B win?

Suppose that B wins x *shillings*, A has then $180 - x$ shillings, and B has $84 + x$ shillings.

Hence
$$180 - x = \tfrac{5}{6}(84 + x);$$
$$1080 - 6x = 420 + 5x,$$
$$11x = 660;$$
$$\therefore \ x = 60.$$

Therefore B wins 60 shillings, or £3.

Example II. A is twice as old as B, ten years ago he was four times as old; what are their present ages?

Let B's age be x years, then A's age is $2x$ years.

Ten years ago their ages were respectively $x - 10$ and $2x - 10$ years; thus we have
$$2x - 10 = 4(x - 10);$$
$$2x - 10 = 4x - 40,$$
$$2x = 30;$$
$$\therefore \ x = 15,$$

so that B is 15 years old, A 30 years.

EXAMPLES IX. b.

1. A has £12 and B has £8; after B has lost a certain sum to A, his money is only three-sevenths of A's; how much did A win?

2. A and B begin to play each with £15; if they play till B's money is four-elevenths of A's, what does B lose?

3. A and B have £28 between them; A gives £3 to B and then finds he has six times as much money as B; how much had each at first?

4. A had three times as much money as B; after giving £3 to B he had only twice as much; what had each at first?

5. A father is four times as old as his son; in 16 years he will only be twice as old; find their ages.

6. A is 20 years older than B, and five years ago A was twice as old as B; find their ages.

7. How old is a man whose age 10 years ago was three-eighths of what it will be in 15 years?

8. A is twice as old as B; 5 years ago he was three times as old; what are their present ages?

9. A father is 24 years older than his son; in 7 years the son's age will be two-fifths of his father's age; what are their present ages?

Example III. A person spent £28. 4s. in buying geese and ducks; if each goose cost 7s., and each duck 3s., and if the total number of birds bought was 108, how many of each did he buy?

In questions of this kind it is of essential importance to have all quantities expressed in the same denomination; in the present instance it will be convenient to express the money in shillings.

Let x be the number of geese, then $108 - x$ is the number of ducks.

Since each goose costs 7 shillings, x geese cost $7x$ shillings.

And since each duck costs 3 shillings, $108 - x$ ducks cost $3(108 - x)$ shillings.

Therefore the amount spent is

$$7x + 3(108 - x) \text{ shillings.}$$

But the question also states that the amount is £28. 4s., that is 564 shillings.

Hence $7x + 3(108 - x) = 564$;

$$7x + 324 - 3x = 564,$$
$$4x = 240,$$
$$\therefore \ x = 60, \text{ the number of geese,}$$

and $108 - x = 48$, the number of ducks.

Note. In all these examples it should be noticed that the unknown quantity x represents a *number* of pounds, ducks, years, etc.; and the student must be careful to avoid beginning a solution with a supposition of the kind, "let $x = A$'s share" or "let $x =$ the ducks," or any statement so vague and inexact.

It will sometimes be found easier not to put x equal to the quantity directly required, but to some other quantity involved in the question: by this means the equation is often simplified.

Example IV. A woman spends 4s. 4½d. in buying eggs, and finds that 9 of them cost as much over one shilling as 15 cost under two shillings; how many eggs did she buy?

Let x be the price of an egg in pence; then 9 eggs cost $9x$ pence, and 15 eggs cost $15x$ pence ;

$$9x - 12 = 24 - 15x,$$
$$24x = 36 ;$$
$$\therefore \ x = 1\tfrac{1}{2}.$$

Thus the price of an egg is 1½d., and the number of eggs $= 52\tfrac{1}{2} \div 1\tfrac{1}{2} = 35$.

10. A sum of £3 is divided between 50 men and women, the men each receiving 1s. 6d. and the women 1s. ; find the number of each sex.

11. The price of 13 yards of cloth is as much less than £1 as the price of 27 yards exceeds £2; find the price per yard.

12. A hundredweight of tea, worth £19. 12s., is made up of two sorts, part worth 4s. a pound and the rest worth 2s. a pound; how much is there of each sort?

13. A man is hired for 60 days on condition that for each day he works he shall receive 7s. 6d., but for each day that he is idle he shall pay 2s. 6d. for his board: at the end he received £6; how many days had he worked?

14. A sum of £2. 11s. is made up of 32 coins, which are either florins or shillings; how many are there of each?

15. A sum of £3. 2s. was paid in half-crowns, florins, and shillings; the number of half-crowns used was four times the number of florins and ten times the number of shillings; how many were there of each?

16. A person buys coffee and tea at 2s. and 3s. a pound respectively; he spends £3. 1s. 9d., and in all gets 24 lbs.; how much of each did he buy?

17. A man sold a horse for a sum of money which was greater by £38 than half the price he paid for it, and gained thereby ten guineas; what did he pay for the horse?

18. Two boys have 240 marbles between them; one arranges his in heaps of 6 each, the other in heaps of 9 each. There are 36 heaps altogether; how many marbles has each?

19. A man's age is four times the combined ages of his two sons, one of whom is three times as old as the other; in 24 years their combined ages will be 12 years less than their father's age; find their respective ages.

20. A sum of money is divided between three persons, A, B, and C, in such a way that A and B have £42 between them, B and C have £45, and C and A have £53; what is the share of each?

21. A person bought a number of oranges for 3s. 9d., and finds that 12 of them cost as much over 5d. as 16 of them cost under 2s. 6d.; how many oranges were bought?

22. By buying eggs at 15 for a shilling and selling them at a dozen for 15d. a man gained 13s. 6d.; find the number of eggs.

23. I bought a certain number of apples at four a penny, and three-fifths of that number at three a penny; by selling them at sixteen for fivepence I gained 4d.; how many apples did I buy?

24. If 8 lbs. of tea and 24 lbs. of sugar cost £1. 12s. 8d., and if 3 lbs. of tea cost as much as 40 lbs. of sugar, find the price of each per pound.

25. Four dozen of port and three dozen of sherry cost £15. 8s.; if a bottle of port costs 1s. 2d. more than a bottle of sherry, find the price of each per dozen.

CHAPTER X.

SIMULTANEOUS EQUATIONS.

65. CONSIDER the equation $2x+5y=23$, which contains *two* unknown quantities.

By transposition we get

$$5y = 23 - 2x ;$$

that is,

$$y = \frac{23 - 2x}{5} \quad \dots\dots\dots\dots\dots\dots(1).$$

From this it appears that for every value we choose to give to x there will be one corresponding value of y. Thus we shall be able to find as many pairs of values as we please which satisfy the given equation.

For instance, if $x=1$, then from (1) we obtain $y = \frac{21}{5}$.

Again, if $x=-2$, then $y = \frac{27}{5}$; and so on.

But if also we have a second equation of the same kind, such as

$$3x + 4y = 24,$$

we have from this

$$y = \frac{24 - 3x}{4} \quad \dots\dots\dots\dots\dots\dots(2).$$

If now we seek values of x and y which satisfy *both* equations, the values of y in (1) and (2) must be identical ;

$$\therefore \quad \frac{23 - 2x}{5} = \frac{24 - 3x}{4} ; \text{ whence } x=4.$$

Substituting this value in the first equation, we have

$$8 + 5y = 23 ; \text{ whence } y=3.$$

Thus, if both equations are to be satisfied by the *same* values of x and y, there is only one solution, namely, $x=4$, $y=3$.

66. DEFINITION. When two or more equations are satisfied by the same values of the unknown quantities they are called **simultaneous equations.**

We proceed to explain the different methods for solving simultaneous equations. In the present chapter we shall confine our attention to the simpler cases in which the unknown quantities are involved in the first degree.

67. Since the two equations are simultaneously true, *any* equation formed by combining them will be satisfied by the values of x and y which satisfy the original equations. Our object will always be to obtain an equation which involves *one only* of the unknown quantities. The process by which we get rid of either of the unknown quantities is called **elimination**, and it must be effected in different ways according to the nature of the equations proposed.

Example 1.　Solve
$$3x + 7y = 27 \dots\dots\dots\dots\dots\dots\dots\dots(1),$$
$$5x + 2y = 16 \dots\dots\dots\dots\dots\dots\dots\dots(2).$$

To eliminate x we multiply (1) by 5 and (2) by 3, so as to make the coefficients of x in both equations equal. This gives
$$15x + 35y = 135,$$
$$15x + 6y = 48 ;$$
subtracting,　　　　　　$29y = 87 ;$
$$\therefore\ y = 3.$$

To find x, substitute this value of y in *either* of the given equations.

Thus from (1)　　　　　$3x + 21 = 27 ;$
$$\therefore\ x = 2, \Big\}$$
and　　　　　　　　　　$y = 3.$

Note. When one of the unknowns has been found, it is immaterial which of the equations we use to complete the solution. Thus, in the present example, if we substitute 3 for y in (2), we have
$$5x + 6 = 16 ;$$
$$\therefore\ x = 2, \text{ as before.}$$

Example 2.　Solve
$$7x + 2y = 47 \dots\dots\dots\dots\dots\dots\dots\dots(1),$$
$$5x - 4y = 1 \dots\dots\dots\dots\dots\dots\dots\dots(2).$$

Here it will be more convenient to eliminate y.

Multiplying (1) by 2,　　$14x + 4y = 94,$
and from (2)　　　　　　$5x - 4y = 1 ;$
adding,　　　　　　　　$19x = 95 ;$
$$\therefore\ x = 5.$$

Substitute this value in (1),
$$\therefore\ 35 + 2y = 47 ;$$
$$\therefore\ y = 6, \Big\}$$
and　　　　　　　　　　$x = 5.$

Note. *Add* when the coefficients of one unknown are equal and *unlike* in sign ; *subtract* when the coefficients are equal and *like* in sign.

Example 3.   Solve $2x = 5y + 1$ ... .(1),

$24 - 7x = 3y$ ....(2),

Here we can eliminate x by substituting in (2) its value obtained from (1).   Thus

$$24 - \frac{7}{2}(5y + 1) = 3y \, ;$$

$$\therefore \; 48 - 35y - 7 = 6y \, ;$$

$$\therefore \; 41 = 41y \, ;$$

$$\therefore \; y = 1, \left.\vphantom{\begin{matrix}1\\1\end{matrix}}\right\}$$

and from (1) $x = 3.$

68.   Any one of the methods given above will be found sufficient; but there are certain arithmetical artifices which will sometimes shorten the work.

Example.   Solve $28x - 23y = 22$ ....(1),

$63x - 55y = 17$ ....(2).

Noticing that 28 and 63 contain a common factor 7, we shall make the coefficients of x in the two equations equal to the *least common multiple* of 28 and 63 if we multiply (1) by 9 and (2) by 4.

Thus $252x - 207y = 198,$

$252x - 220y = 68 \, ;$

subtracting, $13y = 130 \, ;$

that is, $y = 10,$

and therefore from (1), $x = 9.$

EXAMPLES X. a.

Solve the equations :

1. $x + y = 19,$
$x - y = 7.$

2. $x - y = 6,$
$x + y = 0.$

3. $x - y = 25,$
$x + y = 13.$

4. $3x + 5y = 50,$
$4x + 3y = 41.$

5. $x + 5y = 18,$
$3x + 2y = 41.$

6. $4x + y = 10,$
$5x + 7y = 47.$

7. $4x + 5y = 4,$
$5x - 3y = 79.$

8. $4x - 3y = 0,$
$7x - 4y = 180.$

9. $2x + 3y = 22,$
$5x + 2y = 0.$

10. $5x = 7y - 21,$
$21x - 9y = 75.$

11. $55x = 33y,$
$10x = 7y - 15.$

12. $5x - 7y = 11,$
$18x = 12y.$

13. $6y - 5x = 11,$
$4x = 7y - 22.$

14. $3x + 10 = 5y,$
$7y = 4x + 13.$

15. $4y = 47 + 3x,$
$5x = 30 - 15y.$

16. $11x + 13y = 7,$
$13x + 11y = 17.$

17. $13x - 17y = 11,$
$29x - 39y = 17.$

18. $19x + 17y = 7,$
$41x + 37y = 17.$

69.   Before proceeding to solve, it will sometimes be necessary to simplify the equations.

Example.   Solve $3x - \dfrac{y-5}{7} = \dfrac{4x-3}{2}$ ....(1),

$$\dfrac{3y+4}{5} - \dfrac{1}{3}(2x-5) = y \quad ....(2).$$

Clear of fractions.   Thus

from (1), $42x - 2y + 10 = 28x - 21$;

∴ $14x - 2y = -31$ ....(3).

From (2), $9y + 12 - 10x + 25 = 15y$;

∴ $10x + 6y = 37$ ....(4).

Eliminating y from (3) and (4), we find that

$$x = -\dfrac{14}{13}.$$

To obtain the value of y, instead of substituting the value of x in one of the given equations, it will be found simpler to go through a second elimination, thus : eliminating x from (3) and (4), we find that

$$y = \dfrac{207}{26}.$$

70.   Simultaneous equations may often be conveniently solved by considering $\dfrac{1}{x}$ and $\dfrac{1}{y}$ as the unknown quantities.

Example.   Solve $\dfrac{8}{x} - \dfrac{9}{y} = 1$ ....(1),

$$\dfrac{10}{x} + \dfrac{6}{y} = 7 \quad ....(2).$$

Multiply (1) by 2 and (2) by 3 ; thus

$$\dfrac{16}{x} - \dfrac{18}{y} = 2,$$

$$\dfrac{30}{x} + \dfrac{18}{y} = 21 ;$$

adding, $\dfrac{46}{x} = 23$;

multiplying up, $46 = 23x$;

∴ $x = 2$;

and by substituting in (1), $y = 3$.

EXAMPLES X. b.

Solve the equations :

1. $2x - y = 4,$
   $\dfrac{x}{2} + \dfrac{y}{4} = 5.$

2. $4x - y = 1,$
   $\dfrac{x}{2} + \dfrac{3y}{7} = 4.$

3. $x + 2y = 13,$
   $\dfrac{2x}{3} - \dfrac{y}{5} = 1.$

4. $\dfrac{3x}{10} + y = 1,$
   $x + 3y = 2.$

5. $x - \dfrac{2y}{3} = 0,$
   $4x - 3y = 1.$

6. $\dfrac{3x}{5} - y = 7,$
   $4x + 5y = 0.$

7. $x - y = 0,$
   $\dfrac{5}{3}x - \dfrac{9}{2}y = 2\tfrac{5}{6}.$

8. $\dfrac{1}{2}(x+3) = 0,$
   $\dfrac{1}{6}x - y = 4\tfrac{1}{2}.$

9. $\dfrac{3}{5}x - 2y = 20.$
   $\dfrac{1}{2}(y+8) = 2.$

10. $3(x-y) + 2(x+y) = 15,\quad 3(x+y) + 2(x-y) = 25.$

11. $3(x+y-5) = 2(y-x),\quad 3(x-y-7) + 2(x+y-2) = 0.$

12. $4(2x-y-6) = 3(3x-2y-5),\quad 2(x-y+1) + 4x = 3y + 4.$

13. $7(2x-y) + 5(3y-4x) + 30 = 0,\quad 5(y-x+3) = 6(y-2x).$

14. $\dfrac{x+4}{5} = \dfrac{y-4}{7} = 2x + y + 4.$

15. $\dfrac{x-12}{4} = \dfrac{y+18}{3} = \dfrac{2x+3y}{2}.$

16. $\dfrac{8}{x} + \dfrac{9}{y} = 7,$
    $\dfrac{6}{x} - \dfrac{1}{y} = 2\tfrac{2}{3}.$

17. $\dfrac{3}{x} + \dfrac{5}{y} = 37,$
    $\dfrac{7}{x} - \dfrac{3}{y} = 13.$

18. $\dfrac{10}{x} - \dfrac{3}{y} = 8,$
    $\dfrac{3}{x} + \dfrac{2}{y} = -3\tfrac{2}{5}.$

71. In order to solve simultaneous equations which contain two unknown quantities we have seen that we must have two equations. Similarly we find that in order to solve simultaneous equations which contain three unknown quantities we must have three equations.

Rule. *Eliminate one of the unknowns from any pair of the equations, and then eliminate the same unknown from another pair. Two equations involving two unknowns are thus obtained, which may be solved by the rules already given. The remaining unknown is then found by substituting in any one of the given equations.*

Example. Solve the equations :
$$x + \tfrac{5}{7}y = z - \tfrac{8}{7}, \quad 4x + 2y - 3z = 0, \quad x - \tfrac{4}{5}(y-z) = 7.$$

Clearing of fractions, and transposing, we have
$$7x + 5y - 7z = -8 \quad\dotfill\quad (1),$$
$$4x + 2y - 3z = 0 \quad\dotfill\quad (2),$$
$$5x - 4y + 4z = 35 \quad\dotfill\quad (3).$$

Choose y as the unknown to be eliminated.

Multiply (2) by 5, $\qquad 20x + 10y - 15z = 0$;

Multiply (1) by 2, $\qquad 14x + 10y - 14z = -16$;

by subtraction, $\qquad\qquad 6x - z = 16$ ....(4).

Multiply (2) by 2, $\qquad 8x + 4y - 6z = 0$;

from (3), $\qquad\qquad\quad 5x - 4y + 4z = 35$;

by addition, $\qquad\qquad 13x - 2z = 35$.

Multiply (4) by 2, $\qquad 12x - 2z = 32$.

by subtraction, $\qquad\qquad\quad x = 3$.

From (4) we find $\qquad\qquad z = 2$,

and from (2), $\qquad\qquad\quad y = -3$.

EXAMPLES X. c.

Solve the equations :

1. $3x - 2y + z = 4$,
   $2x + 3y - z = 3$,
   $x + y + z = 8$.

2. $3x + 4y - 6z = 16$,
   $4x + y - z = 24$,
   $x - 3y - 2z = 1$.

3. $7x - 4y - 3z = 0$,
   $5x - 3y + 2z = 12$,
   $3x + 2y - 5z = 0$.

4. $4x + 3y - z = 9$,
   $9x - y + 5z = 16$,
   $x + 4y - 3z = 2$.

5. $3y - 6z - 5x = 4$, $\quad 2z - 3x - y = 8$, $\quad x - 2y + 2z + 2 = 0$.

6. $3y + 2z + 5x = 21$, $\quad 8x - 3z + y = 3$, $\quad 2z + 2x - 3y = 39$.

7. $\dfrac{1}{2}x + y + \dfrac{1}{2}z = \dfrac{1}{2}$,

   $x + 2y + \dfrac{1}{3}z = \dfrac{1}{3}$,

   $x + y - 9z = 1$.

8. $\dfrac{1}{2}x - \dfrac{1}{4}y = 5 - \dfrac{1}{6}z$,

   $\dfrac{1}{6}x - \dfrac{1}{3}y = 3 - \dfrac{1}{6}z$,

   $2y + 7 = \dfrac{1}{4}(z - x)$.

9. $\dfrac{1}{3}x + \dfrac{1}{4}(y + z) = 1\dfrac{2}{3}$, $\quad 4x + \dfrac{1}{2}(z - y) = 11$, $\quad \dfrac{1}{3}(z - 4x) = y$.

10. $2x - \dfrac{1}{5}(z - 2y) = 2$, $\quad \dfrac{1}{3}(x + y) = \dfrac{1}{7}(3 - z)$, $\quad x = 4y + 3z$.

11. $\dfrac{x}{3} - \dfrac{y}{2} = y + \dfrac{z}{2} = x + y + z + 2 = 0$.

CHAPTER XI.

PROBLEMS LEADING TO SIMULTANEOUS EQUATIONS.

72. In the Examples discussed in the last chapter we have seen that it is essential to have as many equations as there are unknown quantities to determine. Consequently in the solution of problems which give rise to simultaneous equations, it will always be necessary that the statement of the question should contain as many independent conditions as there are quantities to be determined.

Example 1. Find two numbers whose difference is 11, and one-fifth of whose sum is 9.

Let x be the greater number, y the less;

then
$$x - y = 11 \quad \ldots \ldots \ldots \ldots \ldots \ldots (1).$$

Also
$$\frac{x+y}{5} = 9,$$

or
$$x + y = 45 \quad \ldots \ldots \ldots \ldots \ldots (2).$$

By addition $2x = 56$; and by subtraction $2y = 34$.

The numbers are therefore 28 and 17.

Example 2. If 15 lbs. of tea and 17 lbs. of coffee together cost £3. 5s. 6d., and 25 lbs. of tea and 13 lbs. of coffee together cost £4. 6s. 2d.; find the price of each per pound.

Suppose a pound of tea to cost x shillings,

and $\ldots \ldots \ldots \ldots$ coffee $\ldots \ldots y \ldots \ldots \ldots$

Then from the question we have
$$15x + 17y = 65\tfrac{1}{2} \quad \ldots \ldots \ldots \ldots (1),$$
$$25x + 13y = 86\tfrac{1}{8} \quad \ldots \ldots \ldots \ldots (2).$$

Multiplying (1) by 5 and (2) by 3, we have
$$75x + 85y = 327\tfrac{1}{2},$$
$$75x + 39y = 258\tfrac{1}{2}.$$

Subtracting,
$$46y = 69,$$
$$\therefore \quad y = 1\tfrac{1}{2}.$$

And from (1),
$$15x + 25\tfrac{1}{2} = 65\tfrac{1}{2};$$

whence
$$15x = 40;$$
$$\therefore \quad x = 2\tfrac{2}{3}.$$

$\therefore$ the cost of a pound of tea is $2\tfrac{2}{3}$ shillings, or 2s. 8d., and the cost of a pound of coffee is $1\tfrac{1}{2}$ shillings, or 1s. 6d.

Example 3.　In a bag containing black and white balls, half the number of white is equal to a third of the number of black; and twice the whole number of balls exceeds three times the number of black balls by four.　How many balls did the bag contain?

Let x be the number of white balls, and y the number of black balls; then the bag contains $x + y$ balls.

We have the following equations:

$$\frac{x}{2} = \frac{y}{3} \quad\dots\dots\dots\dots\dots\dots\dots\dots\dots\dots(1),$$

$$2(x + y) = 3y + 4 \quad\dots\dots\dots\dots\dots\dots\dots\dots(2).$$

Substituting from (1) in 2, we obtain

$$\frac{4y}{3} + 2y = 3y + 4 \; ;$$

whence $\qquad\qquad y = 12 ;$

and from (1), $\qquad\quad x = 8.$

Thus there are 8 white and 12 black balls.

73.　In a problem involving *the digits of a number* the student should carefully notice the way in which the value of a number is algebraically expressed in terms of its digits.

Consider a number of three digits such as 435; its value is $4 \times 100 + 3 \times 10 + 5$.　Similarly a number whose digits beginning from the left are x, y, z

$$= x \text{ hundreds} + y \text{ tens} + z \text{ units}$$

$$= 100x + 10y + z.$$

Example.　A certain number of two digits is three times the sum of its digits, and if 45 be added to it the digits will be reversed; find the number.

Let x be the digit in the tens' place, y the digit in the units' place; then the number will be represented by $10x + y$, and the number formed by reversing the digits will be represented by $10y + x$.

Hence we have the two equations

$$10x + y = 3(x + y) \quad\dots\dots\dots\dots\dots\dots(1),$$

and $\qquad\qquad 10x + y + 45 = 10y + x \quad\dots\dots\dots\dots\dots(2).$

From (1), $\qquad\qquad 7x = 2y ;$

from (2), $\qquad\qquad y - x = 5.$

From these equations we obtain $x = 2, y = 7$.

Thus the number is 27.

EXAMPLES XI.

1. Find two numbers whose sum is 54, and whose difference is 12.

2. The sum of two numbers is 97, and their difference is 51 ; find the numbers.

3. One-fifth of the difference of two numbers is 3, and one-third of their sum is 17 ; find the numbers.

4. One-sixth of the sum of two numbers is 14, and half their difference is 13 ; find the numbers.

5. Four sheep and seven cows are worth £131, while three cows and five sheep are worth £66. What is the value of each animal?

6. A farmer bought 7 horses and 9 cows for £330. He could have bought 10 horses and 5 cows for the same money ; find the price of each animal.

7. Twice A's age exceeds three times B's age by 2 years ; if the sum of their ages is 61 years, how old are they?

8. Half A's age exceeds a quarter of B's age by 1 year, and three quarters of B's age exceeds A's by 11 years ; find the age of each.

9. In eight hours C walks 3 miles more than D does in 6 hours, and in seven hours D walks 9 miles more than C does in six hours ; how many miles does each walk per hour?

10. In 9 hours a coach travels one mile more than a train does in 2 hours, but in three hours the train travels 2 miles more than the coach does in 13 hours ; find the rate of each per hour.

11. A bill of £3. 8s. is paid with half-crowns and shillings, and three times the number of half-crowns exceeds twice the number of shillings by 8 ; how many of each are used?

12. A bill of £1. 18s. is paid with shillings and sixpences, and five times the number of sixpences exceeds seven times the number of shillings by 6 ; how many of each are used?

13. Forty-six tons of goods are to be carried in carts and waggons and it is found that this will require 10 waggons and 14 carts, or else 13 waggons and 9 carts ; how many tons can each waggon and each cart carry?

14. A sum of £7. 5s. is given to 17 boys and 15 girls ; the same amount could have been given to 13 boys and 20 girls ; find how much each boy and each girl receives.

15. A certain number of two digits is seven times the sum of the digits, and if 36 be taken from the number the digits will be reversed ; find the number.

16. A certain number of two digits is four times the sum of the digits, and if 27 be added to the number the digits will be reversed ; find the number.

17. A certain number between 10 and 100 is six times the sum of the digits, and the number exceeds the number formed by reversing the digits by 9 ; find the number.

18. The digits of a number between 10 and 100 are equal to each other, and the number exceeds 5 times the sum of the digits by 8 ; find the number.

19. A man has £100 in sovereigns, half-crowns, and shillings ; the number of the coins is 852, and their weight is 235 ounces. If a sovereign weighs $\frac{1}{4}$ oz., a half-crown $\frac{1}{2}$ oz., and a shilling $\frac{1}{5}$ oz., find how many of each kind of the coins he has.

20. A man has £5 worth of silver in half-crowns, shillings, and threepenny pieces. He has in all 70 coins. If he changed the threepenny pieces for halfpence, and half the shillings for sixpences, he would then have 180 coins. How many of each had he at first?

21. Divide £100 between 3 men, 5 women, 4 boys, and 3 girls, so that each man shall have as much as a woman and a girl, each woman as much as a boy and a girl, and each boy half as much as a man and a girl.

[*Chapter XVII. will furnish further practice in Problems.*]

MISCELLANEOUS EXAMPLES I.

The following Exercise consists of Miscellaneous Examples for revision on all the rules hitherto explained. The questions are selected from recent Examination Papers set by the Science and Art Department in Mathematics, Stage I.

1. If $x=2$ and $y=-\frac{1}{2}$, find the numerical values of

$$3(x+y)-2(x-y)\ ; \quad x^4y^6\ ; \quad x^3y^5\ ; \quad \text{and} \quad \frac{x^3-y^3}{x-y}.$$

2. Subtract $\quad 6x+1-2(x+5y)-\{2-(x+2y-1)\}$

from $\quad 2(2-y)-7x-2\{1-3y+6(y-x)\}.$

3. Find by how much $\quad (x^2-3x+1)^2$

is greater than $\quad x(x-1)(x-2)(x-3).$

4. Multiply together $x^2-3x+2,\ x^2-2x-3,$ and x^2+5x+6 ; and divide $\quad 7x^3-23x^2y+7xy^2-3y^3$ by $x-3y.$

5. Given $x=\frac{1}{3},\ y=-\frac{3}{2},\ z=-4$; find the values of

$$6xy-4yz+2zx\ ; \quad \frac{9x}{8y^2}+\frac{24y}{z^2}+\frac{z}{36x^2}\ ; \quad \frac{3x}{y+z}+\frac{z}{x+y}.$$

6. Solve the equations :

 (i.) $2x+3y=3x+2y=25$; (ii.) $2x=9-3y,$
                $5y=24-6x.$

7. Divide $5a^4-18a^3x+5a^2x^2-x^4$ by $x^2+3ax-a^2$.

8. If $x=3$, and $4y=-1$, find the numerical values of
     $4(x-4y^3)$; $(x-2y)(x+3y)$; x^3y^5.

9. Subtract $3\{x+2y-\frac{2}{3}(y-4)\}-4(3x-y+2)$
from $3x+2-[7y-4x-\{4(x-2)-3(2y+1)\}]$.

10. Multiply $x^4+2x^3-8x-16$ by x^2-2x+4 ;
and divide $x^5-5x^2y^3+5xy^4-y^5$ by $x^2-2xy+y^2$.

11. Given $2x=1$, $3y=-4$, $2z=7$; find the numerical values of the following expressions :

$$\frac{x}{3y}-\frac{y}{2z} ;\qquad x^2y^3+2y^2z-3yz^2 ;\qquad \frac{1}{x+y}+\frac{2}{y+z}.$$

12. Simplify

  $(a-b)(b+c)(c+a)+(b-c)(c+a)(a+b)+(c-a)(a+b)(b+c)$;

and find its value, when $a=1$, $b=3$, $c=-2$.

13. Multiply $2x^3-3x^2y+4xy^2-5y^3$ by $2x^2+3xy+4y^2$; and find the value of the product when $x=-1$, $y=\cdot1$.

14. Simplify the expression

  $(ac-b^2)(ce-d^2)+(ae-c^2)(bd-c^2)-(ad-bc)(be-ca)$.

15. Find the value of

$$.\ \frac{(ac-bc)(a+b)+bc(c-a)-ca(a-b)}{(b-c)(c-a)(a+b)},$$

when $a=1$, $b=3$, $c=4$.

16. Simplify $2x^2-3\left(\frac{y^2}{2}-x^2\right)-\frac{1}{2}\{4x^2-\frac{1}{2}(24x^2+5y^2)\}$,

and divide the simplified expression by $x+\frac{y}{6}$.

17. Simplify the following expressions, and find their product :
     $5x-(2y-x-1)-2\{3y-2(x+y)\}$,
and $3x-4y-2(x-2y)+\frac{3}{2}\{2x-1-3(x-y)\}$.

18. Solve the equations :

 (i.) $\frac{x}{3}+5=\frac{2y}{3}$,   (ii.) $\frac{x+y}{3}=x-7$,

    $y-x=\frac{x}{3}$; $\frac{x}{3}=y+1$.

CHAPTER XII.

RESOLUTION INTO FACTORS.

74. DEFINITION. When an algebraical expression is the product of two or more expressions each of these latter quantities is called a **factor** of it, and the determination of these quantities is called the **resolution** of the expression into its factors.

In this chapter we shall explain the principal rules by which the resolution of expressions into their component factors may be effected.

75. When each of the terms which compose an expression is divisible by a common factor, the expression may be simplified by dividing each term separately by this factor, and enclosing the quotient within brackets; the common factor being placed outside as a coefficient.

Example 1. The terms of the expression $3a^2 - 6ab$ have a common factor $3a$;
$$\therefore \; 3a^2 - 6ab = 3a(a - 2b).$$

Example 2. $5a^2bx^3 - 15abx^2 - 20b^3x^2 = 5bx^2(a^2x - 3a - 4b^2)$.

EXAMPLES XII. a.

Resolve into factors :

1. $x^2 + ax$.   **2.** $2a^2 - 3a$.   **3.** $a^3 - a^2$.   **4.** $a^3 - a^2b$.

5. $3m^2 - 6mn$.   **6.** $p^2 + 2p^2q$.   **7.** $x^5 - 5x^2$.   **8.** $y^2 + xy$.

9. $5a^2 - 25a^2b$.   **10.** $12x + 48x^2y$.   **11.** $10c^3 - 25c^4d$.

12. $27 - 162x$.   **13.** $x^2y^2z^2 + 3xy$.   **14.** $17x^2 - 51x$.

15. $2a^3 - a^2 + a$.   **16.** $3x^3 + 6a^2x^2 - 3a^3x$.

17. $7p^2 - 7p^3 + 14p^4$.   **18.** $4b^5 + 6a^2b^3 - 2b^2$.

19. $x^3y^3 - x^2y^2 + 2xy$.   **20.** $26a^3b^5 + 39a^4b^2$.

76. An expression may be resolved into factors if the terms can be arranged in groups which have a compound factor common.

Example 1.   Resolve into factors $x^2 - ax + bx - ab$.

Noticing that the first two terms contain a common factor x, and the last two terms a common factor b, we enclose the first two terms in one bracket, and the last two in another.   Thus

$$
\begin{aligned}
x^2 - ax + bx - ab &= (x^2 - ax) + (bx - ab) \\
&= x(x - a) + b(x - a) \\
&= (x - a) \text{ taken } x \text{ times } plus \ (x - a) \text{ taken } b \text{ times} \\
&= (x - a) \text{ taken } (x + b) \text{ times} \\
&= (x - a)(x + b).
\end{aligned}
$$

Example 2.   Resolve into factors $12a^2 + bx^2 - 4ab - 3ax^2$.

$$
\begin{aligned}
12a^2 + bx^2 - 4ab - 3ax^2 &= (12a^2 - 4ab) - (3ax^2 - bx^2) \\
&= 4a(3a - b) - x^2(3a - b) \\
&= (3a - b)(4a - x^2).
\end{aligned}
$$

EXAMPLES XII. b.

Resolve into factors :

1. $x^2 + xy + xz + yz$.
2. $x^2 - xz + xy - yz$.
3. $a^2 + 2a + ab + 2b$.
4. $a^2 + ac + 4a + 4c$.
5. $2a + 2x + ax + x^2$.
6. $3q - 3p + pq - p^2$.
7. $am - bm - an + bn$.
8. $ab - by - ay + y^2$.
9. $pq + qr - pr - r^2$.
10. $2mx + nx + 2my + ny$.
11. $ax - 2ay - bx + 2by$.
12. $2a^2 + 3ab - 2ac - 3bc$.
13. $ac^2 + b + bc^2 + a$.
14. $ac^2 - 2a - bc^2 + 2b$.
15. $a^3 - a^2 + a - 1$.
16. $2x^3 + 3 + 2x + 3x^2$.
17. $a^2x - aby + 2ax - 2by$.
18. $axy + bcxy - az - bcz$.
19. $7x^3 - 4x^2 - 21x + 12$.
20. $3x^3 - 12ax^2 - 2a^2x + 8a^3$.

Factors of Trinomial Expressions.

77. Before proceeding to the next case of resolution into factors, we draw the students' attention to the way in which, in forming the product of two binomials, the coefficients of the different terms combine so as to give a trinomial result.

Thus

$$(x+5)(x+3)=x^2+8x+15 \quad\dots\dots\dots\dots(1),$$
$$(x-5)(x-3)=x^2-8x+15 \quad\dots\dots\dots\dots(2),$$
$$(x+5)(x-3)=x^2+2x-15 \quad\dots\dots\dots\dots(3),$$
$$(x-5)(x+3)=x^2-2x-15 \quad\dots\dots\dots\dots(4).$$

We now propose to consider the converse problem : namely, the resolution of a trinomial expression, similar to those which occur on the right-hand side of the above identities, into its component binomial factors.

By examining the above results, we notice that :

(i.) The first term of both the factors is x.

(ii.) The product of the second terms of the two factors is equal to the third term of the trinomial ; e.g. in (2) we see that 15 is the product of -5 and -3 ; and in (3) we see that -15 is the product of $+5$ and -3.

(iii.) The algebraic sum of the second terms of the two factors is equal to the coefficient of x in the trinomial ; e.g. in (4) the sum of -5 and $+3$ gives -2, the coefficient of x in the trinomial.

The application of these laws will be easily understood from the following examples.

Example 1. Resolve into factors $x^2 + 11x + 24$.

The second terms of the factors must be such that their product is $+24$, and their sum $+11$. It is clear that they must be $+8$ and $+3$.

$$\therefore \ x^2 + 11x + 24 = (x + 8)(x + 3).$$

Example 2. Resolve into factors $x^2 - 10x + 24$.

The second terms of the factors must be such that their product is $+24$, and their sum -10. Hence they must *both* be *negative*, and it is easy to see that they must be -6 and -4.

$$\therefore \ x^2 - 10x + 24 = (x - 6)(x - 4).$$

Example 3. Resolve into factors $x^2 - 11ax + 10a^2$.

The second terms of the factors must be such that their product is $+10a^2$, and their sum $-11a$. Hence they must be $-10a$ and $-a$.

$$\therefore \ x^2 - 11ax + 10a^2 = (x - 10a)(x - a).$$

EXAMPLES XII. c.

Resolve into factors :

1. $x^2 + 3x + 2$.	**2.** $y^2 + 5y + 6$.	**3.** $y^2 + 7y + 12$.
4. $a^2 - 3a + 2$.	**5.** $a^2 - 6a + 8$.	**6.** $b^2 - 5b + 6$.
7. $b^2 + 13b + 42$.	**8.** $b^2 - 13b + 40$.	**9.** $z^2 - 13z + 36$.
10. $x^2 - 15x + 56$.	**11.** $x^2 - 15x + 54$.	**12.** $z^2 + 15z + 44$.

Resolve into factors :

13. $b^2 - 12b + 36.$ **14.** $a^2 + 15a + 56.$ **15.** $a^2 - 12a + 27.$

16. $x^2 + 9x + 20.$ **17.** $x^2 - 10x + 9.$ **18.** $x^2 - 16x + 64.$

19. $y^2 - 23y + 102.$ **20.** $y^2 - 24y + 95.$ **21.** $y^2 + 54y + 729.$

22. $a^2 + 10ab + 21b^2.$ **23.** $a^2 + 12ab + 11b^2.$ **24.** $a^2 - 23ab + 132b^2.$

25. $m^4 + 8m^2 + 7.$ **26.** $m^4 + 9m^2n^2 + 14n^4.$ **27.** $x^2y^2 - 5xy + 6.$

28. $a^2b^2 - 15ab + 54.$ **29.** $13 + 14y + y^2.$ **30.** $216 - 35a + a^2.$

78. Next consider a case where the third term of the trinomial is negative.

Example 1. Resolve into factors $x^2 + 2x - 35.$

The second terms of the factors must be such that their product is -35, and their *algebraical sum* $+2$. Hence they must have *opposite* signs, and the greater of them must be *positive* in order to give its sign to their sum.

The required terms are therefore $+7$ and -5.

$$\therefore \quad x^2 + 2x - 35 = (x + 7)(x - 5).$$

Example 2. Resolve into factors $x^2 - 3x - 54.$

The second terms of the factors must be such that their product is -54, and their *algebraical sum* -3. Hence they must have *opposite* signs, and the greater of them must be *negative* in order to give its sign to their sum.

The required terms are therefore -9 and $+6$.

$$\therefore \quad x^2 - 3x - 54 = (x - 9)(x + 6).$$

EXAMPLES XII. d.

Resolve into factors :

1. $x^2 + x - 2.$ **2.** $x^2 - x - 6.$ **3.** $x^2 - x - 20.$

4. $y^2 + 4y - 12.$ **5.** $y^2 + 4y - 21.$ **6.** $y^2 - 5y - 36.$

7. $a^2 + 8a - 33.$ **8.** $a^2 - 13a - 30.$ **9.** $a^2 + a - 132.$

10. $b^2 - 12b - 45.$ **11.** $b^2 + 14b - 51.$ **12.** $b^2 + 10b - 39.$

13. $m^2 - m - 56.$ **14.** $m^2 - 5m - 84.$ **15.** $m^2 + m - 56.$

16. $p^2 - 8p - 65.$ **17.** $p^2 + 3p - 108.$ **18.** $p^2 + p - 110.$

19. $x^2 + 2x - 48.$ **20.** $x^2 - 7x - 120.$ **21.** $x^2 - x - 132.$

22. $y^4 + 13y^2 - 48.$ **23.** $y^2 + 4xy - 96x^2.$ **24.** $y^2 + 7xy - 98x^2.$

25. $a^4 + a^2b^2 - 72b^4.$ **26.** $a^2 + ab - 240b^2.$ **27.** $a^2b^2 - 5ab - 14.$

28. $a^2b^2 - 2abc - 35c^2.$ **29.** $96 - 4b - b^2.$ **30.** $72 + b - b^2.$

79. We proceed now to the resolution into factors of trinomial expressions when the coefficient of the highest power is not unity.

By observing the manner in which, in ordinary multiplication, the terms of the product are formed, we may write down the following results :

$$(3x+2)(x+4)=3x^2+14x+8 \dots \dots (1),$$
$$(3x-2)(x-4)=3x^2-14x+8 \dots \dots (2),$$
$$(3x+2)(x-4)=3x^2-10x-8 \dots \dots (3),$$
$$(3x-2)(x+4)=3x^2+10x-8 \dots \dots (4).$$

Here we see, as before, that

(i.) If the third term of the trinomial is positive, then the second terms of its factors have both the same sign, and this sign is the same as that of the middle term of the trinomial.

(ii.) If the third term of the trinomial is negative, then the second terms of its factors have opposite signs.

Now consider in detail the result $3x^2-14x+8=(3x-2)(x-4)$.

The first term $3x^2$ is the product of $3x$ and x.

The third term $+8 \dots \dots -2$ and -4.

The middle term $-14x$ is the result of adding together the two products $3x \times -4$ and $x \times -2$.

Again, consider the result $3x^2-10x-8=(3x+2)(x-4)$.

The first term $3x^2$ is the product of $3x$ and x.

The third term $-8 \dots \dots +2$ and -4.

The middle term $-10x$ is the result of adding together the two products $3x \times -4$ and $x \times 2$; and its sign is negative because the greater of these two products is negative.

The above observations lead us to the following method.

Example 1.   Resolve into factors $7x^2-19x-6$.

Write down $(7x \quad 3)(x \quad 2)$ for a first trial, noticing that 3 and 2 must have opposite signs.   These factors give $7x^2$ and -6 for the first and third terms.   But since $7 \times 2 - 3 \times 1 = 11$, the combination fails to give the correct coefficient of the middle term.

Next try $(7x \quad 2)(x \quad 3)$.

Since $7 \times 3 - 2 \times 1 = 19$, these factors will be correct if we insert the signs so that the negative shall predominate.

Thus $\qquad 7x^2-19x-6=(7x+2)(x-3)$.

[Verify by mental multiplication.]

Example 2.   Resolve into factors $14x^2 + 29x - 15$ ....(1),

$$14x^2 - 29x - 15 ....(2).$$

In each case we may write down $(7x\ \ 3)(2x\ \ 5)$ as a first trial, noticing that 3 and 5 must have opposite signs.

And since $7 \times 5 - 3 \times 2 = 29$, we have only now to insert the proper signs in each factor.

In (1) the positive sign must predominate,

in (2) the negative...........................

Therefore $14x^2 + 29x - 15 = (7x - 3)(2x + 5).$

$14x^2 - 29x - 15 = (7x + 3)(2x - 5).$

Example 3.   Resolve into factors $5x^2 + 17x + 6$ ....(1),

$$5x^2 - 17x + 6 ....(2).$$

In (1) we notice that the factors which give 6 are both positive.

In (2) ....negative.

And therefore for (1) we may write $(5x +\ \ )(x +\ \ )$.

(2)....................$(5x -\ \ )(x -\ \ )$.

And, since $5 \times 3 + 1 \times 2 = 17$, we see that

$$5x^2 + 17x + 6 = (5x + 2)(x + 3).$$

$$5x^2 - 17x + 6 = (5x - 2)(x - 3).$$

EXAMPLES XII. e.

Resolve into factors :

1. $2a^2 + 3a + 1.$ 2. $3a^2 + 4a + 1.$ 3. $4a^2 + 5a + 1.$

4. $2a^2 + 5a + 2.$ 5. $3a^2 + 10a + 3.$ 6. $2a^2 + 7a + 3.$

7. $5a^2 + 7a + 2.$ 8. $2a^2 + 9a + 10.$ 9. $2a^2 + 7a + 6.$

10. $2x^2 + 9x + 4.$ 11. $2x^2 + 5x - 3.$ 12. $3x^2 + 5x - 2.$

13. $3y^2 + y - 2.$ 14. $3y^2 - 7y - 6.$ 15. $2y^2 + 9y - 5.$

16. $2b^2 - 5b - 3.$ 17. $6b^2 + 7b - 3.$ 18. $2b^2 + b - 15.$

19. $4m^2 + 5m - 6.$ 20. $4m^2 - 4m - 3.$ 21. $6m^2 - 7m - 3.$

22. $4x^2 - 8xy - 5y^2.$ 23. $6x^2 - 7xy + 2y^2.$ 24. $6x^2 - 13xy + 2y^2.$

25. $12a^2 - 17ab + 6b^2.$ 26. $6a^2 - 5ab - 6b^2.$ 27. $6a^2 + 35ab - 6b^2.$

28. $2 - 3y - 2y^2.$ 29. $3 + 23y - 8y^2.$ 30. $8 + 18y - 5y^2.$

31. $4 + 17x - 15x^2$ 32. $6 - 13a + 6a^2.$ 33. $28 - 31b - 5b^2.$

The Difference of Two Squares.

80. By multiplying $a+b$ by $a-b$ we obtain the identity

$$(a+b)(a-b)=a^2-b^2,$$

from which we see that *the difference of the squares of any two quantities is equal to the product of the sum and the difference of the two quantities.*

Thus any expression which is the difference of two squares may at once be resolved into factors.

Example. Resolve into factors $25x^2-16y^2$.

$$25x^2-16y^2=(5x)^2-(4y)^2$$
$$=(5x+4y)(5x-4y).$$

The intermediate step may usually be omitted.

Example. $1-49c^6=(1+7c^3)(1-7c^3)$.

The difference of the squares of two numerical quantities is sometimes conveniently found by the aid of the formula

$$a^2-b^2=(a+b)(a-b).$$

Example. $(329)^2 \; (171)^2=(329+171)(329-171)$
$$=500 \times 158$$
$$=79000.$$

EXAMPLES XII. f.

Resolve into factors :

1. a^2-9. **2.** a^2-49. **3.** a^2-81. **4.** x^2-25.

5. $64-x^2$. **6.** $81-4x^2$. **7.** $4y^2-1$. **8.** y^2-9a^2.

9. $4y^2-25$. **10.** $9y^2-49x^2$. **11.** $4m^2-81$. **12.** $36a^2-1$.

13. $9a^2-25b^2$. **14.** $121-16y^2$. **15.** $25-c^4$.

16. $49a^4-100b^2$. **17.** $4p^2q^2-81$. **18.** $a^4b^4c^2-9$.

19. x^6-4a^4. **20.** x^4-25z^4. **21.** $a^{10}-p^2q^4$.

22. $16a^{16}-9b^6$. **23.** $25x^{12}-4$. **24.** $a^6b^8c^4-9x^2$.

Find by factors the value of

25. $(39)^2-(31)^2$. **26.** $(51)^2-(49)^2$. **27.** $(1001)^2-1$.

28. $(82)^2-(18)^2$. **29.** $(275)^2-(225)^2$. **30.** $(936)^2-(64)^2$.

The Sum or Difference of Two Cubes.

81. If we divide a^3+b^3 by $a+b$ the quotient is a^2-ab+b^2 : and if we divide a^3-b^3 by $a-b$ the quotient is a^2+ab+b^2.

We have therefore the following identities :

$$a^3+b^3=(a+b)(a^2-ab+b^2) ;$$
$$a^3-b^3=(a-b)(a^2+ab+b^2).$$

These results are very important, and enable us to resolve into factors any expression which can be written as the sum or the difference of two cubes.

Example 1.　　$8x^3-27y^3=(2x)^3-(3y)^3$
$$=(2x-3y)(4x^2+6xy+9y^2).$$

Note. The middle term $6xy$ is the *product* of $2x$ and $3y$.

Example 2.　　$64a^3+1=(4a)^3+(1)^3$
$$=(4a+1)(16a^2-4a+1).$$

We may usually omit the intermediate step and write down the factors at once.

Examples.　$343a^6-27x^3=(7a^2-3x)(49a^4+21a^2x+9x^2).$
$8x^9+729=(2x^3+9)(4x^6-18x^3+81).$

EXAMPLES XII. g.

Resolve into factors :

1. $a^3-b^3.$　　2. $a^3+b^3.$　　3. $1+x^3.$　　4. $1-y^3.$
5. $8x^3+1.$　　6. $x^3-8z^3.$　　7. $a^3+27b^3.$　　8. $x^3y^3-1.$
9. $1-8a^3.$　10. $b^3-8.$　11. $27+x^3.$　12. $64-p^3.$
13. $125a^3+1.$　14. $216-b^3.$　15. $x^3y^3+343.$
16. $1000x^3+1.$　17. $512a^3-1.$　18. $a^3b^3c^3-27.$
19. $8x^3-343.$　20. $x^3+216y^3.$　21. $x^6-27z^3.$
22. $m^3-1000n^6.$　23. $a^3-729b^3.$　24. $125a^6+512b^3.$

Harder Cases of Resolution into Factors.

82. We shall now give some harder applications of the foregoing rules, followed by a miscellaneous exercise in which all the processes of this chapter will be illustrated.

Example 1. Resolve into factors $(a+2b)^2-16x^2.$

This expression, being *the difference between two squares*, is resolved into factors by the rule of Art. 80.

$$\therefore (a+2b)^2-16x^2=(a+2b+4x)(a+2b-4x).$$

If the factors contain like terms they should be collected so as to give the result in its simplest form.

Example 2. $(3x+7y)^2 - (2x-3y)^2$
$$= \{(3x+7y) + (2x-3y)\} \{(3x+7y) - (2x-3y)\}$$
$$= (3x+7y+2x-3y)(3x+7y-2x+3y)$$
$$= (5x+4y)(x+10y).$$

83. By suitably grouping together the terms, compound expressions can often be expressed as the difference of two squares, and so be resolved into factors.

Example 1. Resolve into factors $9a^2 - c^2 + 4cx - 4x^2$.
$$9a^2 - c^2 + 4cx - 4x^2 = 9a^2 - (c^2 - 4cx + 4x^2)$$
$$= (3a)^2 - (c-2x)^2$$
$$= (3a+c-2x)(3a-c+2x).$$

Example 2. Resolve into factors $2bd - a^2 - c^2 + b^2 + d^2 + 2ac$.

Here the terms $2bd$ and $2ac$ suggest the proper preliminary arrangement of the expression. Thus
$$2bd - a^2 - c^2 + b^2 + d^2 + 2ac = b^2 + 2bd + d^2 - a^2 + 2ac - c^2$$
$$= b^2 + 2bd + d^2 - (a^2 - 2ac + c^2)$$
$$= (b+d)^2 - (a-c)^2$$
$$= (b+d+a-c)(b+d-a+c).$$

84. Sometimes an expression may be resolved into more than two factors.

Example 1. Resolve into factors, $32a^5b - 162ab^5$.
$$32a^5b - 162ab^5 = 2ab(16a^4 - 81b^4)$$
$$= 2ab(4a^2 + 9b^2)(4a^2 - 9b^2)$$
$$= 2ab(4a^2 + 9b^2)(2a + 3b)(2a - 3b).$$

Example 2. Resolve into factors $x^6 - y^6$.
$$x^6 - y^6 = (x^3 + y^3)(x^3 - y^3)$$
$$= (x+y)(x^2 - xy + y^2)(x-y)(x^2 + xy + y^2).$$

Note. When an expression can be arranged either as the difference of two squares, or as the difference of two cubes, it will be found simplest to first use the rule for the difference of two squares.

85. The following case is important.

Example. Resolve into factors $x^4 + x^2y^2 + y^4$.
$$x^4 + x^2y^2 + y^4 = (x^4 + 2x^2y^2 + y^4) - x^2y^2$$
$$= (x^2 + y^2)^2 - (xy)^2$$
$$= (x^2 + y^2 + xy)(x^2 + y^2 - xy)$$
$$= (x^2 + xy + y^2)(x^2 - xy + y^2).$$

68 ALGEBRA. [CHAP.

86. The student should verify by actual multiplication the following identity. [See Art. 41, Ex. 3.]

$$a^3 + b^3 + c^3 - 3abc = (a+b+c)(a^2+b^2+c^2-bc-ca-ab).$$

Similarly

$$x^3 - 8y^3 + 125z^3 + 30xyz = x^3 + (-2y)^3 + (5z)^3 - 3x(-2y)(5z)$$
$$= (x - 2y + 5z)(x^2 + 4y^2 + 25z^2 + 10yz - 5zx + 2xy).$$

EXAMPLES XII.

Resolve into two or more factors :

1. $(x+y)^2 - z^2$.
2. $(x-y)^2 - z^2$.
3. $(a+2b)^2 - c^2$.
4. $(a+3c)^2 - 1$.
5. $(2x-1)^2 - a^2$.
6. $a^2 - (b+c)^2$.
7. $4a^2 - (b-1)^2$.
8. $9 - (a+x)^2$.
9. $(2a-3b)^2 - c^2$.
10. $(18x+y)^2 - (17x-y)^2$.
11. $(6a+3)^2 - (5a-4)^2$.
12. $4a^2 - (2a-3b)^2$.
13. $x^2 - (2b-3c)^2$.
14. $(x+y)^2 - (m-n)^2$.
15. $(3x+2y)^2 - (2x-3y)^2$.
16. $a^2 - 2ax + x^2 - 4b^2$.
17. $x^2 + a^2 + 2ax - z^2$.
18. $1 - a^2 - 2ab - b^2$.
19. $12xy + 25 - 4x^2 - 9y^2$.
20. $c^2 - a^2 - b^2 + 2ab$.
21. $x^2 - 2x + 1 - m^2 - 4mn - 4n^2$.
22. $x^4 + y^4 - z^4 - a^4 + 2x^2y^2 - 2a^2z^2$.
23. $(m+n+p)^2 - (m-n+p)^2$.
24. $a^4 + a^2 + 1$.
25. $a^4b^4 - 16$.
26. $256x^4 - 81y^4$.
27. $16a^4b^2 - b^6$.
28. $64m^7 - mn^6$.
29. $x^4 - x^4y^4$.
30. $a^2b^5 - 81a^2b$.
31. $400a^2x - x^3$.
32. $1 - 729y^6$.
33. $216b^6 + a^3b^3$.
34. $250z^3 + 2$.
35. $1029 - 3x^3$.
36. $6x^3y^2 + 15x^2y^2 - 36xy^2$.
37. $2m^8n^4 - 7m^4n^6 - 4n^8$.
38. $98x^4 - 7x^2y^2 - y^4$.
39. $a^2b^2 - a^2 - b^2 + 1$.
40. $x^3 - 2x^2 - x + 2$.
41. $(a+b)^3 + 1$.
42. $a^2x^3 - 8a^2y^3 - 4b^2x^3 + 32b^2y^3$.
43. $2p - 3q + 4p^2 - 9q^2$.
44. $119 + 10m - m^2$.
45. $24a^2b^2 - 30ab^3 - 36b^4$.
46. $240x^2 + x^6y^4 - x^{10}y^8$.
47. $x^4 + 4x^2 + 16$.
48. $x^4 + y^4 - 7x^2y^2$.
49. $a^4 - 18a^2b^2 + b^4$.
50. $x^8 + x^4 + 1$.
51. $(a+b)^4 - c^4$.
52. $(c+d)^3 + (c-d)^3$.
53. $a^2 - b^2 + c(a-b)$.
54. $(a+b)^2 - ac - bc$.
55. $a^2 - 9b^2 + (a-3b)^2$.
56. $4(x-y)^3 - 25(x-y)$.
57. $a^3 + b^3 + 8c^3 - 6abc$.
58. $a^3 - 27b^3 + c^3 + 9abc$.
59. $a^3 + 8c^3 + 1 - 6ac$.
60. $8a^3 + 27b^3 + c^3 - 18abc$.

Converse Use of Factors.

87. The actual processes of multiplication and division can often be partially or wholly avoided by a skilful use of factors.

For example, the formula for resolving into factors the difference of two squares enables us to write down at once the product of the sum and the difference of two quantities.

Example 1.   Multiply $2a + 3b - c$ by $2a - 3b + c$.

These expressions may be arranged thus :
$$2a + (3b - c) \text{ and } 2a - (3b - c).$$
Hence the product $= \{2a + (3b - c)\} \{2a - (3b - c)\}.$
$$= (2a)^2 - (3b - c)^2.$$
$$= 4a^2 - (9b^2 - 6bc + c^2)$$
$$= 4a^2 - 9b^2 + 6bc - c^2.$$

Example 2.   Find the product of
$$x + 2, \; x - 2, \; x^2 - 2x + 4, \; x^2 + 2x + 4.$$

Taking the first factor with the third, and the second with the fourth,

the product $= \{(x + 2)(x^2 - 2x + 4)\} \{(x - 2)(x^2 + 2x + 4)\}$
$$= (x^3 + 8)(x^3 - 8)$$
$$= x^6 - 64.$$

EXAMPLES XII. k.

Employ factors to obtain the product of

1. $a - b + c, \; a - b - c.$ **2.** $2x - y + z, \; 2x + y + z.$

3. $1 + 2x - x^2, \; 1 - 2x - x^2.$ **4.** $c^2 + 3c + 2, \; c^2 - 3c - 2.$

5. $a + b - c + d, \; a + b + c - d.$ **6.** $p - q + x - y, \; p - q - x + y.$

Find, by factors, the continued product of

7. $(a - b)^2, \; (a + b)^2, \; (a^2 + b^2)^2.$

8. $(1 - x)^3, \; (1 + x)^3, \; (1 + x^2)^3.$

9. $a^2 - 4a + 3, \; a^2 - a - 2, \; a^2 + 5a + 6.$

10. $3 - y, \; 3 + y, \; 9 - 3y + y^2, \; 9 + 3y + y^2.$

11. $1 + c + c^2, \; 1 - c + c^2, \; 1 - c^2 + c^4.$

12. Divide $a^3(a + 2)(a^2 - a - 56)$ by $a^2 + 7a$, employing factors.

13. Divide $3x^2(x + 4)(x^2 - 9)$ by $x^2 + x - 12.$

14. Divide the product of $2a^2 + 11a - 21$ and $3a^2 - 20a - 7$ by $a^2 - 49.$

15. Divide $x^6 - 7x^3 - 8$ by $(x + 1)(x^2 + 2x + 4).$

CHAPTER XIII.

HIGHEST COMMON FACTOR.

88. DEFINITION. The **highest common factor** of two or more algebraical expressions is the *expression of highest dimensions* (Art. 22) which divides each of them without remainder.

The abbreviation H.C.F. is sometimes used for *highest common factor*.

Note. The term *greatest common measure* is sometimes used instead of *highest common factor* ; but this usage is incorrect, for in Algebra our object is to find the factor of *highest dimensions* which is common to two or more expressions, and we are not concerned with the *numerical* values of the expressions or their divisors. The term *greatest common measure* ought to be confined solely to arithmetical quantities, for it can easily be shewn by trial that the algebraical highest common factor is not always the greatest common measure. [See Hall and Knight's *Elementary Algebra*, Art. 145.]

89. In the case of *simple expressions* the highest common factor can be written down by inspection.

Example. The highest common factor of a^3b^4, $a^2b^6c^2$, a^4b^7c is a^2b^4; for a^2 is the highest power of a that will divide the given expressions, and b^4 is the highest power of b that will divide them ; while c is not a *common* factor.

90. If the expressions have numerical coefficients, find by Arithmetic their greatest common measure, and prefix it as a coefficient to the algebraical highest common factor.

Example. The highest common factor of $21a^4x^3y$, $35a^2x^4y$, $28a^3xy$ is $7a^2xy$; for it consists of the product of

(1) The greatest common measure of 35, 28, and 21 ;

(2) The highest power of each letter which divides every one of the given expressions.

91. An analogous method will enable us readily to find the highest common factor of *compound* expressions which are given as the product of factors, or which can be easily resolved into factors.

Example 1. Find the highest common factor of
$$3a^2 + 9ab, \quad a^3 - 9ab^2, \quad a^3 + 6a^2b + 9ab^2.$$

Resolving each expression into its factors, we have
$$3a^2 + 9ab = 3a(a + 3b),$$
$$a^3 - 9ab^2 = a(a + 3b)(a - 3b),$$
$$a^3 + 6a^2b + 9ab^2 = a(a + 3b)(a + 3b) ;$$
therefore the H.C.F. is $a(a + 3b)$.

Example 2. Find the H.C.F. of $x(a - x)^2$, $a(a - x)^3$, $2ax(a - x)^5$.
The H.C.F. is $(a - x)^2$, for it contains the highest power of the compound factor $a - x$, which is common to the given expressions.

Example 3. Find the highest common factor of
$$ax^2 + 2a^2x + a^3, \quad 2ax^2 - 4a^2x - 6a^3, \quad 3(ax + a^2)^2.$$

Here $ax^2 + 2a^2x + a^3 = a(x^2 + 2ax + a^2) = a(x + a)^2,$
$$2ax^2 - 4a^2x - 6a^3 = 2a(x^2 - 2ax - 3a^2) = 2a(x + a)(x - 3a),$$
$$3(ax + a^2)^2 = 3\{a(x + a)\}^2 = 3a^2(x + a)^2 ;$$
$$\therefore \quad \text{H.C.F.} = a(x + a).$$

EXAMPLES XIII. a.

Find the highest common factor of

1. $a^2x^3y, \quad b^3xy^4, \quad cx^4y^2.$
2. $12a^3bc^2, \quad 18ab^2c^3.$
3. $15x^2y, \quad 60x^5y^2z^3, \quad 25x^3z^4.$
4. $17xy^2z, \quad 51xy^2z^2, \quad 34x^2yz.$
5. $77a^3b^5c^2, \quad 33a^2b^3c^5, \quad ab^2c^6.$
6. $x^2 - y^2, \quad x^2 - xy.$
7. $3(a - b)^3, \quad a^2 - 2ab + b^2.$
8. $3a^3 - 2a^2b, \quad 3a^2 - 2ab.$
9. $9a^2 - 4b^2, \quad 6a^2 + 4ab.$
10. $x^6 - x^4y^2, \quad x^3y^2 + x^2y^3.$
11. $a^2x^3(a - x)^3, \quad 2a^2x^2(a - x)^2.$
12. $2x^2 - 8x + 8, \quad (x - 2)^3.$
13. $x^2y^2 - y^4, \quad xy^2 + y^3, \quad xy - y^2.$
14. $x^3y^3 - y^6, \quad y^2(xy - y^2)^2.$
15. $(a^2 - ax)^2, \quad (ax - x^2)^3.$
16. $(abc - bc^2)^2, \quad (a^2c - ac^2)^2.$
17. $x^3 - x^2 - 42x, \quad x^4 - 49x^2.$
18. $(x^3 - 5x^2)^2, \quad x^5 - 8x^4 + 15x^3.$
19. $2x^2 - 9x + 4, \quad 3x^2 \quad 7x - 20.$
20. $3c^4 + 5c^3 - 12c^2, \quad 6c^5 + 7c^4 - 20c^3.$
21. $4m^4 - 9m^2, \quad 6m^3 - 5m^2 - 6m, \quad 6m^4 + 5m^3 - 6m^2.$
22. $3a^4x^3 - 8a^3x^3 + 4a^2x^3, \quad 3a^5x^2 - 11a^4x^2 + 6a^3x^2,$
     $3a^4x^3 + 16a^3x^3 - 12a^2x^3.$

92. We shall now work out examples to illustrate the process of finding the highest common factor of expressions which cannot be readily resolved into factors. The method is analogous to that used in Arithmetic for finding the greatest common measure of two numbers. For a proof of the rule the reader may consult Hall and Knight's *Elementary Algebra*, Arts. 146, 147; but we may here conveniently *enunciate* two principles, which the student should bear in mind in reading the examples which follow.

I. *If an expression contain a certain factor, any multiple of the expression is divisible by that factor.*

II. *If two expressions have a common factor, it will divide their sum and their difference; and also the sum and the difference of any multiples of them.*

Example. Find the highest common factor of

$$4x^3 - 3x^2 - 24x - 9 \text{ and } 8x^3 - 2x^2 - 53x - 39.$$

x	$4x^3 - 3x^2 - 24x - 9$	$8x^3 - 2x^2 - 53x - 39$	2
	$4x^3 - 5x^2 - 21x$	$8x^3 - 6x^2 - 48x - 18$	
$2x$	$2x^2 - \ 3x - 9$	$4x^2 - \ 5x - 21$	2
	$2x^2 - \ 6x$	$4x^2 - \ 6x - 18$	
3	$3x - 9$	$x - \ 3$	
	$3x - 9$		

Therefore the H.C.F. is $x - 3$.

Explanation. First arrange the given expressions according to descending or ascending powers of x. The expressions so arranged having their first terms of the same order, we take for divisor that whose highest power has the smaller coefficient. Arrange the work in parallel columns as above. When the first remainder $4x^2 - 5x - 21$ is made the divisor we put the quotient x to the *left* of the dividend. Again, when the second remainder $2x^2 - 3x - 9$ is in turn made the divisor, the quotient 2 is placed to the *right*; and so on. As in Arithmetic, the last divisor $x - 3$ is the highest common factor required.

93. This method is only useful to determine the *compound* factor of the highest common factor. Simple factors of the given expressions must be first removed from them, and the highest common factor of these, if any, must be reserved and multiplied into the *compound* factor given by the rule.

Example. Find the highest common factor of

$$24x^4 - 2x^3 - 60x^2 - 32x \text{ and } 18x^4 - 6x^3 - 39x^2 - 18x.$$

We have $24x^4 - 2x^3 - 60x^2 - 32x = 2x(12x^3 - x^2 - 30x - 16)$,

and $18x^4 - 6x^3 - 39x^2 - 18x = 3x(6x^3 - 2x^2 - 13x - 6)$.

Also $2x$ and $3x$ have the common factor x. Removing the simple factors $2x$ and $3x$, and *reserving* their common factor x, we continue as in Art. 92.

$$
\begin{array}{r|l|l|r}
2x & 6x^3 - 2x^2 - 13x - 6 & 12x^3 - x^2 - 30x - 16 & 2 \\
   & 6x^3 - 8x^2 - 8x   & 12x^3 - 4x^2 - 26x - 12 & \\
\cline{2-2}\cline{3-3}
2 & 6x^2 - 5x - 6   & 3x^2 - 4x - 4 & x \\
   & 6x^2 - 8x - 8   & 3x^2 + 2x   & \\
\cline{2-2}\cline{3-3}
   & 3x + 2   & -6x - 4 & -2 \\
   &   & -6x - 4 & \\
\end{array}
$$

Therefore the H.C.F. is $x(3x + 2)$.

94. So far the process of Arithmetic has been found exactly applicable to the algebraical expressions we have considered. But in many cases certain modifications of the arithmetical method will be found necessary. These will be more clearly understood if it is remembered that, at every stage of the work, the remainder must contain as a factor of itself the highest common factor we are seeking. [See Art. 92, I. & II.]

Example 1. Find the highest common factor of

$$3x^3 - 13x^2 + 23x - 21 \text{ and } 6x^3 + x^2 - 44x + 21.$$

$$
\begin{array}{l|l|r}
3x^3 - 13x^2 + 23x - 21 & 6x^3 + x^2 - 44x + 21 & 2 \\
                        & 6x^3 - 26x^2 + 46x - 42 & \\
\cline{2-2}
                        & 27x^2 - 90x + 63 & \\
\end{array}
$$

Here on making $27x^2 - 90x + 63$ a divisor, we find that it is not contained in $3x^3 - 13x^2 + 23x - 21$ with an *integral* quotient. But noticing that $27x^2 - 90x + 63$ may be written in the form $9(3x^2 - 10x + 7)$, and also bearing in mind that every remainder in the course of the work contains the H.C.F., we conclude that the H.C.F. we are seeking is contained in $9(3x^2 - 10x + 7)$. But the two original expressions have no *simple* factors, therefore their H.C.F. can have none. We may therefore *reject* the factor 9 and go on with divisor $3x^2 - 10x + 7$.

Resuming the work, we have

$$
\begin{array}{r|l}
x & 3x^3 - 13x^2 + 23x - 21 \\
  & 3x^3 - 10x^2 + \ 7x \\
-1 & \ \ - 3x^2 + 16x - 21 \\
  & \ \ - 3x^2 + 10x - \ 7 \\ \hline
  & \quad 2\ )\ 6x - 14 \\ \cline{2-2}
  & \quad \quad 3x - \ 7
\end{array}
\qquad
\begin{array}{l|l}
3x^2 - 10x + 7 & x \\
3x^2 - \ 7x & \\ \cline{1-1}
\ \ - 3x + 7 & -1 \\
\ \ - 3x + 7 & \\ \hline
\end{array}
$$

Therefore the highest common factor is $3x - 7$.

The factor 2 has been removed on the same grounds as the factor 9 above.

95. Sometimes the process is more convenient when the expressions are arranged in ascending powers.

Example. Find the highest common factor of

$$3 - 4a - 16a^2 - 9a^3 \ \dots\dots\dots\dots\dots\dots\dots(1),$$

and

$$4 - 7a - 19a^2 - 8a^3 \ \dots\dots\dots\dots\dots\dots\dots(2).$$

As the expressions stand we cannot begin to divide one by the other without using a fractional quotient. The difficulty may be obviated by *introducing* a suitable factor, just as in the last case we found it useful to remove a factor when we could no longer proceed with the division in the ordinary way. The given expressions have no common *simple* factor, hence their H.C.F. cannot be affected if we multiply either of them by any simple factor.

Multiply (1) by 4 and use (2) as a divisor:

$$
\begin{array}{r|l}
 & 4 - \ 7a - \ 19a^2 - \ \ 8a^3 \\
 & 5 \\ \cline{2-2}
4 & 20 - 35a - \ 95a^2 - \ 40a^3 \\
 & 20 - 28a - \ 48a^2 \\ \cline{2-2}
 & \ \ - 7a - \ 47a^2 - \ 40a^3 \\
 & \ \ - 5 \\ \cline{2-2}
7a & \ \ 35a + 235a^2 + 200a^3 \\
 & \ \ 35a - \ 49a^2 - \ 84a^3 \\ \cline{2-2}
 & 284a^2\ |\ 284a^2 + 284a^3 \\ \cline{2-2}
 & \quad\quad 1 + a
\end{array}
\qquad
\begin{array}{l|l}
12 - 16a - 64a^2 - 36a^3 & 3 \\
12 - 21a - 57a^2 - 24a^3 & \\ \cline{1-1}
a\ |\ 5a - \ 7a^2 - 12a^3 & \\ \cline{1-1}
5 - \ 7a \ - 12a^2 & 5 \\
5 + \ 5a & \\ \cline{1-1}
\ \ - 12a - 12a^2 & -12a \\
\ \ - 12a \ - 12a^2 & \\
\end{array}
$$

Therefore the H.C.F. is $1 + a$.

After the first division the factor a is removed as explained in Art. 94; then the factor 5 is introduced because the first term of $4 - 7a - 19a^2 - 8a^3$ is not divisible by the first term of $5 - 7a - 12a^2$. At the next stage a factor -5 is introduced, and finally the factor $284a^2$ is removed.

96. From the last two examples it appears that we may multiply or divide either of the given expressions, or any of the remainders which occur in the course of the work, by any factor which does not divide both of the given expressions.

EXAMPLES XIII. b.

Find the highest common factor of

1. $2x^3 + 3x^2 + x + 6, \; 2x^3 + x^2 + 2x + 3.$

2. $2y^3 - 9y^2 + 9y - 7, \; y^3 - 5y^2 + 5y - 4.$

3. $2x^3 + 8x^2 - 5x - 20, \; 6x^3 - 4x^2 - 15x + 10.$

4. $a^3 + 3a^2 - 16a + 12, \; a^3 + a^2 - 10a + 8.$

5. $6x^3 - x^2 - 7x - 2, \; 2x^3 - 7x^2 + x + 6.$

6. $q^3 - 3q + 2, \; q^3 - 5q^2 + 7q - 3.$

7. $a^4 + a^3 - 2a^2 + a - 3, \; 5a^3 + 3a^2 - 17a + 6.$

8. $3y^4 - 3y^3 - 15y^2 - 9y, \; 4y^5 - 16y^4 - 44y^3 - 24y^2.$

9. $15x^4 - 15x^3 + 10x^2 - 10x, \; 30x^5 + 120x^4 + 20x^3 + 80x^2.$

10. $2m^4 + 7m^3 + 10m^2 + 35m, \; 4m^4 + 14m^3 - 4m^2 - 6m + 28.$

11. $3x^4 - 9x^3 + 12x^2 - 12x, \; 6x^3 - 6x^2 - 15x + 6.$

12. $2a^5 - 4a^4 - 6a, \; a^5 + a^4 - 3a^3 - 3a^2.$

13. $x^3 + 4x^2 - 2x - 15, \; x^3 - 21x - 36.$

14. $9a^4 + 2a^2x^2 + x^4, \; 3a^4 - 8a^3x + 5a^2x^2 - 2ax^3.$

15. $2 - 3a + 5a^2 - 2a^3, \; 2 - 5a + 8a^2 - 3a^3.$

16. $3x^2 - 5x^3 - 15x^4 - 4x^5, \; 6x - 7x^2 - 29x^3 - 12x^4.$

17. $6 - 8a - 32a^2 - 18a^3, \; 20 - 35a - 95a^2 - 40a^3.$

18. $9x^2 - 15x^3 - 45x^4 - 12x^5, \; 42x - 49x^2 - 203x^3 - 84x^4.$

19. $3x^5 - 5x^3 + 2, \; 2x^5 - 5x^2 + 3.$

20. $4x^5 - 6x^3 - 28x, \; 6x^4 + 10x^3 - 17x^2 - 35x - 14.$

CHAPTER XIV.

Lowest Common Multiple.

97. Definition. The **lowest common multiple** of two or more algebraical expressions is *the expression of lowest dimensions* which is divisible by each of them without remainder.

The abbreviation L.C.M. is sometimes used instead of the words *lowest common multiple.*

98. In the case of *simple expressions*, the lowest common multiple can be written down by inspection, as follows :

Example. The lowest common multiple of $21a^4x^3y$, $35a^2x^4y$, and $28a^3xy$ is $420a^4x^4y$; for it consists of the product of

(1) the lowest common multiple of 21, 35, and 28 ;

(2) the lowest power of each letter which is divisible by every power of that letter occurring in the given expressions.

99. The lowest common multiple of compound expressions which are given as the product of factors, or which can be easily resolved into factors, can be found in a similar way.

Example 1. The lowest common multiple of $6x^2(a-x)^2$, $8a^3(a-x)^3$, and $12ax(a-x)^5$ is $24a^3x^2(a-x)^5$;

for it consists of the product of

(1) the L.C.M. of the numerical coefficients ;

(2) the lowest power of each factor which is divisible by every power of that factor occurring in the given expressions.

Example 2. Find the lowest common multiple of
$$(yz^2 - xyz)^2, \quad y^2(xz^2 - x^3), \quad z^4 + 2xz^3 + x^2z^2.$$

Resolving each expression into its factors, we have
$$(yz^2 - xyz)^2 = \{yz(z-x)\}^2 = y^2z^2(z-x)^2,$$
$$y^2(xz^2 - x^3) = y^2x(z^2 - x^2) = xy^2(z-x)(z+x),$$
$$z^4 + 2xz^3 + x^2z^2 = z^2(z^2 + 2xz + x^2) = z^2(z+x)^2.$$

Therefore the L.C.M. is $xy^2z^2(z+x)^2(z-x)^2$.

EXAMPLES XIV. a.

Find the lowest common multiple of

1. xy^2, $3yz^2$, $2zx^2$.   2. $27a^3$, $81b^3$, $18a^2b^5$.

3. $5ax^6$, $6cy$, $7a^2x^3c^5z$.   4. $15a^2b^3$, $20ax^2y$, $30x^2$.

5. $66a^2b^3cx^4$, $55ab^5xy^3z$, $121x^3yz^3$.   6. a^2, $a^3 - a^2$.

7. $4m^2$, $6m^3 - 8m^2$.   8. $b^2 + b$, $b^3 - b$.

9. $x^2 - 4$, $x^3 + 8$.   10. $9a^2b - b$, $6a^2 + 2a$.

11. $m^2 - 5m + 6$, $m^2 + 5m - 14$.   12. $y^2 + 3y^3$, $y^3 - 9y^5$.

13. $x^3 + 27y^3$, $x^2 + xy - 6y^2$.   14. $c^2 - 3cx - 18x^2$, $c^2 - 8cx + 12x^2$.

15. $a^2 - 4a - 5$, $a^2 - 8a + 15$, $a^3 - 2a^2 - 3a$.

16. $2x^2 - 4xy - 16y^2$, $x^2 - 6xy + 8y^2$, $3x^2 - 12y^2$.

17. $3x^3 - 12a^2x$, $4x^2 + 16ax + 16a^2$.   18. $a^5c - a^3c^3$, $(a^2c + ac^2)^2$.

19. $(a^2x - 2ax^2)^2$, $(2ax - 4x^2)^2$.   20. $(2a - a^2)^3$, $4a^2 - 4a^3 + a^4$.

21. $2x^2 - x - 3$, $(2x - 3)^2$, $4x^2 - 9$.

22. $2x^2 - 7x - 4$, $6x^2 - 7x - 5$, $x^3 - 8x^2 + 16x$.

23. $10x^2y^2(x^3 - y^3)$, $15y^4(x - y)^3$, $12x^3y(x - y)(x^2 - y^2)$.

24. $2x^2 + x - 6$, $7x^2 + 11x - 6$, $(7x^2 - 3x)^2$.

25. $6a^3 - 7a^2x - 3ax^2$, $10a^2x - 11ax^2 - 6x^3$, $10a^2 - 21ax - 10x^2$.

100. When the given expressions are such that their factors cannot be determined by inspection, they must be resolved by finding the highest common factor.

Example. Find the lowest common multiple of
$$2x^4 + x^3 - 20x^2 - 7x + 24 \text{ and } 2x^4 + 3x^3 - 13x^2 - 7x + 15.$$

The highest common factor is $x^2 + 2x - 3$.

By division, we obtain
$$2x^4 + x^3 - 20x^2 - 7x + 24 = (x^2 + 2x - 3)(2x^2 - 3x - 8).$$
$$2x^4 + 3x^3 - 13x^2 - 7x + 15 = (x^2 + 2x - 3)(2x^2 - x - 5).$$

Therefore the L.C.M. is $(x^2 + 2x - 3)(2x^2 - 3x - 8)(2x^2 - x - 5)$.

EXAMPLES XIV. b.

Find the lowest common multiple of

1. $x^3 - 2x^2 - 13x - 10$ and $x^3 - x^2 - 10x - 8$.

2. $y^3 + 3y^2 - 3y - 9$ and $y^3 + 3y^2 - 8y - 24$.

3. $m^3 + 3m^2 - m - 3$ and $m^3 + 6m^2 + 11m + 6$.

4. $2x^4 - 2x^3 + x^2 + 3x - 6$ and $4x^4 - 2x^3 + 3x - 9$.

5. Find the highest common factor and the lowest common multiple of $(x - x^2)^3$, $(x^2 - x^3)^2$, $x^3 - x^4$.

6. Find the lowest common multiple of $(a^4 - a^2x^2)^2$, $(a^2 + ax)^3$, $(ax - x^2)^2$.

7. Find the highest common factor and lowest common multiple of $6x^2 + 5x - 6$ and $6x^2 + x - 12$; and shew that the product of the H.C.F. and L.C.M. is equal to the product of the two given expressions.

8. Find the highest common factor and the lowest common multiple of $a^2 + 5ab + 6b^2$, $a^2 - 4b^2$, $a^3 - 3ab^2 + 2b^3$.

9. Find the lowest common multiple of $1 - x^2 - x^4 + x^5$ and $1 + 2x + x^2 - x^4 - x^5$.

10. Find the highest common factor of $(a^3 - 4ab^2)^2$, $(a^3 + 2a^2b)^3$, $(a^2x + 2abx)^2$.

11. Find the highest common factor and the lowest common multiple of $(3a^2 - 2ax)^2$, $2a^2x(9a^2 - 4x^2)$, $6a^3x - 13a^2x^2 + 6ax^3$.

12. Find the lowest common multiple of $x^3 + x^2y + xy^2$, $x^3y - y^4$, $x^5y + x^3y^3 + xy^5$.

MISCELLANEOUS EXAMPLES II.

The following Miscellaneous Examples on Factors, Highest Common Factor, and Lowest Common Multiple have been selected from Examination Papers set by the Science and Art Department in Mathematics, Stage I.

Resolve into their elementary factors :

1. $x^2 + 17x - 18$.

2. $x^2 - x - 2$.

3. $x^2 + x - 2$.

4. $2x^2 + 5x - 12$.

5. $x^3 - 8$.

6. $x^3 + x^2y - xy^2 - y^3$.

7. $x^2 + y^2 - z^2 + 2xy$.

8. $2x^3 + 3ax^2 - 2a^2x - 3a^3$.

9. $x^2 - 6xy + 9y^2 - z^2$.

10. $xy + 4x - 9y - 36$.

11. $(1 + xy)^2 - (x + y)^2$.

12. $a^2 - b^2 - c^2 + d^2 + 2(bc - ad)$.

13. $x^3 + 2x^2 - 9x - 18$.

14. $x^2 - y^2 - 6x + 6y$.

15. $x^4 + 4x^2y^2 + 16y^4$.

Write down the following expressions in factors, and find their least common multiple :

16. $15x^3 - 15,\ 6x^2 + 12x - 18,\ 10x^2 - 90.$

17. $4x^2 + 8x - 12,\ 9x^2 - 9x - 54,\ 6x^4 - 30x^2 + 24.$

18. $x^2 - 8x + 12,\ 3x^2 - 20x + 12,\ 3x^3 - 2x^2 - 12x + 8.$

19. $2x^2 + 3x - 2,\ (5x - 7)^2 - (x - 5)^2,\ 2x^3 - x^2 - 8x + 4.$

Find the greatest common measure of

20. $45x^3y + 3x^2y^2 - 9xy^3 + 6y^4$ and $54x^2y - 24y^3.$

21. $3x^4 + 7x^3 + 2x^2 - 31x - 35$ and $15x^4 + 2x^3 - 34x^2 - 34x - 21.$

Find the greatest common measure of the following expressions, and hence find the factors of the expressions :

22. $x^4 - 5x^3 - 6x^2 + 35x - 7$ and $3x^3 - 23x^2 + 43x - 8.$

23. $x^3 + 3x^2 - 25x + 21$ and $2x^3 - 9x^2 + 10x - 3.$

24. $4x^4 - 4x^3 + x^2 - 1$ and $2x^3 + 5x^2 - 2x + 3.$

25. $2x^4 - 3x^3 + 4x^2 - 9x - 6$ and $4x^3 - 9x^2 + 12x - 27.$

26. Multiply $a^3 - x^3$ by $a^2 - x^2$, and divide the product by $(a - x)^2.$

27. Find the product of $x^2 + 5x + \frac{25}{2}$ and $x^2 - 5x + \frac{25}{2}.$

28. Subtract $(x^2 - 7x + 13)^2$ from $(x^2 + 7x - 13)^2.$

29. Obtain $(x^3 + 3x + 15)^2 - (x^3 - 3x + 15)^2$ in its simplest form, and find its value when $2x = -5.$

30. Shew that the following expression is the product of three factors :
$$3a(2a^2 + b^2) + 9a^2b + b(a + b)(2a + b).$$

CHAPTER XV.

FRACTIONS.

101. IN Arithmetic the fraction $\frac{3}{8}$ is defined as that which is obtained when the unit is divided into *eight* equal parts of which *three* are taken. It is then shewn that *the same result is given by dividing* 3 *units by* 8.

It is convenient for algebraical purposes to adopt the latter view of a fraction, and to define the fraction $\frac{a}{b}$ as *the quotient obtained when* a *is divided by* b, whatever values a and b may have.

When a and b represent positive whole numbers, the fraction $\frac{a}{b}$ indicates *that the unit has been divided into* b *equal parts of which* a *have been taken.*

The same general rules apply to algebraical as to arithmetical fractions. For proofs of these rules the reader is referred to Hall and Knight's *Elementary Algebra*, Chapters XIX. and XXI.

Reduction to Lowest Terms.

102. The value of a fraction is not altered if we multiply or divide both numerator and denominator by the same quantity. [*Elementary Algebra*, Art. 150.]

Rule. **To reduce a fraction to its lowest terms:** *divide numerator and denominator by every factor which is common to both, that is, by their highest common factor.*

Examples. (1) $\dfrac{6a^2c}{9ac^2} = \dfrac{3ac \times 2a}{3ac \times 3c} = \dfrac{2a}{3c}.$

(2) $\dfrac{7x^2yz}{28x^3yz^2} = \dfrac{7x^2yz \times 1}{7x^2yz \times 4xz} = \dfrac{1}{4xz}.$

(3) $\dfrac{24a^3c^2x^2}{18a^3x^2 - 12a^2x^3} = \dfrac{6a^2x^2 \times 4ac^2}{6a^2x^2(3a - 2x)} = \dfrac{4ac^2}{3a - 2x}.$

(4) $\dfrac{6x^2 - 8xy}{9xy - 12y^2} = \dfrac{2x(3x - 4y)}{3y(3x - 4y)} = \dfrac{2x}{3y}.$

Note. Dividing numerator and denominator of a fraction by a common factor is called *cancelling* that factor. The beginner should be careful not to begin cancelling until he has expressed both numerator and denominator in the most convenient form, by resolution into factors where necessary.

EXAMPLES XV. a.

Reduce to lowest terms :

1. $\dfrac{14xy^3}{21x^2z^3}.$

2. $\dfrac{15k^2p^3m^4}{25k^3pm^2}.$

3. $\dfrac{27a^4b^3x^2}{45a^3b^4x^4}.$

4. $\dfrac{42x^2y^2z^2}{210x^3y^2z}.$

5. $\dfrac{3x^2}{6x^2 - 2xy}.$

6. $\dfrac{3ab + b^2}{6a^2b^2 + 2ab^3}.$

7. $\dfrac{5x^2yz^2}{5x^2y + 10x^2z}.$

8. $\dfrac{2x^2y^2 - 8}{3x^2y + 6x}.$

9. $\dfrac{x^2 + 4x}{x^2 + x - 12}.$

10. $\dfrac{7a^2x - 7a^2c}{5cx^2 - 10c^2x + 5c^3}.$

11. $\dfrac{x^2 + x - 30}{5x^2 + 30x}.$

12. $\dfrac{(2a + b)^2}{4a^3 - ab^2}.$

13. $\dfrac{a^3 + b^3}{a^2 - ab - 2b^2}.$

14. $\dfrac{2c^2 + 5cd - 3d^2}{c^2 + 6cd + 9d^2}.$

15. $\dfrac{x^2 - 4x - 21}{3x^2 + 10x + 3}.$

16. $\dfrac{2x^2 + x - 3}{2x^2 + 11x + 12}.$

17. $\dfrac{3x^3 - 24}{4a^2 + 4a - 24}.$

18. $\dfrac{18a^3 + 6a^2x + 2ax^2}{27a^3 - x^3}.$

103. When the factors of the numerator and denominator cannot be determined by inspection, the fraction may be reduced to its lowest terms by dividing both numerator and denominator by the highest common factor, which may be found by the rules given in Chap. XIII.

Example. Reduce to lowest terms $\dfrac{3x^3 - 13x^2 + 23x - 21}{15x^3 - 38x^2 - 2x + 21}.$

The H.C.F. of numerator and denominator is $3x - 7$.

Dividing numerator and denominator by $3x - 7$, we obtain as respective quotients $x^2 - 2x + 3$ and $5x^2 - x - 3$.

Thus $\dfrac{3x^3 - 13x^2 + 23x - 21}{15x^3 - 38x^2 - 2x + 21} = \dfrac{(3x - 7)(x^2 - 2x + 3)}{(3x - 7)(5x^2 - x - 3)} = \dfrac{x^2 - 2x + 3}{5x^2 - x - 3}.$

104. If either numerator or denominator can readily be resolved into factors we may use the following method.

Example. Reduce to lowest terms $\dfrac{x^3 + 3x^2 - 4x}{7x^3 - 18x^2 + 6x + 5}.$

The numerator $= x(x^2 + 3x - 4) = x(x + 4)(x - 1)$.

Of these factors the only one which can be a common divisor is $x - 1$. Hence, arranging the denominator so as to shew $x - 1$ as a factor,

the fraction $= \dfrac{x(x + 4)(x - 1)}{7x^2(x - 1) - 11x(x - 1) - 5(x - 1)}$

$= \dfrac{x(x + 4)(x - 1)}{(x - 1)(7x^2 - 11x - 5)} = \dfrac{x(x + 4)}{7x^2 - 11x - 5}.$

EXAMPLES XV. b.

Reduce to lowest terms :

1. $\dfrac{x^3 - x^2 + 2x - 2}{3x^4 + 7x^2 + 2}$.

2. $\dfrac{a^3 + a + 2}{a^3 - 4a^2 + 5a - 6}$.

3. $\dfrac{y^3 - 2y^2 - 2y - 3}{3y^3 + 4y^2 + 4y + 1}$.

4. $\dfrac{m^3 - m^2 - 2m}{m^3 - m^2 - m - 2}$.

5. $\dfrac{a^3 - 2ab^2 + 21b^3}{a^3 - 4a^2b - 21ab^2}$.

6. $\dfrac{9x^3 - a^2x - 2a^3}{3x^3 - 10ax^2 - 7a^2x - 4a^3}$.

7. $\dfrac{5x^3 - 4x - 1}{2x^3 - 3x^2 + 1}$.

8. $\dfrac{c^3 + 2c^2d - 12cd^2 - 9d^3}{2c^3 + 6c^2d - 28cd^2 - 24d^3}$.

9. $\dfrac{x^4 - 21x + 8}{8x^4 - 21x^3 + 1}$.

10. $\dfrac{y^5 + 6y^4 + 2y^3 - 9y^2}{y^4 + 7y^3 + 3y^2 - 11y}$.

11. $\dfrac{1 - x^2 + 6x^3}{2 - x + 9x^3}$.

12. $\dfrac{2 - 5x - 4x^2 + 3x^3}{4 + 4x + 9x^2 + 4x^3 - 5x^4}$.

Multiplication and Division of Fractions.

105. Rule. To multiply together two or more fractions: *multiply the numerators to form a new numerator, and the denominator to form a new denominator.*

[Hall and Knight's *Elementary Algebra*, Art. 157.]

Thus $\dfrac{a}{b} \times \dfrac{c}{d} = \dfrac{ac}{bd}$.

Similarly, $\dfrac{a}{b} \times \dfrac{c}{d} \times \dfrac{e}{f} = \dfrac{ace}{bdf}$;

and so for any number of fractions.

In practice the application of this rule is modified by removing in the course of the work factors which are common to numerator and denominator.

Example 1. $\dfrac{2a}{3b} \times \dfrac{5x^2}{2a^2b} \times \dfrac{3b^2}{2x} = \dfrac{2a \times 5x^2 \times 3b^2}{3b \times 2a^2b \times 2x} = \dfrac{5x}{2a}$,

this result being obtained by cancelling out like *factors* in numerator and denominator.

Example 2.   Simplify $\dfrac{2a^2+3a}{4a^3} \times \dfrac{4a^2-6a}{12a+18}$.

The expression $= \dfrac{a(2a+3)}{4a^3} \times \dfrac{2a(2a-3)}{6(2a+3)} = \dfrac{2a-3}{12a}$,

by cancelling those factors which are common to both numerator and denominator.

106. Rule. To divide one fraction by another: *invert the divisor, and proceed as in multiplication.* [*Elem. Alg.* Art. 158.]

Thus
$$\frac{a}{b} \div \frac{c}{d} = \frac{a}{b} \times \frac{d}{c} = \frac{ad}{bc}.$$

Example 1. $\dfrac{7a^3}{4x^3y^2} \times \dfrac{6c^3x}{5ab^2} \div \dfrac{28a^2c^2}{15b^2xy^2}$

$$= \frac{7a^3}{4x^3y^2} \times \frac{6c^3x}{5ab^2} \times \frac{15b^2xy^2}{28a^2c^2} = \frac{9c}{8x},$$

since all the other factors cancel one another.

Example 2.   Simplify $\dfrac{6x^2-ax-2a^2}{ax-a^2} \times \dfrac{x-a}{9x^2-4a^2} \div \dfrac{2x+a}{3ax+2a^2}$.

The expression $= \dfrac{6x^2-ax-2a^2}{ax-a^2} \times \dfrac{x-a}{9x^2-4a^2} \times \dfrac{3ax+2a^2}{2x+a}$

$$= \frac{(3x-2a)(2x+a)}{a(x-a)} \times \frac{x-a}{(3x+2a)(3x-2a)} \times \frac{a(3x+2a)}{2x+a}$$

$$= 1,$$

since all the factors cancel each other.

EXAMPLES XV. c.

Simplify

1. $\dfrac{3ab^2}{5b^3c} \times \dfrac{15b^2c^2}{9a^2b}$.

2. $\dfrac{a^2m}{b^2y} \times \dfrac{2cd^2}{3ab} \times \dfrac{9my}{4m^2}$.

3. $\dfrac{4a^2b}{9xy} \times \dfrac{3p^2q^2}{8a^2b^2} \div \dfrac{pq}{x^2y^2}$.

4. $\dfrac{2a^3p^2}{5ax^4} \times \dfrac{10b^2}{4x^2} \div \dfrac{b^2p^2}{3x^6}$.

5. $\dfrac{x^2-1}{x^2+3x} \times \dfrac{2x^3+6x^2}{x^2+x}$.

6. $\dfrac{ab+2}{4a^2-12ab} \times \dfrac{a^2b-3ab^2}{a^2b^2-4}$.

7. $\dfrac{2c^2+3cd}{4c^2-9d^2} \div \dfrac{c+d}{2cd-3d^2}$.

8. $\dfrac{5y-10y^2}{12y^2+6y^3} \div \dfrac{1-2y}{2y+y^2}$.

9. $\dfrac{b^2-5b}{3b-4a} \times \dfrac{9b^2-16a^2}{b^2-25}$.

10. $\dfrac{x^2+9x+20}{x^2+5x+4} \div \dfrac{x^2+7x+10}{x^2+3x+2}$.

11. $\dfrac{y^2-y-12}{y^2-16} \times \dfrac{y^2-2y-24}{y^2+6y+9}$.

12. $\dfrac{a^3+27}{a^2+9a+14} \div \dfrac{a^2-4a-21}{a^2-49}$.

13. $\dfrac{2a^2-3a-2}{a^2-a-6} \times \dfrac{3a^2-8a-3}{3a^2-5a-2}$.

14. $\dfrac{b^3+125}{5b^2+24b-5} \times \dfrac{25b^2-1}{b^3-5b^2+25b}$.

15. $\dfrac{2p^2+4p}{p^2-9} \times \dfrac{p^2-5p+6}{p^2-5p} \times \dfrac{p^2-2p-15}{p^2-4}$.

16. $\dfrac{64a^2b^2-1}{x^2-x-56} \times \dfrac{x^2-49}{8a^3b-a^2} \div \dfrac{x-7}{a^2x-8a^2}$.

17. $\dfrac{4x^2+4x-15}{x^2+2x-48} \times \dfrac{x+8}{2x^2-15x+18} \div \dfrac{2x^2+5x}{(x-6)^2}$.

18. $\dfrac{a^2+8ab-9b^2}{a^2+6ab-27b^2} \times \dfrac{a^2-7ab+12b^2}{a^3-b^3} \times \dfrac{a^3+a^2b+ab^2}{a^2-3ab-4b^2}$.

19. $\dfrac{ax^2-16a^3}{x^2-ax-30a^2} \times \dfrac{x^2+ax-20a^2}{ax^2+9a^2x+20a^3} \div \dfrac{x^2-8ax+16a^2}{x^2+8ax+15a^2}$.

20. $\dfrac{(a-b)^2-c^2}{a^2-ab+ac} \times \dfrac{a^2+ab+ac}{(a-c)^2-b^2} \times \dfrac{(a+b)^2-c^2}{(a+b+c)^2}$.

Addition and Subtraction of Fractions.

107. To find the algebraical sum of a number of fractions we must, as in Arithmetic, first reduce them to a common denominator. For this purpose it is usually most convenient to take the *lowest* common denominator.

Rule. To reduce fractions to their lowest common denominator: *find the L.C.M. of the given denominators, and take it for the common denominator; divide it by the denominator of the first fraction, and multiply the numerator of this fraction by the quotient so obtained; and do the same with all the other given fractions.*

Example. Express with lowest common denominator

$$\frac{5x}{2a(x-a)} \quad \text{and} \quad \frac{4a}{3x(x^2-a^2)}.$$

The lowest common denominator is $6ax(x-a)(x+a)$.

We must therefore multiply the numerators by $3x(x+a)$ and $2a$ respectively.

Hence the equivalent fractions are

$$\frac{15x^2(x+a)}{6ax(x-a)(x+a)} \quad \text{and} \quad \frac{8a^2}{6ax(x-a)(x+a)}.$$

xv.]   FRACTIONS.   85

108. We may now enunciate the rule for the addition or subtraction of fractions.

Rule. To add or subtract fractions : *reduce them to the lowest common denominator ; find the algebraical sum of the numerators, and retain the common denominator.*

Thus $\dfrac{a}{b} + \dfrac{c}{d} = \dfrac{ad+bc}{bd}$,

and $\dfrac{a}{b} - \dfrac{c}{d} = \dfrac{ad-bc}{bd}$.   [*Elem. Alg.*, Art. 165.]

Example 1. Simplify $\dfrac{5x}{3} + \dfrac{3}{4}x - \dfrac{7x}{6}$.

The least common denominator is 12.

The expression $= \dfrac{20x + 9x - 14x}{12} = \dfrac{15x}{12} = \dfrac{5x}{4}$.

Example 2. Simplify $\dfrac{3ab}{5x} - \dfrac{ab}{2x} - \dfrac{ab}{10x}$.

The expression $= \dfrac{6ab - 5ab - ab}{10x} = \dfrac{0}{10x} = 0$.

When no denominator is expressed the denominator 1 may be understood.

Example 3. $3x - \dfrac{a^2}{4y} = \dfrac{3x}{1} - \dfrac{a^2}{4y} = \dfrac{12xy - a^2}{4y}$.

If a fraction is not in its lowest terms it should be simplified before combining it with other fractions.

Example 4. $\dfrac{ax}{2} - \dfrac{x^2y}{3xy} = \dfrac{ax}{2} - \dfrac{x}{3} = \dfrac{3ax - 2x}{6}$.

Example 5. Find the value of $\dfrac{x-2y}{xy} + \dfrac{3y-a}{ay} - \dfrac{3x-2a}{ax}$.

The lowest common denominator is axy.

Thus the expression $= \dfrac{a(x-2y) + x(3y-a) - y(3x-2a)}{axy}$

$= \dfrac{ax - 2ay + 3xy - ax - 3xy + 2ay}{axy}$

$= 0$,

since the terms in the numerator destroy each other.

Note. To ensure accuracy the beginner is recommended to use brackets as in the first line of work above.

EXAMPLES XV. d.

Find the value of

1. $\dfrac{a}{4} - \dfrac{a}{8} + \dfrac{a}{12}$.

2. $\dfrac{2x}{3} - \dfrac{x}{6} + \dfrac{9x}{12}$.

3. $\dfrac{a}{xy} + \dfrac{2a}{yz} - \dfrac{3a}{xz}$.

4. $\dfrac{xy}{5x} - \dfrac{2y}{3} + \dfrac{4y}{8}$.

5. $2 + \dfrac{a}{b} - \dfrac{b^2}{ab}$.

6. $a + \dfrac{b}{p^2 q} - \dfrac{c}{q^2}$.

7. $\dfrac{a-2}{3} + \dfrac{a-1}{2} + \dfrac{a+5}{6}$.

8. $\dfrac{2b-1}{5} + \dfrac{b-3}{2} - \dfrac{7b+3}{10}$.

9. $-\dfrac{2x-1}{5} + \dfrac{3x-1}{7} - \dfrac{x-2}{35}$.

10. $\dfrac{2x-5}{x} - \dfrac{x-4}{x} - \dfrac{x^2-4x}{3x^2}$.

11. $-\dfrac{a+x}{2a} + \dfrac{a+2x}{3a} \cdot \dfrac{x-5a}{6a}$.

12. $\dfrac{2a^2-5a}{a} - \dfrac{a^3+3a^2}{a^2} + \dfrac{9a^3-a^4}{a^3}$.

13. $-\dfrac{x-y}{y} + \dfrac{x+y}{x} - \dfrac{6xy-4x^2}{3xy}$.

14. $\dfrac{ab-bc}{2bc} - \dfrac{a}{3c} - \dfrac{2a^2-ab}{2ab}$.

15. $\dfrac{2ay-xy+4x}{2xy} - 1 - \dfrac{a}{2x}$.

16. $\dfrac{a^2-ab}{a^2b} - \dfrac{b-c}{bc} - \dfrac{2c^2-ac}{c^2a}$.

109. We shall now consider the addition and subtraction of fractions whose denominators are compound expressions. *The lowest common multiple of the denominators should always be written down by inspection when possible.*

Example 1. Simplify $\dfrac{2x-3a}{x-2a} - \dfrac{2x-a}{x-a}$.

The lowest common denominator is $(x-2a)(x-a)$.

Hence, multiplying the numerators by $x-a$ and $x-2a$ respectively, we have

$$\text{the expression} = \frac{(2x-3a)(x-a) - (2x-a)(x-2a)}{(x-2a)(x-a)}$$

$$= \frac{2x^2-5ax+3a^2 - (2x^2-5ax+2a^2)}{(x-2a)(x-a)}$$

$$= \frac{2x^2-5ax+3a^2-2x^2+5ax-2a^2}{(x-2a)(x-a)}$$

$$= \frac{a^2}{(x-2a)(x-a)}.$$

Note. In finding the value of such an expression as

$$-(2x-a)(x-2a),$$

the beginner should first express the product in brackets, and then remove the brackets, as we have done. After a little practice he will be able to take both steps together.

Example 2. Find the value of $\dfrac{3x+2}{x^2-16}+\dfrac{x-5}{(x+4)^2}.$

The lowest common denominator is $(x-4)(x+4)^2$.

Hence the expression $=\dfrac{(3x+2)(x+4)+(x-5)(x-4)}{(x-4)(x+4)^2}$

$=\dfrac{3x^2+14x+8+x^2-9x+20}{(x-4)(x+4)^2}$

$=\dfrac{4x^2+5x+28}{(x-4)(x+4)^2}.$

If a fraction is not in its lowest terms, it should be simplified before it is combined with other fractions.

Example 3. Simplify $\dfrac{a^2+2a}{a^2+a-2}-\dfrac{a}{2a+2}-\dfrac{a^2+a+2}{2a^2-2}.$

The expression $=\dfrac{a(a+2)}{(a+2)(a-1)}-\dfrac{a}{2(a+1)}-\dfrac{a^2+a+2}{2(a^2-1)}$

$=\dfrac{a}{a-1}-\dfrac{a}{2(a+1)}-\dfrac{a^2+a+2}{2(a^2-1)}$

$=\dfrac{2a(a+1)-a(a-1)-(a^2+a+2)}{2(a^2-1)}$

$=\dfrac{2a-2}{2(a^2-1)}=\dfrac{2(a-1)}{2(a^2-1)}=\dfrac{1}{a+1}.$

Sometimes the work will be simplified by first combining two of the fractions, instead of finding the lowest common multiple of all the denominators at once.

Example 4. Simplify $\dfrac{3}{8(a-x)}-\dfrac{1}{8(a+x)}-\dfrac{a-2x}{4(a^2+x^2)}.$

Taking the first two fractions together,

the expression $=\dfrac{3(a+x)-(a-x)}{8(a^2-x^2)}-\dfrac{a-2x}{4(a^2+x^2)}$

$=\dfrac{a+2x}{4(a^2-x^2)}-\dfrac{a-2x}{4(a^2+x^2)}$

$=\dfrac{(a+2x)(a^2+x^2)-(a-2x)(a^2-x^2)}{4(a^4-x^4)}$

$=\dfrac{a^3+2a^2x+ax^2+2x^3-(a^3-2a^2x-ax^2+2x^3)}{4(a^4-x^4)}$

$=\dfrac{4a^2x+2ax^2}{4(a^4-x^4)}=\dfrac{ax(2a+x)}{2(a^4-x^4)}.$

EXAMPLES XV. e.

Find the value of

1. $\dfrac{1}{a-2}+\dfrac{1}{a-3}$.

2. $\dfrac{1}{x-4}-\dfrac{1}{x-2}$.

3. $\dfrac{a}{x-a}-\dfrac{b}{x-b}$.

4. $\dfrac{a-x}{a+x}+\dfrac{a+x}{a-x}$.

5. $\dfrac{x}{x-1}-\dfrac{x^2}{x^2-1}$.

6. $\dfrac{3a}{a^2-4}-\dfrac{1}{a+2}$.

7. $\dfrac{x^2}{x^2-4y^2}+\dfrac{x-2y}{x+2y}$.

8. $\dfrac{3a}{2x(x-a)}-\dfrac{2a}{3x(x+a)}$.

9. $\dfrac{5}{x-2}-\dfrac{4x}{(x-2)(x+1)}$.

10. $\dfrac{1}{y^2-2y-3}+\dfrac{3(y+2)}{y^2-y-6}$.

11. $\dfrac{1}{1-a}+\dfrac{a}{(1-a)^2}$.

12. $\dfrac{3}{x+y}-\dfrac{2x}{(x+y)^2}$.

13. $\dfrac{2x+y}{x^2-y^2}-\dfrac{2x-y}{(x+y)^2}$.

14. $\dfrac{b+c}{b^2-2bc+c^2}-\dfrac{b-2c}{b^2-c^2}$.

15. $\dfrac{x}{xy-y^2}-\dfrac{xy}{x^3-x^2y}$.

16. $\dfrac{4a^2-b^2}{2ab-b^2}-\dfrac{4a}{2a+b}$.

17. $\dfrac{x^2}{x^3+1}-\dfrac{1}{x+1}$.

18. $\dfrac{2b-4}{b^3+8}+\dfrac{1}{b+2}$.

19. $\dfrac{x^2-2y^2}{x^2+xy+y^2}+\dfrac{x^2y^2-2y^4}{x^3y-y^4}$.

20. $\dfrac{1}{a^2-3a+2}+\dfrac{1}{a^2+3a-10}$.

21. $x+2-\dfrac{x-2}{x-1}$.

22. $4+\dfrac{a-6}{2+a}-2a$.

23. $\dfrac{1}{x^2}+\dfrac{x^2}{x+1}-\dfrac{1}{x}$.

24. $\dfrac{1}{x}-\dfrac{2}{x-2}+\dfrac{1}{x-4}$.

25. $\dfrac{6}{2x-1}-\dfrac{3}{2x+1}-\dfrac{2-3x}{4x^2-1}$.

26. $\dfrac{1+2a}{3-3a}-\dfrac{3a^2+2a}{2-2a^2}+1$.

27. $\dfrac{2x}{9-6x}+\dfrac{5}{6+4x}-\dfrac{4x^2-9x}{27-12x^2}$.

28. $\dfrac{1}{x-a}+\dfrac{2a}{(x-a)^2}+\dfrac{a^2}{(x-a)^3}$.

29. $\dfrac{2}{(a+1)^2}-\dfrac{a-3}{(a+1)^4}+\dfrac{2}{(a+1)^3}$.

30. $\dfrac{1}{2y^2-y-3}-\dfrac{1}{2y^2+y-1}$.

31. $\dfrac{5}{4+3x-x^2}-\dfrac{2}{3+4x+x^2}$.

32. $\dfrac{1}{z(z-1)}+\dfrac{1}{z(z+1)}-\dfrac{2}{z^2-1}$.

33. $\dfrac{2}{(x-2)^2}-\dfrac{x}{x^2+4}+\dfrac{1}{x-2}$.

34. $\dfrac{y-2}{(y-3)(y-4)}-\dfrac{2(y-3)}{(y-2)(y-4)}+\dfrac{y-4}{(y-2)(y-3)}$.

35. $\dfrac{1}{1-x} - \dfrac{2+x}{(1-x)(2-x)} + \dfrac{2+3x+3x^2}{(1-x)(2-x)(3+x)}$.

36. $\dfrac{2}{x^2-5xy+6y^2} - \dfrac{3}{x^2-xy-6y^2} + \dfrac{1}{x^2-4y^2}$.

37. $\dfrac{5a}{6(a^2-1)} - \dfrac{a+3}{2(a^2+2a-3)} + \dfrac{a+1}{3a^2+6a+3}$.

38. $\dfrac{a}{a-b} - \dfrac{b^2}{a^2+ab+b^2} - \dfrac{a^3+b^3}{a^3-b^3}$.   **39.** $\dfrac{3(6-x)}{x^3+27} + \dfrac{x-3}{x^2-3x+9} - \dfrac{1}{x+3}$.

40. $\dfrac{1}{(x-y)^2} - \dfrac{1}{x^2+2xy+y^2} - \dfrac{4xy}{x^4-2x^2y^2+y^4}$.

41. $\dfrac{x}{(x-a)^2} - \dfrac{a}{x^2-a^2} - \dfrac{ax}{(x-a)^3}$.

42. $\dfrac{1}{2-x} + \dfrac{1}{2+x} - \dfrac{3}{4+x^2}$.   **43.** $\dfrac{x}{4(1+x)} - \dfrac{x}{4(1-x)} + \dfrac{3}{2(1+x^2)}$.

44. $\dfrac{3}{2m-4} - \dfrac{3}{2m+4} - \dfrac{2}{3m^2+12}$.   **45.** $\dfrac{a}{a-b} - \dfrac{b}{a+b} - \dfrac{b^2}{a^2+b^2}$.

46. $\dfrac{x-3}{x-4} = \dfrac{x-1}{x-2} - \dfrac{1}{(x-2)^2}$.   **47.** $\dfrac{x-3}{x-6} - \dfrac{x-6}{x-3} + \dfrac{x-3}{x} - \dfrac{x}{x-3}$.

Changes of Sign in Addition of Fractions.

110. An algebraical fraction has been defined in Art. 101 as the quotient obtained by *dividing the numerator by the denominator.* Hence, by Art. 39, it follows from the Rule of Signs as applied to division that

(1) *If the signs of* BOTH *numerator and denominator of a fraction be changed, the sign of the whole fraction will be unchanged.*

(2) *If the sign of the* NUMERATOR ALONE *be changed, the sign of the whole fraction will be changed.*

(3) *If the sign of the* DENOMINATOR ALONE *be changed, the sign of the whole fraction will be changed.*

Example 1. $\dfrac{b-a}{y-x} = \dfrac{-(b-a)}{-(y-x)} = \dfrac{-b+a}{-y+x} = \dfrac{a-b}{x-y}$.

Example 2. $\dfrac{x-x^2}{2y} = -\dfrac{-x+x^2}{2y} = -\dfrac{x^2-x}{2y}$.

Example 3. $\dfrac{3x}{4-x^2} = -\dfrac{3x}{-4+x^2} = -\dfrac{3x}{x^2-4}$.

111.   The following examples illustrate an important application of the foregoing principles.

Example 1.   Simplify $\dfrac{a}{x+a} + \dfrac{2x}{x-a} + \dfrac{a(3x-a)}{a^2-x^2}$.

Here it is evident that the lowest common denominator of the first two fractions is $x^2 - a^2$, therefore it will be convenient to alter the sign of the denominator in the third fraction.

Thus the expression $= \dfrac{a}{x+a} + \dfrac{2x}{x-a} - \dfrac{a(3x-a)}{x^2-a^2}$

$$= \frac{a(x-a) + 2x(x+a) - a(3x-a)}{x^2-a^2}$$

$$= \frac{ax - a^2 + 2x^2 + 2ax - 3ax + a^2}{x^2-a^2}$$

$$= \frac{2x^2}{x^2 - a^2}.$$

Example 2.   Simplify $\dfrac{1}{(a-b)(a-c)} + \dfrac{1}{(b-c)(b-a)} + \dfrac{1}{(c-a)(c-b)}$.

Here in finding the L.C.M. of the denominators it must be observed that there are not *six* different compound factors to be considered; for three of them differ from the other three only in sign.

Thus $(a-c) = -(c-a),$

$(b-a) = -(a-b),$

$(c-b) = -(b-c).$

Hence, replacing the second factor in each denominator by its equivalent, we may write the expression in the form

$$-\frac{1}{(a-b)(c-a)} - \frac{1}{(b-c)(a-b)} - \frac{1}{(c-a)(b-c)}.$$

Now the L.C.M. is $(b-c)(c-a)(a-b)$;

and the expression $= \dfrac{-(b-c) - (c-a) - (a-b)}{(b-c)(c-a)(a-b)}$

$$= \frac{-b+c-c+a-a+b}{(b-c)(c-a)(a-b)}$$

$$= 0.$$

Note.   In examples of this kind it will be found convenient to arrange the expressions **cyclically**, that is, so that a is followed by b, b by c, and c by a.

112. If the sign of each of *two* factors in a product is changed, the sign of the product is unaltered ; thus

$$(a-x)(b-x)=\{-(x-a)\}\{-(x-b)\}=(x-a)(x-b).$$

Similarly, $(a-x)^2=(x-a)^2.$

In other words, in the simplification of fractions we may change the sign of each of *two* factors in a denominator without altering the sign of the fraction ; thus

$$\frac{1}{(b-a)(c-b)}=\frac{1}{(a-b)(b-c)}.$$

113. The arrangement adopted in the following example is worthy of notice.

Example. Simplify $\dfrac{1}{a-x}-\dfrac{1}{a+x}-\dfrac{2x}{a^2+x^2}-\dfrac{4x^3}{a^4+x^4}.$

Here it should be evident that the first two denominators give L.C.M. a^2-x^2, which readily combines with a^2+x^2 to give L.C.M. a^4-x^4, which again combines with a^4+x^4 to give L.C.M. a^8-x^8. Hence it will be convenient to proceed as follows :

$$\text{The expression}=\frac{a+x-(a-x)}{a^2-x^2}-\ldots\ldots-\ldots$$

$$=\frac{2x}{a^2-x^2}-\frac{2x}{a^2+x^2}-$$

$$=\frac{4x^3}{a^4-x^4}-\frac{4x^3}{a^4+x^4}$$

$$=\frac{8x^7}{a^8-x^8}.$$

EXAMPLES XV. f.

Find the value of

1. $\dfrac{5}{1+2x}-\dfrac{3x}{1-2x}+\dfrac{4-13x}{4x^2-1}.$　　**2.** $\dfrac{10}{9-a^2}-\dfrac{2}{3+a}+\dfrac{1}{a-3}.$

3. $\dfrac{5a}{6(a^2-1)}+\dfrac{1}{2(1-a)}+\dfrac{1}{3(a+1)}.$　　**4.** $\dfrac{2y}{2y-3}-\dfrac{5}{6y+9}+\dfrac{12y+8}{27-12y^2}.$

5. $\dfrac{x+a}{x-a}-\dfrac{x-a}{x+a}+\dfrac{4ax}{a^2-x^2}.$　　**6.** $\dfrac{a}{a-b}-\dfrac{b}{a+b}+\dfrac{b}{b-a}.$

7. $\dfrac{a}{x^2-x}+\dfrac{a}{x-x^3}-\dfrac{a}{x^2-1}.$　　**8.** $\dfrac{x^6-x^3y^3}{y^6-x^6}+\dfrac{x^3y^3}{x^3y^3-y^6}.$

9. $\dfrac{1}{(y-2)(y-3)}+\dfrac{2}{(y-1)(3-y)}+\dfrac{1}{(y-1)(y-2)}.$

10. $\dfrac{a}{(x-a)(a-b)}-\dfrac{b}{(x-b)(a-b)}+\dfrac{x}{(a-x)(b-x)}.$

11. $\dfrac{2}{x-1}+\dfrac{3}{(1-x)^2}-\dfrac{1}{2x-1}.$ **12.** $\dfrac{1}{a-b}-\dfrac{a}{(a-b)^2}-\dfrac{ab}{(b-a)^3}.$

13. $\dfrac{a+c}{(a-b)(x-a)}-\dfrac{b+c}{(b-a)(b-x)}.$ **14.** $\dfrac{x-z}{(x-y)(a-x)}-\dfrac{y-z}{(y-x)(y-a)}.$

15. $\dfrac{a+b}{b}-\dfrac{2a}{a+b}+\dfrac{a^3-a^2b}{b(b^2-a^2)}.$ **16.** $\dfrac{1}{a+x}+\dfrac{1}{a-2x}-\dfrac{1}{x-a}+\dfrac{1}{2x+a}.$

17. $\dfrac{3}{a+x}-\dfrac{1}{3x+a}+\dfrac{3}{x-a}+\dfrac{1}{a-3x}.$

18. $\dfrac{x}{(x-y)(x-z)}+\dfrac{y}{(y-z)(y-x)}+\dfrac{z}{(z-x)(z-y)}.$

19. $\dfrac{a}{(b-c)(b-a)}+\dfrac{b}{(c-a)(c-b)}+\dfrac{c}{(a-b)(a-c)}.$

20. $\dfrac{y-z}{(x-y)(x-z)}+\dfrac{z-x}{(y-z)(y-x)}+\dfrac{x-y}{(z-x)(z-y)}.$

21. $\dfrac{1+p}{(p-q)(p-r)}+\dfrac{1+q}{(q-r)(q-p)}+\dfrac{1+r}{(r-p)(r-q)}.$

22. $\dfrac{1}{4(x+a)}-\dfrac{1}{4(a-x)}+\dfrac{x}{2(x^2-a^2)}+\dfrac{x^3}{a^4-x^4}.$

23. $\dfrac{1}{2a^3(a+x)}-\dfrac{1}{2a^3(x-a)}+\dfrac{1}{a^2(a^2+x^2)}+\dfrac{2a^4}{x^8-a^8}.$

24. $\dfrac{a}{a^2-b^2}-\dfrac{b}{a^2+b^2}+\dfrac{a^3+b^3}{b^4-a^4}+\dfrac{ab}{(a+b)(a^2+b^2)}.$

25. $\dfrac{1}{x-2}+\dfrac{2}{(2+x)^2}+\dfrac{2}{(2-x)^2}-\dfrac{1}{x+2}.$

Simplification of Complex Fractions.

114. DEFINITION. A fraction whose numerator and denominator are whole numbers is called a **Simple Fraction.**

A fraction of which the numerator or denominator is itself a fraction is called a **Complex Fraction.**

Thus $\quad \dfrac{a}{\frac{b}{c}}, \ \dfrac{\frac{a}{b}}{x}, \ \dfrac{\frac{a}{b}}{\frac{c}{d}}$ are Complex Fractions.

In the last of these types the outside quantities, a and d, are sometimes referred to as the *extremes*, while the two middle quantities, b and c, are called the *means*.

Instead of using the horizontal line to separate numerator and denominator, it is sometimes convenient to write complex fractions in the forms

$$a\Big/\frac{b}{c}, \quad \frac{a}{b}\Big/x, \quad \frac{a}{b}\Big/\frac{c}{d}$$

115. An algebraical fraction has been defined in Art. 101 as the result obtained by *dividing the numerator by the denominator.*

Thus $\qquad \dfrac{\frac{a}{b}}{\frac{c}{d}} = \dfrac{a}{b} \div \dfrac{c}{d} = \dfrac{a}{b} \times \dfrac{d}{c} = \dfrac{ad}{bc}.$

The student should notice the following particular cases, and should be able to write down the results readily.

$$\frac{1}{\frac{a}{b}} = 1 \div \frac{a}{b} = 1 \times \frac{b}{a} = \frac{b}{a}.$$

$$\frac{a}{\frac{1}{b}} = a \div \frac{1}{b} = a \times b = ab.$$

116. The following examples illustrate the simplification of complex fractions.

Example 1. Simplify $\dfrac{x + \dfrac{a^2}{x}}{x - \dfrac{a^4}{x^3}}.$

The expression $= \left(x + \dfrac{a^2}{x}\right) \div \left(x - \dfrac{a^4}{x^3}\right) = \dfrac{x^2 + a^2}{x} \div \dfrac{x^4 - a^4}{x^3}$

$\qquad\qquad = \dfrac{x^2 + a^2}{x} \times \dfrac{x^3}{x^4 - a^4} = \dfrac{x^2}{x^2 - a^2}.$

Example 2. Simplify $\dfrac{\dfrac{3}{a}+\dfrac{a}{3}-2}{\dfrac{a}{6}+\dfrac{1}{2}-\dfrac{3}{a}}$.

Here the reduction may be simply effected by multiplying the fractions above and below by $6a$, which is the L.C.M. of the denominators.

Thus the expression $= \dfrac{18+2a^2-12a}{a^2+3a-18}$

$$=\frac{2(a^2-6a+9)}{(a+6)(a-3)}=\frac{2(a-3)}{a+6}.$$

Example 3. Simplify $\dfrac{\dfrac{a^2+b^2}{a^2-b^2}-\dfrac{a^2-b^2}{a^2+b^2}}{\dfrac{a+b}{a-b}-\dfrac{a-b}{a+b}}$.

The numerator $= \dfrac{(a^2+b^2)^2-(a^2-b^2)^2}{(a^2+b^2)(a^2-b^2)}=\dfrac{4a^2b^2}{(a^2+b^2)(a^2-b^2)}$;

similarly the denominator $= \dfrac{4ab}{(a+b)(a-b)}$.

Hence the fraction $= \dfrac{4a^2b^2}{(a^2+b^2)(a^2-b^2)} \div \dfrac{4ab}{(a+b)(a-b)}$

$$=\frac{4a^2b^2}{(a^2+b^2)(a^2-b^2)} \times \frac{(a+b)(a-b)}{4ab}$$

$$=\frac{ab}{a^2+b^2}.$$

117. In the case of fractions like the following, called **Continued Fractions**, we begin from the lowest fraction, and simplify step by step.

Example. Find the value of $\dfrac{1}{4-\dfrac{3}{2+\dfrac{x}{1-x}}}$.

The expression $= \dfrac{1}{4-\dfrac{3}{\dfrac{2-2x+x}{1-x}}}=\dfrac{1}{4-\dfrac{3(1-x)}{2-x}}=\dfrac{1}{\dfrac{8-4x-3+3x}{2-x}}$

$$=\frac{1}{\dfrac{5-x}{2-x}}=\frac{2-x}{5-x}.$$

EXAMPLES XV. g.

Find the value of

1. $\dfrac{1}{x+\dfrac{y}{z}}$.

2. $\dfrac{a}{b-\dfrac{c}{d}}$.

3. $\dfrac{1-a}{\dfrac{1}{a^2}-1}$.

4. $\dfrac{b}{\dfrac{1}{1-a}}$.

5. $\dfrac{\dfrac{a}{x}-\dfrac{x}{a}}{\dfrac{1}{x}-\dfrac{1}{a}}$.

6. $\dfrac{\dfrac{1}{x^2}-\dfrac{1}{y^2}}{\dfrac{y}{x}-\dfrac{x}{y}}$.

7. $\dfrac{a-\dfrac{b}{d}}{\dfrac{a}{b}-\dfrac{1}{d}}$.

8. $\dfrac{\dfrac{p}{q}-r}{\dfrac{1}{p^1q}-\dfrac{r}{p^2}}$.

9. $\dfrac{a+\dfrac{6}{a}-5}{1+\dfrac{8}{a^2}-\dfrac{6}{a}}$.

10. $\dfrac{y-3+\dfrac{y^2}{3}}{y-\dfrac{9}{y}+3}$.

11. $\dfrac{\dfrac{1}{n}-\dfrac{3}{n^2}-\dfrac{4}{n^3}}{n-\dfrac{16}{n}}$.

12. $\dfrac{x-2+\dfrac{6}{x+3}}{x-4+\dfrac{12}{x+3}}$.

13. $\dfrac{b-2-\dfrac{6}{b+3}}{b-4+\dfrac{6}{b+3}}$.

14. $\dfrac{\dfrac{a}{b^2}-\dfrac{b}{a^2}}{\dfrac{1}{a^2}+\dfrac{1}{ab}+\dfrac{1}{b^2}}$.

15. $\dfrac{\dfrac{c+d}{c-d}-\dfrac{c-d}{c+d}}{\dfrac{c+d}{c-d}+\dfrac{c-d}{c+d}}$.

16. $\dfrac{a-\dfrac{a-b}{1-ab}}{1-\dfrac{a(a-b)}{1-ab}}$.

17. $\dfrac{\dfrac{x+3}{7}-\dfrac{x+3}{x+4}}{\dfrac{x-3}{4}+\dfrac{x-3}{x-1}}$.

18. $1+\dfrac{1}{1+\dfrac{1}{a}}$.

19. $x+\dfrac{1}{x-\dfrac{1}{x}}$.

20. $2-\dfrac{3}{4-\dfrac{c}{d}}$.

21. $\dfrac{x}{1+\dfrac{x}{1-\dfrac{1}{x}}}$.

22. $\dfrac{1}{x+\dfrac{1}{x+\dfrac{2}{x}}}$.

23. $\dfrac{1}{1-\dfrac{1}{1-\dfrac{1}{y}}}$.

24. $\dfrac{1-x^2}{2-\dfrac{x}{1-\dfrac{1}{1+x}}}$.

25. $\dfrac{y}{1-\dfrac{1-y}{1-\dfrac{y^2}{2-y}}}$.

26. $\dfrac{a}{b-\dfrac{c}{d-\dfrac{e}{f}}}$.

27. $\dfrac{x^2-1}{2x^2-\dfrac{4x^2-1}{1+\dfrac{x}{x-1}}}$.

28. $\dfrac{3a-2c}{3a-2c-\dfrac{3a}{1-\dfrac{3(a-c)}{3a-2c}}}$.

Miscellaneous Fractions.

118. The following exercise contains miscellaneous examples which illustrate most of the processes connected with fractions.

EXAMPLES XV. h.

Simplify the following fractions :

1. $\dfrac{1-x^3}{1+2x+2x^2+x^3}.$

2. $\dfrac{12x^2+x-1}{1-8x+16x^2} \div \dfrac{1+6x+9x^2}{16x^2-1}.$

3. $\dfrac{a+b}{a-b}+\dfrac{4ab}{b^2-a^2}.$

4. $\dfrac{a+b}{a^2-ab-2b^2}-\dfrac{2a}{a^2-4b^2}.$

5. $\dfrac{x^3-1}{x-1}-\dfrac{x^4+x^2+1}{x^2+x+1}.$

6. $\dfrac{(x+y)^2}{x-y}-\dfrac{(x-y)^2}{x+y}.$

7. $\dfrac{abx^2-acx+bxy-cy}{ax^2+xy-ax-y}.$

8. $\dfrac{1}{x}\left(\dfrac{a}{a-x}-\dfrac{a}{a+3x}\right)-\dfrac{3}{a+3x}.$

9. $\dfrac{2}{x^2-6x+8}+\dfrac{3}{x^2-11x+28}+\dfrac{5}{x^2-9x+14}.$

10. $\dfrac{3a-\dfrac{1}{3a}}{3a+1}\times\dfrac{a}{3a-1}.$

11. $\dfrac{\dfrac{x^2+a^2}{x}-a}{a^3+x^3}+\dfrac{\dfrac{1}{x}}{x+a}.$

12. $\dfrac{1}{1-\dfrac{x}{x-1}}-\dfrac{1}{\dfrac{x}{x+1}-1}.$

13. $\dfrac{2x^2-\dfrac{7x^2-27}{x-1}}{3x^2-\dfrac{3(x^2-27x+54)}{1-x}}.$

14. $\dfrac{cd(a^2+b^2)+ab(c^2+d^2)}{cd(a^2-b^2)+ab(c^2-d^2)}.$

15. $\left(\dfrac{x}{1+x^2}\times\dfrac{1+x}{x^2}\right)-\dfrac{1}{x^2}.$

16. $\dfrac{1}{x^3-3ax^2+4a^3}-\dfrac{1}{x^3-ax^2-4a^2x+4a^3}.$

17. $\dfrac{\dfrac{a^2+4}{2}-a}{\dfrac{2}{a}-1}\times\dfrac{a^2-4}{a^3+8}.$

18. $\dfrac{\dfrac{2}{a^2}\left(4a^2-\dfrac{1}{9}\right)}{\dfrac{1}{a}+6}+\dfrac{1}{3}.$

19. $\dfrac{2x^3+x^2-3x}{35x^2+24x-35}\times\dfrac{5x^2-8x-21}{x^3+7x^2-8x}\div\dfrac{2x^2-3x-9}{7x^2+51x-40}.$

20. $\dfrac{q+r-p}{(p-q)(p-r)}+\dfrac{r+p-q}{(q-r)(q-p)}+\dfrac{p+q-r}{(r-p)(r-q)}.$

21. $\left\{ \left(\dfrac{x+y}{x-y} + \dfrac{x-y}{x+y} \right) \div \left(\dfrac{x+y}{x-y} - \dfrac{x-y}{x+y} \right) \right\} - \dfrac{x^3 + x^2y + xy^2 + y^3}{2x^2y + 2xy^2}.$

22. $\dfrac{a^2 - (b-c)^2}{(c+a)^2 - b^2} + \dfrac{b^2 - (c-a)^2}{(a+b)^2 - c^2} + \dfrac{c^2 - (a-b)^2}{(b+c)^2 - a^2}.$

23. $\dfrac{a^3 - 1}{a^2 + a - 6} \div \left[\dfrac{a^2 - 4a + 3}{a^2 - 4a + 4} \div \left\{ \dfrac{a^2 - 9}{a^4 + a^2 + 1} \div \dfrac{a^2 - a - 2}{a^3 + 1} \right\} \right].$

24. $\left(\dfrac{1}{2a} + \dfrac{1}{2a - x} \right) \left(\dfrac{1}{3a} - \dfrac{1}{3a - x} \right) - \dfrac{x^2 - 4ax}{6a^2(x - 2a)(x - 3a)}.$

25. $\dfrac{4ab^2}{2a^4 + 32b^4} + \dfrac{1}{8a + 16b} - \dfrac{a}{4a^2 + 16b^2} - \dfrac{1}{8(2b - a)}.$

26. $\dfrac{3b^2 + b}{6b^2 - 1 - b} + \dfrac{2b - 7}{1 - 2b} + \dfrac{2b^2 - 3b}{4b^2 - 8b + 3} + 3.$

27. $\dfrac{1}{\left(1 - \dfrac{y}{x} \right)\left(1 - \dfrac{z}{x} \right)} + \dfrac{1}{\left(1 - \dfrac{z}{y} \right)\left(1 - \dfrac{x}{y} \right)} + \dfrac{1}{\left(1 - \dfrac{x}{z} \right)\left(1 - \dfrac{y}{z} \right)}.$

28. $\dfrac{\dfrac{x^4 + x^3y + x^2y^2}{(x^2 - y^2)^3} \times \left(1 + \dfrac{y}{x} \right)^2}{\left(1 - \dfrac{y^3}{x^3} \right) \div \left(\dfrac{y}{x^2} + \dfrac{1}{x} \right)}.$

29. $\dfrac{m^2 + \dfrac{1}{m^2} + 1}{m^2 - \dfrac{1}{m^4}} - \dfrac{m^3 + m}{\dfrac{1}{m} - m^3}.$

MISCELLANEOUS EXAMPLES III.

The following Miscellaneous Examples on Fractions have been selected from Examination Papers set by the Science and Art Department in Mathematics, Stage I.

1. Simplify $\dfrac{x^2 - 8x + 7}{x^2 - 6x - 7}.$

Reduce to their simplest forms :

2. $\dfrac{1}{a-2} + \dfrac{1}{a+2} - \dfrac{4}{a^2 - 4}.$

3. $\dfrac{(m-a)(m+a) + n(2m+n)}{m+n+a}.$

4. $\dfrac{x+2y}{x-2y} - \dfrac{x-2y}{x+2y}.$

5. $\left(1 - \dfrac{2x}{1+2x} \right) \left(1 + \dfrac{2x}{1-2x} \right).$

6. Reduce to its lowest terms $\dfrac{2x^4 - 3x^3 - 47x^2 - 45x + 18}{3x^4 - 13x^3 - 45x^2 + 94x - 24}$.

Simplify the following expressions :

7. $\dfrac{1}{x-1} - \dfrac{x}{x^2+x+1}$.

8. $\dfrac{2}{y+3x} - \dfrac{1}{y-3x} + \dfrac{9x}{y^2-9x^2}$.

9. $\left(\dfrac{x-y}{x+y} + \dfrac{x+y}{x-y}\right) \div \left(\dfrac{x}{y} + \dfrac{y}{x}\right)$.

10. $\dfrac{1}{1-\dfrac{x}{y}} + \dfrac{\dfrac{x}{y}}{1-\dfrac{y}{x}}$.

Reduce the following expressions to their simplest forms :

11. $\dfrac{x-a}{\dfrac{1}{a}-\dfrac{1}{b}} \times \dfrac{a-b}{1-\dfrac{a}{x}}$.

12. $\left(\dfrac{1}{2} + \dfrac{1}{3a}\right) \div \left(9a - \dfrac{4}{a}\right)$.

13. $\dfrac{1+\dfrac{a}{b}}{\dfrac{b}{a}-1} \div \dfrac{a^2-b^2}{1-\dfrac{2b}{a}+\dfrac{b^2}{a^2}}$.

14. $\dfrac{\dfrac{1}{a}}{\dfrac{1}{a}-\dfrac{1}{b}} \div \dfrac{(a+b)^2-4ab}{\dfrac{1}{a^2}+\dfrac{1}{b^2}-\dfrac{2}{ab}}$.

Reduce the following fractions to their lowest terms :

15. $\dfrac{x^4 - 13x^2 + 36}{x^4 - x^3 - 7x^2 + x + 6}$.

16. $\dfrac{x^8 + x^4y^4 + y^8}{x^6 - y^6}$.

Simplify the following expressions :

17. $\dfrac{\dfrac{c}{a}}{\left(1-\dfrac{b}{a}\right)\left(1-\dfrac{a}{c}\right)} - \dfrac{\dfrac{a}{c}}{\left(\dfrac{b}{a}-1\right)\left(\dfrac{a}{c}-1\right)}$.

18. $\left(1 - \dfrac{x+1}{x^2-4x+5}\right)\left(1 - \dfrac{2x-4}{x^2-2x+1}\right)$.

Reduce to their simplest forms :

19. $\left\{1 - \dfrac{4xy}{(x+y)^2}\right\} \div \dfrac{x^2-y^2}{(x+y)^3}$.

20. $\dfrac{2x}{1-\dfrac{x^2}{y^2}} + \dfrac{y}{1-\dfrac{y}{x}}$.

21. $\left(x^2 - \dfrac{2x}{x-1}\right) \times \dfrac{x^3-1}{x^2-1} \div (x^3+x^2+x)$.

CHAPTER XVI.

HARDER EQUATIONS.

119. SOME of the equations in this chapter will serve as a useful exercise for revision of the methods already explained; but we also add others presenting more difficulty, the solution of which will often be facilitated by some special artifice.

The following examples worked in full will sufficiently illustrate the most useful methods.

Example 1. Solve $\dfrac{6x-3}{2x+7} = \dfrac{3x-2}{x+5}$.

Clearing of fractions, we have

$$(6x-3)(x+5) = (3x-2)(2x+7),$$
$$6x^2 + 27x - 15 = 6x^2 + 17x - 14 ;$$
$$\therefore\ 10x = 1 ;$$
$$\therefore\ x = \frac{1}{10}.$$

Note. By a simple reduction many equations can be brought to the form in which the above equation is given. When this is the case, the necessary simplification is readily completed by multiplying up, or "multiplying across", as it is sometimes called.

Example 2. Solve $\dfrac{8x+23}{20} - \dfrac{5x+2}{3x+4} = \dfrac{2x+3}{5} - 1$.

Multiplying by 20, we have

$$8x + 23 - \frac{20(5x+2)}{3x+4} = 8x + 12 - 20.$$

By transposition,
$$31 = \frac{20(5x+2)}{3x+4}.$$

Multiplying across,
$$93x + 124 = 20(5x+2),$$
$$84 = 7x ;$$
$$\therefore\ x = 12.$$

120. When two or more fractions have the same denominator, they should be taken together and simplified.

Example 1. Solve $\dfrac{24-5x}{x-2}+\dfrac{8x-49}{4-x}=\dfrac{28}{x-2}-13.$

By transposition, we have

$$\frac{8x-49}{4-x}+13=\frac{28-(24-5x)}{x-2};$$

$$\therefore\ \frac{3-5x}{4-x}=\frac{4+5x}{x-2}.$$

Multiplying across, we have

$$3x-5x^2-6+10x=16-4x+20x-5x^2;$$

that is, $-3x=22;$

$$\therefore\ x=-\frac{22}{3}.$$

Example 2. Solve $\dfrac{x-8}{x-10}+\dfrac{x-4}{x-6}=\dfrac{x-5}{x-7}+\dfrac{x-7}{x-9}.$

This equation might be solved by at once clearing of fractions, but the work would be laborious. The solution will be much simplified by proceeding as follows.

The equation may be written in the form

$$\frac{(x-10)+2}{x-10}+\frac{(x-6)+2}{x-6}=\frac{(x-7)+2}{x-7}+\frac{(x-9)+2}{x-9};$$

whence we have

$$1+\frac{2}{x-10}+1+\frac{2}{x-6}=1+\frac{2}{x-7}+1+\frac{2}{x-9};$$

which gives

$$\frac{1}{x-10}+\frac{1}{x-6}=\frac{1}{x-7}+\frac{1}{x-9}.$$

Transposing,

$$\frac{1}{x-10}-\frac{1}{x-7}=\frac{1}{x-9}-\frac{1}{x-6};$$

$$\therefore\ \frac{3}{(x-10)(x-7)}=\frac{3}{(x-9)(x-6)}.$$

Hence, since the numerators are equal, the denominators must be equal;

that is, $(x-10)(x-7)=(x-9)(x-6),$

$$x^2-17x+70=x^2-15x+54;$$

$$\therefore\ 16=2x;$$

$$\therefore\ x=8.$$

121. The following example illustrates the method of solving simple equations with decimal coefficients.

Example. $\dfrac{(\cdot 3x - 2)(\cdot 3x - 1)}{\cdot 2x - 1} - \dfrac{1}{6}(\cdot 3x - 2) = \cdot 4x - 2.$

Multiplying through by 10, the equation may be written

$$\frac{(3x - 20)(3x - 10)}{2x - 10} - \frac{1}{6}(3x - 20) = 4x - 20.$$

Multiplying by $6(x - 5)$, we have

$$3(3x - 20)(3x - 10) - (x - 5)(3x - 20) = 6(4x - 20)(x - 5),$$

or, $\quad 27x^2 - 270x + 600 - 3x^2 + 35x - 100 = 24x^2 - 240x + 600 \,;$

$$\therefore \; 5x = 100 \,;$$
$$\therefore \quad x = 20.$$

Sometimes it is found more simple to work entirely in decimals, as illustrated in Art. 60.

EXAMPLES XVI. a.

Solve the following equations :

1. $\dfrac{3}{5x - 9} = \dfrac{1}{4x - 10}.$

2. $\dfrac{7}{6x - 17} = \dfrac{3}{4x - 13}.$

3. $\dfrac{7}{9} = \dfrac{3 - 4x}{4 - 5x}.$

4. $\dfrac{1}{6 - 5x} + \dfrac{4}{17x + 3} = 0.$

5. $\dfrac{5x - 8}{x - 4} = \dfrac{5x + 14}{x + 7}.$

6. $\dfrac{8x - 1}{6x + 2} = \dfrac{4x - 3}{3x - 1}.$

7. $\dfrac{22x - 12}{8x - 5} = 2 + \dfrac{3x + 7}{4x + 8}.$

8. $\dfrac{9x - 22}{2x - 5} - \dfrac{3x - 5}{2x - 7} = 3.$

9. $\dfrac{8x - 19}{4x - 10} - \dfrac{1}{2} = \dfrac{3x - 4}{2x + 1}.$

10. $\dfrac{7x + 2}{3(x - 1)} = \dfrac{1}{3} + \dfrac{6x - 1}{3x + 1}.$

11. $\dfrac{x - 5}{2} + \dfrac{2x - 1}{3x + 2} = \dfrac{5x - 1}{10} - 1\tfrac{2}{5}.$

12. $\dfrac{5x - 17}{13 - 4x} + \dfrac{2x - 11}{14} - \dfrac{23}{42} = \dfrac{3x - 7}{21}.$

13. $x - \dfrac{4x - 3}{7x + 4} - \dfrac{1 - 9x}{6} = \dfrac{4x + 3}{8} - \dfrac{1}{24} + 2x.$

14. $\dfrac{3}{x + 1} - \dfrac{2\tfrac{1}{3}}{x + 2} = \dfrac{1}{x + 3} - \dfrac{1}{3x + 6}.$

15. $\dfrac{3\tfrac{1}{2}}{x - 4} - \dfrac{18}{3x - 18} = \dfrac{7}{4x - 16} - \dfrac{4}{x - 6}.$

Solve the following equations :

16. $\dfrac{1}{x+6}+\dfrac{1}{3x+12}=\dfrac{3}{2x+10}-\dfrac{1}{6(x+4)}.$

17. $\dfrac{x-1}{x-2}-\dfrac{x-5}{x-6}=\dfrac{x-3}{x-4}-\dfrac{x-7}{x-8}.$

18. $\dfrac{1}{x-9}+\dfrac{1}{x-17}=\dfrac{1}{x-11}+\dfrac{1}{x-15}.$

19. $\dfrac{1}{2x-1}+\dfrac{1}{2x-7}=\dfrac{1}{2x-3}+\dfrac{1}{2x-5}.$

20. $\dfrac{x-1}{x-2}-\dfrac{x}{x-1}=\dfrac{x-4}{x-5}-\dfrac{x-3}{x-4}.$

21. $\dfrac{5x-64}{x-13}-\dfrac{4x-55}{x-14}=\dfrac{2x-11}{x-6}-\dfrac{x-6}{x-7}.$

22. $\dfrac{5x+31}{x+6}-\dfrac{2x+9}{x+5}=\dfrac{x-6}{x-5}+\dfrac{2x-13}{x-6}.$

23. $\dfrac{12x+1}{3x-1}+\dfrac{5}{1-9x^2}=\dfrac{11+12x}{1+3x}.$

24. $\dfrac{5x^2}{x^2-9}-\dfrac{x+3}{x-3}=5-\dfrac{x-3}{x+3}.$ **25.** $\dfrac{2x-3}{\cdot3x-\cdot4}=\dfrac{\cdot4x-\cdot6}{\cdot06x-\cdot07}.$

26. $\dfrac{\cdot3x-1}{\cdot5x-\cdot4}=\dfrac{\cdot5+1\cdot2x}{2x-\cdot1}.$ **27.** $\dfrac{1-1\cdot4x}{\cdot2+x}=\dfrac{\cdot7(x-1)}{\cdot1-\cdot5x}.$

28. $(2x+1\cdot5)(3x-2\cdot25)=(2x-1\cdot125)(3x+1\cdot25).$

29. $\dfrac{x-2}{\cdot05}-\dfrac{x-4}{\cdot0625}=56.$ **30.** $\cdot08\dot3(x-\cdot625)=\cdot0\dot9(x-\cdot59375).$

Literal Equations.

122. In the equations we have discussed hitherto the co-efficients have been numerical quantities. When equations involve *literal* coefficients, these are supposed to be known, and will appear in the solution.

Example 1.   Solve $(x+a)(x+b)-c(a+c)=(x-c)(x+c)+ab.$

Multiplying out, we have

$$x^2+ax+bx+ab-ac-c^2=x^2-c^2+ab;$$

whence $$ax+bx=ac,$$
$$(a+b)x=ac;$$
$$\therefore \quad x=\dfrac{ac}{a+b}.$$

Example 2.   Solve $\dfrac{a}{x-a} - \dfrac{b}{x-b} = \dfrac{a-b}{x-c}$.

Simplifying the left side, we have

$$\frac{a(x-b) - b(x-a)}{(x-a)(x-b)} = \frac{a-b}{x-c},$$

$$\frac{(a-b)x}{(x-a)(x-b)} = \frac{a-b}{x-c};$$

$$\frac{x}{(x-a)(x-b)} = \frac{1}{x-c}.$$

Multiplying across, $x^2 - cx = x^2 - ax - bx + ab,$

$$ax + bx - cx = ab,$$

$$(a+b-c)x = ab;$$

$$\therefore \quad x = \frac{ab}{a+b-c}.$$

Example 3.   Solve the simultaneous equations :

$$ax - by = c \quad\dots\dots\dots\dots\dots\dots\dots \quad (1),$$

$$px + qy = r \quad\dots\dots\dots\dots\dots\dots\dots \quad (2).$$

To eliminate y, multiply (1) by q and (2) by b ;

thus $aqx - bqy = cq,$

$$bpx + bqy = br.$$

By addition, $(aq + bp)x = cq + br$;

$$\therefore \quad x = \frac{cq + br}{aq + bp}.$$

We might obtain y by substituting this value of x in *either* of the equations (1) or (2); but y is more conveniently found by eliminating x, as follows.

Multiplying (1) by p and (2) by a, we have

$$apx - bpy = cp,$$

$$apx + aqy = ar.$$

By subtraction, $(aq + bp)y = ar - cp$;

$$y = \frac{ar - cp}{aq + bp}.$$

EXAMPLES XVI. b.

Solve the following equations:

1. $ax + b^2 = a^2 - bx.$

2. $x^2 - a^2 = (2a - x)^2.$

3. $a^2(a - x) + abx = b^2(x - b).$

4. $(b + 1)(x + a) = (b - 1)(x - a).$

5. $a(x + b) - b^2 = a^2 - b(a - x).$

6. $c^2 x - d^3 = d^2 x + c^3.$

7. $a(x - a) + b(x - b) + c(x - c) = 2(ab + bc + ca).$

8. $\dfrac{a^2}{x} - b = \dfrac{b^2}{x} + a.$

9. $\dfrac{x}{2a} = \dfrac{x}{b} + \dfrac{1}{b^2} - \dfrac{1}{4a^2}.$

10. $x + (x - a)(x - b) + a^2 + b^2 = b + x^2 - a(b - 1).$

11. $\dfrac{2x - a}{b} - \dfrac{3x - b}{a} = \dfrac{3a^2 - 8b^2}{ab}.$

12. $\dfrac{a - x}{a - b} - \dfrac{b - x}{a + b} = \dfrac{a^2 + b^2}{a^2 - b^2}.$

13. $\dfrac{ax - b}{c} + \dfrac{bx - c}{a} = \dfrac{a - cx}{b}.$

14. $\dfrac{x + a - b}{x + b + c} = \dfrac{x + b - c}{x + a + b}.$

15. $p(p - x) - \dfrac{p}{q}(x - q)^2 - p(p - q) + pq\left(\dfrac{x}{q} - 1\right)^2 = 0.$

16. $\dfrac{ax}{a + c} + 2ac - a^2 = \dfrac{cx}{a - c} - c^2.$

17. $\dfrac{x + m}{x - n} + \dfrac{x + n}{x - m} = 2.$

Solve the following simultaneous equations:

18. $x - y = a + b,$
$ax + by = 0.$

19. $cx - dy = c^2 + d^2,$
$x + y = 2c.$

20. $ax = by,$
$x - y = c.$

21. $\dfrac{x}{2} + \dfrac{y}{3} = a + b,$
$\dfrac{x}{a} + \dfrac{y}{b} = 5.$

22. $\dfrac{a}{x} - \dfrac{b}{y} = 0,$
$\dfrac{x}{a} + \dfrac{y}{b} = 2.$

23. $\dfrac{x + y}{x - y} = \dfrac{a}{b},$
$\dfrac{x - y}{a + b} = 2b.$

24. $\dfrac{x}{b} - \dfrac{y}{a} = \dfrac{a}{b} + \dfrac{b}{a},$
$a(a + x) = b(b - y).$

25. $\dfrac{x + y}{p} - \dfrac{x - y}{q} = 0.$
$\dfrac{x - y}{2p} + \dfrac{x + y}{2q} = p^2 + q^2.$

26. $\dfrac{2x - b}{a} = \dfrac{2y + a}{b} = \dfrac{3x + y}{a + 2b}.$

27. $\dfrac{ax + by}{bx + ay} = \dfrac{1}{2} = \dfrac{a^2 - b^2}{bx + ay}.$

MISCELLANEOUS EXAMPLES IV.

The following Miscellaneous Simple Equations have been selected from Examination Papers set by the Science and Art Department in Mathematics, Stage I.

Solve the equations :

1. $x - \dfrac{3}{5} - \dfrac{5(x-2)}{4} = \dfrac{3}{2}\left(x - \dfrac{1}{10}\right).$

2. $\dfrac{7x - 28}{3} - 3\tfrac{1}{4} + \dfrac{4x - 21}{7} = x - 7\tfrac{1}{4} - \dfrac{9 - 7x}{8}.$

3. $7\left(x + \dfrac{1}{3}\right) - 5x\left(\dfrac{1}{3x} + \dfrac{1}{2\cdot}\right) = 4.$ 4. $\dfrac{1}{x-3} - \dfrac{1}{x-7} = \dfrac{1}{x-2} - \dfrac{1}{x-6}.$

5. $\left(2 - 3x\right)\left(4 - \dfrac{5}{x}\right) = \left(3 - 2x\right)\left(6 - \dfrac{7}{x}\right).$

6. $\dfrac{1}{3}(0{\cdot}75 - x) + \dfrac{1}{5}(0{\cdot}47 + 2x) = \left(3 - \dfrac{1}{15}\right)x.$

7. $x - \dfrac{2x - 0{\cdot}3}{0{\cdot}7} = \dfrac{5 - x}{0{\cdot}35}.$ 8. $\dfrac{x}{2} - \dfrac{0{\cdot}05x - 7{\cdot}5}{0{\cdot}6} = \dfrac{0{\cdot}25x + 3{\cdot}8}{0{\cdot}3}.$

Solve the following literal equations :

9. $a(x - a) + b(x - b) + 2ab = 0.$ 10. $\dfrac{b}{a} - \dfrac{dx}{c} = \dfrac{ax}{b} - \dfrac{c}{d}.$

11. $(ax + b)(bx - a) = a(bx^2 - a).$ 12. $\dfrac{x + a}{a} + \dfrac{x}{x - a} = \dfrac{x - a}{a}.$

Solve the following simultaneous equations :

13. $0{\cdot}5x + 0{\cdot}07y = 0{\cdot}93,$
    $0{\cdot}03x - 0{\cdot}4y = 0{\cdot}46.$

14. $3{\cdot}4x - 0{\cdot}02y = 0{\cdot}01,$
    $x + 0{\cdot}2y = 0{\cdot}6.$

15. $2x - 7y = 0{\cdot}2,$
    $0{\cdot}2x + 0{\cdot}15y = 1{\cdot}72.$

16. $0{\cdot}5x - 0{\cdot}6y - 1{\cdot}38,$
    $\dfrac{x}{3} + \dfrac{y}{0{\cdot}5} = 2.$

17. $(x + y)^2 - (x - y)^2 = 352,$
    $x(y + 5) = 143.$

18. $ax + by = bx - ay = c.$

19. The expression $ax^2 + bx - 30$ is equal to 240 when $x = 5$, and is equal to 100 when $x = -2$; find the values of a and b.

CHAPTER XVII.

HARDER PROBLEMS.

123. IN previous chapters we have given collections of problems which lead to simple equations. We add here a few examples of somewhat greater difficulty.

Example 1. If the numerator of a fraction is increased by 2 and the denominator by 1, it becomes equal to $\frac{5}{8}$; and if the numerator and denominator are each diminished by 1, it becomes equal to $\frac{1}{2}$: find the fraction.

Let x be the numerator of the fraction, y the denominator; then the fraction is $\frac{x}{y}$.

From the first supposition,

$$\frac{x+2}{y+1}=\frac{5}{8} \dots\dots\dots\dots\dots\dots\dots\dots (1),$$

from the second,

$$\frac{x-1}{y-1}=\frac{1}{2} \dots\dots\dots\dots\dots\dots\dots\dots (2).$$

From the first equation, $8x-5y=-11$,
and from the second, $2x-y=1$;
whence $x=8$, $y=15$.

Thus the fraction is $\frac{8}{15}$.

Example 2. At what time between 4 and 5 o'clock will the minute-hand of a watch be 13 minutes in advance of the hour-hand?

Let x denote the required number of minutes after 4 o'clock; then, as the minute-hand travels twelve times as fast as the hour-hand, the hour-hand will move over $\frac{x}{12}$ minute-divisions in x minutes.

At 4 o'clock the minute-hand is 20 divisions behind the hour-hand, and finally the minute-hand is 13 divisions in advance; therefore the minute-hand moves over $20+13$, or 33 divisions more than the hour-hand.

Hence $x=\frac{x}{12}+33$,

$$\frac{11}{12}x=33;$$

$$\therefore \quad x=36.$$

Thus the time is 36 minutes past 4.

If the question be asked as follows: "At what *times* between 4 and 5 o'clock will there be 13 minutes between the two hands?" we must also take into consideration the case when the minute-hand is 13 divisions *behind* the hour-hand. In this case the minute-hand gains $20 - 13$, or 7 divisions.

$$\therefore \quad x = \frac{x}{12} + 7; \qquad \text{whence} \quad x = 7\frac{7}{11}.$$

Therefore the *times* are $7\frac{7'}{11}$ past 4, and 36' past 4.

Example 3. A grocer buys 15 lbs. of figs and 28 lbs. of currants for £1. 1s. 8d.; by selling the figs at a loss of 10 per cent., and the currants at a gain of 30 per cent., he clears 2s. 6d. on his outlay; how much per pound did he pay for each?

Let x, y denote the number of pence in the price of a pound of figs and currants respectively; then the outlay is

$$15x + 28y \text{ pence.}$$

Therefore $\qquad 15x + 28y = 260. \quad\text{...........................}(1).$

The loss upon the figs is $\frac{1}{10} \times 15x$ pence, and the gain upon the currants is $\frac{3}{10} \times 28y$ pence; therefore the total gain is

$$\frac{42y}{5} - \frac{3x}{2} \text{ pence;}$$

$$\therefore \quad \frac{42y}{5} - \frac{3x}{2} = 30;$$

that is, $\qquad 28y - 5x = 100. \quad\text{...........................}(2).$

From (1) and (2) we find that $x = 8$, and $y = 5$; that is, the figs cost 8d. a pound, and the currants cost 5d. a pound.

Example 4. Two persons A and B start simultaneously from two places, c miles apart, and walk in the same direction. A travels at the rate of p miles an hour, and B at the rate of q miles; how far will A have walked before he overtakes B?

Suppose A has walked x miles, then B has walked $x - c$ miles. A walking at the rate of p miles an hour will travel x miles in $\frac{x}{p}$ hours; and B will travel $x - c$ miles in $\frac{x-c}{q}$ hours; these two times being equal, we have

$$\frac{x}{p} = \frac{x-c}{q},$$

$$qx = px - pc;$$

whence $x = \frac{pc}{p-q}. \quad \therefore A$ has travelled $\frac{pc}{p-q}$ miles.

Example 5. A train travelled a certain distance at a uniform rate. Had the speed been 6 miles an hour more, the journey would have occupied 4 hours less; and had the speed been 6 miles an hour less, the journey would have occupied 6 hours more. Find the distance.

Let the speed of the train be x miles per hour, and let the time occupied be y hours; then the distance traversed will be represented by xy miles.

On the first supposition the speed per hour is $x+6$ miles, and the time taken is $y-4$ hours. In this case the distance traversed will be represented by $(x+6)(y-4)$ miles.

On the second supposition the distance traversed will be represented by $(x-6)(y+6)$ miles.

All these expressions for the distance must be equal;

$$\therefore \ xy = (x+6)(y-4) = (x-6)(y+6).$$

From these equations we have

$$xy = xy + 6y - 4x - 24,$$

or $6y - 4x = 24$ ....(1) ;

and $xy = xy - 6y + 6x - 36,$

or $6x - 6y = 36$ ....(2).

From (1) and (2) we obtain $x=30$, $y=24$.
Hence the distance is 720 miles.

EXAMPLES XVII.

1. If the numerator of a fraction is increased by 5 it reduces to $\frac{2}{3}$, and if the denominator is increased by 9 it reduces to $\frac{1}{3}$: find the fraction.

2. Find a fraction such that it reduces to $\frac{3}{5}$ if 7 be subtracted from its denominator, and reduces to $\frac{3}{4}$ on subtracting 3 from its numerator.

3. If unity is taken from the denominator of a fraction it reduces to $\frac{1}{3}$; if 3 is added to the numerator it reduces to $\frac{4}{7}$: required the fraction.

4. Find a fraction which becomes $\frac{3}{4}$ on adding 5 to the numerator and subtracting 1 from the denominator, and reduces to $\frac{1}{3}$ on subtracting 4 from the numerator and adding 7 to the denominator.

5. If 9 is added to the numerator a certain fraction will be increased by $\frac{1}{3}$; if 6 is taken from the denominator the fraction reduces to $\frac{2}{3}$; required the fraction.

6. At what time between 9 and 10 o'clock are the hands of a watch together?

7. When are the hands of a clock 8 minutes apart between the hours of 5 and 6?

8. At what time between 10 and 11 o'clock is the hour-hand six minutes ahead of the minute-hand?

9. At what time between 1 and 2 o'clock are the hands of a watch in the same straight line?

10. When are the hands of a clock at right angles between the hours of 5 and 6?

11. At what times between 12 and 1 o'clock are the hands of a watch at right angles?

12. A person buys 20 yards of cloth and 25 yards of canvas for £1. 17s. 6d. By selling the cloth at a gain of 15 per cent. and the canvas at a gain of 20 per cent. he clears 6s. 3d. : find the price of each per yard.

13. A dealer spends £695 in buying horses at £25 each and cows at £20 each ; through disease he loses 20 per cent. of the horses and 25 per cent. of the cows. By selling the animals at the price he gave for them he receives £540 ; find how many of each kind he bought.

14. The population of a certain district is 33000, of whom 835 can neither read nor write. These consist of 2 per cent. of all the males and 3 per cent. of all the females : find the number of males and females.

15. Two persons C and D start simultaneously from two places a miles apart, and walk to meet each other ; if C walks p miles per hour, and D one mile per hour faster than C, how far will D have walked when they meet?

16. A can walk a miles per hour faster than B ; supposing that he gives B a start of c miles, and that B walks n miles per hour, how far will A have walked when he overtakes B?

17. A, B, C start from the same place at the rates of a, $a+b$, $a+2b$ miles an hour respectively. B starts n hours after A, how long after B must C start in order that they may overtake A at the same instant, and how far will they then have walked?

18. Find the distance between two towns when by increasing the speed 7 miles per hour a train can perform the journey in 1 hour less, and by reducing the speed 5 miles per hour can perform the journey in 1 hour more.

19. A person buys a certain quantity of land. If he had bought 7 acres more each acre would have cost £4 less, and if each acre had cost £18 more he would have obtained 15 acres less : how much did he pay for the land?

20. A can walk half a mile per hour faster than B, and three quarters of a mile per hour faster than C. To walk a certain distance C takes three quarters of an hour more than B, and two hours more than A : find their rates of walking per hour.

MISCELLANEOUS EXAMPLES V.

The following Miscellaneous Problems have been selected from Examination Papers set by the Science and Art Department in Mathematics, Stage I.

1. Oranges are bought for half-a-crown a hundred ; some are sold at 3s. 6d. a hundred, and the rest at 2s. 10½d. a hundred : the same profit is made, as if they had all been sold at 3s. 1½d. a hundred. Of a thousand oranges sold, how many fetch 3s. 6d. a hundred ?

2. A man has 1000 apples for sale ; at first he sells so as to gain at the rate of 50 per cent. on the cost price ; when he has done this for a time, the sale falls off, so he sells the remainder for what he can get, and finds that by doing so he loses at the rate of 10 per cent. If his total gain is at the rate of 29 per cent., how many apples did he sell for what he could get ?

3. A sum of £23. 14s. is to be divided between A, B, and C : if B gets 20 per cent. more than A, and 25 per cent. more than C, how much does each get ?

4. The sides of a triangle ABC are together 61 miles long ; BC is ⅚th of AB, and three miles longer than CA : find the length of the sides severally.

5. A market-woman spent 10s. 10d. in buying eggs, some at two a penny and others at 3 for twopence : she sold them for a sovereign, thereby gaining ½d. on each egg. How many of each kind did she buy ?

6. The circumference of the large wheel of a bicycle (1885) is 3½ times that of the small wheel : the small wheel makes 10 turns more than the large wheel in running 21 yards. Find the circumference of each wheel.

7. Find two numbers whose sum is 39, and whose difference equals a third part of the greater.

8. Divide 279 into two parts such that one-third of the first part is less by 15 than one-fifth of the second part.

9. Find a number such that, when diminished by 3, one-fourth of the remainder may be greater by 2 than one-fifth of the original number.

10. A person spent 9s. in buying apples at the rate of 7d. a dozen, and oranges at the rate of 20 a shilling. If he had bought two-thirds as many apples and twice as many oranges, he would have had to pay 13s. 4d. How many of each did he buy ?

11. Find the fraction which is equal to $\frac{3}{7}$ when 10 is added to its numerator, and which is equal to $\frac{1}{3}$ when 4 is subtracted from its denominator.

12. A man has 81 coins, some of them crowns and the rest shillings : if he exchanged the crowns for florins, and the shillings for half-crowns, he would neither gain nor lose. How many crowns had he?

13. If 10 yards of silk and 7 yards of satin cost £5. 6s. 4d., and if 3 yards of the satin cost as much as 4 yards of the silk, find the price of a yard of each.

14. A man buys oranges at 6d. a dozen, and an equal number at ninepence a score : he sells them at ninepence a dozen, and makes a profit of 5s. 6d. How many oranges did he buy?

15. A bill of £13. 12s. 6d. was paid with 40 coins, some of them half-crowns, and the others half-sovereigns ; find the number used of each kind.

16. Find the fraction which becomes equal to $\frac{1}{2}$ if its denominator be decreased by 6, and becomes equal to $\frac{3}{4}$ if its numerator and denominator be each increased by 124.

17. A tradesman sends in a bill for £12, part of which is for labour, and the other part for materials. If the charge for labour had been twice what it was, and the charge for materials one-third of what it was, the amount of the bill would still have been £12. What was the charge for labour, and what for materials?

18. A certain number consisting of two digits, exceeds four times the sum of its digits by 3 : if the number be increased by 18, the result is the same as if the number formed by reversing the digits were diminished by 18. Find the number.

19. A number is divided into two parts : the difference between the parts is 5, and two-thirds of the smaller part is less than three-fourths of the larger part by 8. Find the number.

20. A man bought two pairs of shoes, and one pair cost him 2s. more than the other pair. If he had paid 3s. less for each pair, they would have cost him four-fifths of what he paid. How many shillings did each pair cost him?

CHAPTER XVIII.

Positive Integral Indices.

124. We have adopted the symbol a^7 as an abbreviation for $a \times a \times a \times a \times a \times a \times a$, that is, for $a \cdot a \cdot a \ldots$ to *seven* factors.

From this it follows that

(1) $\quad a^7 \times a^3 = (a \cdot a \cdot a \cdot a \cdot a \cdot a \cdot a) \times (a \cdot a \cdot a)$
$$= a \cdot a \cdot a \ldots \text{ to } (7+3) \text{ factors}$$
$$= a^{7+3} = a^{10}.$$

(2) $\quad a^7 \div a^3 = (a \cdot a \cdot a \cdot a \cdot a \cdot a \cdot a) \div (a \cdot a \cdot a)$
$$= \frac{a \cdot a \cdot a \cdot a \cdot a \cdot a \cdot a}{a \cdot a \cdot a}$$
$$= a \cdot a \cdot a \ldots \text{ to } (7-3) \text{ factors}$$
$$= a^{7-3} = a^4.$$

(3) $\quad (a^7)^3 = a^7 \times a^7 \times a^7 = a^{7+7+7}$
$$= a^{7 \times 3} = a^{21}.$$

125. Of these important rules we shall now give general proofs for all cases when the indices are positive whole numbers.

DEFINITION. When m is a *positive integer*, a^m stands for the product of m factors each equal to a.

Rule I. *To prove that* $a^m \times a^n = a^{m+n}$, *when* m *and* n *are positive integers.*

By definition, $a^m = a \cdot a \cdot a \ldots\ldots$ to m factors ;
$$a^n = a \cdot a \cdot a \ldots\ldots \text{ to } n \text{ factors} ;$$
$\therefore \ a^m \times a^n = (a \cdot a \cdot a \ldots \text{ to } m \text{ factors}) \times (a \cdot a \cdot a \ldots \text{ to } n \text{ factors})$
$$= a \cdot a \cdot a \ldots \text{ to } m+n \text{ factors}$$
$$= a^{m+n}, \text{ by definition.}$$

Cor. If p is also a positive integer, then
$$a^m \times a^n \times a^p = a^{m+n+p} ;$$
and so for any number of factors.

Rule II. *To prove that* $a^m \div a^n = a^{m-n}$, *when* m *and* n *are positive integers, and* m $>$ n.

$$a^m \div a^n = \frac{a^m}{a^n} = \frac{a \,.\, a \,.\, a \ldots \text{ to } m \text{ factors}}{a \,.\, a \,.\, a \ldots \text{ to } n \text{ factors}}$$

$$= a \,.\, a \,.\, a \ldots \text{ to } m - n \text{ factors}$$

$$= a^{m-n}.$$

Rule III. *To prove that* $(a^m)^n = a^{mn}$, *when* m *and* n *are positive integers.*

$$(a^m)^n = a^m \,.\, a^m \,.\, a^m \ldots \text{ to } n \text{ factors}$$

$$= (a \,.\, a \,.\, a \ldots \text{ to } m \text{ factors})(a \,.\, a \,.\, a \ldots \text{ to } m \text{ factors})\ldots\ldots$$

the bracket being repeated n times,

$$= a \,.\, a \,.\, a \ldots \text{ to } mn \text{ factors}$$

$$= a^{mn}.$$

Examples.   (1) $(a^5)^3 \times (a^3)^5 \div (a^7)^4 = a^{15} \times a^{15} \div a^{28}$

$$= a^{15+15-28} = a^2.$$

(2) $\left\{ \dfrac{(8a^3)^2}{49b^2} \times \dfrac{(7b)^3}{(2a)^6} \right\}^2 \div \left(\dfrac{7b}{a} \right)^3 = \left\{ \dfrac{2^6 a^6 \times 7^3 b^3}{7^2 b^2 \times 2^6 a^6} \right\}^2 \div \left(\dfrac{7b}{a} \right)^3$

$$= (7b)^2 \div \left(\dfrac{7b}{a} \right)^3 = \dfrac{a^3}{7b}.$$

(3) $\dfrac{81 x^{3m-n} \times 27 x^{m+6n}}{(9x^{m+n})^4} = \dfrac{3^{4+3} \,.\, x^{4m+5n}}{3^8 \,.\, x^{4m+4n}}$

$$= \dfrac{x^{4m+5n-4m-4n}}{3} = \dfrac{x^n}{3}.$$

126.   It is evident from the Rule of Signs that

(1) no *even* power of *any* quantity can be *negative*;

(2) any *odd* power of a quantity will have *the same sign* as the quantity itself.

Note. It is especially worthy of notice that the *square* of every expression, whether positive or negative, is *positive*.

Examples.   (1) $(-x^3)^2 = (-x^3)(-x^3) = +(x^3)^2 = x^6.$

(2) $(-a^5)^3 = (-a^5)(-a^5)(-a^5) = -(a^5)^3 = -a^{15}.$

(3) $(-3a^3)^4 = (-3)^4 (a^3)^4 = 81 a^{12}.$

(4) $(b-a)^4 \times (a-b)^2 = (a-b)^4 \times (a-b)^2 = (a-b)^6.$

(5) $(b-a)^5 \div (a-b)^3 = -(a-b)^5 \div (a-b)^3 = -(a-b)^2.$

EXAMPLES XVIII.

Reduce to their simplest forms :

1. $a^3 \times a^5 \times a^7.$ **2.** $a^{10} \times a^2 \div a^4.$

3. $\dfrac{x^{10} \times 16c^6}{8c^8}.$ **4.** $(a^3)^3 \times (x^2)^2 \div (x^2)^5.$

5. $\dfrac{(a^3b^2c)^2}{(ab^2c^3)^3} \times \dfrac{b^2c^7}{a^2}.$ **6.** $\dfrac{(3x^4)^2 \times (2x^2)^4}{4x^{10} \times 6^2}.$

7. $\left(\dfrac{3a^3b}{c}\right)^3 \times \dfrac{c^2}{(9a^4b)^2} \div \dfrac{3ab}{c}.$ **8.** $\left\{\dfrac{125b^5}{2a^2} \times \dfrac{(4a^3)^2}{(5ab)^3}\right\}^2 \div \dfrac{2^5}{(a^2b)^2}.$

Find the values of the following expressions, when $a=2$, $b=-\frac{1}{2}$, $c=-3$, $d=\frac{1}{3}$, $m=5$, $n=3$:

9. $81a^5b^4d^3.$ **10.** $-\dfrac{243a^3b^2d}{4c^3}.$

11. $27a^m \times b^n \div (3cd)^3.$ **12.** $\left(\dfrac{a}{n}\right)^{a-c} \times (-3c)^n \div \left(\dfrac{1}{b}\right)^m.$

Simplify the following expressions :

13. $\dfrac{a^{3p} \cdot a^{8p} \cdot a^{7p}}{a^{9p}}.$ **14.** $\dfrac{4c^{2p-q} \cdot 8x^{p+q} \cdot 16x^{3q}}{(8x^{p+q})^3}.$

15. $\left\{\dfrac{a^{m-n} \times b^{m+n}}{a^{m+n} \cdot b^{m-n}}\right\}^m \div (ab^2)^{mn}.$ **16.** $\dfrac{3^2 \times (27)^4}{9^5 \times 81}.$

17. $\dfrac{2^{6m} \times 4^{3m} \times 8^{2m}}{2^{18m+1}}.$ **18.** $\dfrac{(a+b)^2 \times (a+b)^5}{(a+b)^3}.$

19. $\dfrac{(a^2-b^2)^4 \times (a+b)^2}{\{(a-b)(a+b)^3\}^2}.$ **20.** $\dfrac{(x-y)^{3m}(x+y)^{2m}}{\{(x^2-y^2)^m\}^3}.$

21. If m and n are positive whole numbers, explain the meaning of A^m; also explain why $A^m \times A^n = A^{m+n}$.

     Simplify $\dfrac{27a^{m+p} \cdot b^{n-q}}{81a^m b^n}.$

22. Simplify $\left(\dfrac{bc}{a}\right)^{m+n} \cdot \left(\dfrac{a}{c}\right)^m \cdot \left(\dfrac{a}{b}\right)^n$; and find its value when $a=7$, $b=3$, $c=2$, $m=2$, $n=1$.

MISCELLANEOUS EXAMPLES VI.

The following collection of Miscellaneous Examples covers all the rules and methods explained in the course of this text-book. The questions are selected from the Examination Papers set by the Science and Art Department, in Mathematics, Stage I.

1. Find the value of

$$(b-c)^3+2(c-a)^3+(a-b)^3-3(b-c)(c-a)(a-b),$$

when $\qquad a=1, \ b=-\frac{1}{2}, \ c=\frac{3}{2}.$

2. Multiply $x^3+ax^2y+axy^2+y^3$ by x^2-xy+y^2; and divide $mx^5+m^2x^4-(2m-3)x^3+(3m+7)x^2+(7m-6)x-14$ by $x^2+mx-2.$

3. Find the G.C.M. and L.C.M. of

$$2x^4+9x^2+5x+12 \ \text{ and } \ 2x^4+4x^3+13x^2+11x+12.$$

4. Simplify the following expressions :

(i.) $\dfrac{\dfrac{x}{x+1}+\dfrac{x}{x-1}}{\dfrac{2}{x^2-1}}-\dfrac{4x-\dfrac{1}{x}}{2+\dfrac{1}{x}};$

(ii.) $\left(\dfrac{1}{x^3-y^3}-\dfrac{1}{x^3+y^3}\right)\div\left(\dfrac{1}{x-y}-\dfrac{1}{x+y}-\dfrac{2y}{x^2+y^2}\right).$

5. Solve the equations :

(i.) $\dfrac{x-3}{2}+\dfrac{x-4}{3}=5;$ (ii.) $\dfrac{x+m}{x-n}+\dfrac{x+n}{x-m}=2;$

(iii.) $\dfrac{5+x}{3}=\dfrac{7+y}{5}=\dfrac{9+x+y}{7}.$

6. The depth of a pond at one end is twice as great as at the other. Eighteen inches of water (in depth) are drained off, and the deep end is then three times as deep as the shallow end. What were the original depths?

7. Find the value of

$$\frac{a+b}{ab}(a^2+b^2-c^2)+\frac{b+c}{bc}(b^2+c^2-a^2)+\frac{c+a}{ca}(c^2+a^2-b^2)$$

when $a=3, \ b=4, \ c=-5.$

8. Divide $(a+1)^2x^3+(a+1)x^2+a^2(a-1)x-a^5$ by $(a+1)x-a^2$; and find the value of the quotient when $a=-\frac{1}{2},$ and $x=-2.$

9. Reduce to its lowest terms $\dfrac{x^4+4x^3-19x^2-46x+120}{x^4-25x^2+144}$.

10. Simplify the following expressions :

(i.) $x+y-\dfrac{9x^2-4y^2}{3x+2y}$;

(ii.) $\left(\dfrac{1}{2a+b}+\dfrac{1}{2a-b}-\dfrac{3a}{4a^2-b^2}\right)\times\dfrac{4a^2+4ab+b^2}{2a-b}$.

11. Solve the equations :

(i.) $\dfrac{x-1}{3}-\dfrac{2x-7}{4}=\dfrac{3}{4}$;

(ii.) $\dfrac{2x+5}{3}=\dfrac{y+4}{2}=\dfrac{2x+2y+9}{6}$;

and (iii.) find the relation between a and b, when

$$\frac{3}{x-a}-\frac{2}{x+a}=\frac{x+b}{x^2-a^2}.$$

12. At what times between 4 and 5 o'clock are the hands of a watch exactly at right angles to one another?

13. Given $6a=1$, $9b=-1$, and $2c-1=0$, find the numerical values of

(i.) $8a^3+27b^3+c^3-18abc$;

(ii.) $\dfrac{a}{b+c}+\dfrac{b}{c+a}+\dfrac{c}{a+b}$.

14. Divide $x^3+8y^3-27z^3+18xyz$ by $x+2y-3z$, and test your answer by substituting $x=5$, $y=-4$, $z=3$, in the dividend, divisor, and quotient.

15. Find the value of $x^3-8y^3+29z^3+18xyz$, when $2y=x+3z$ and $z=5$.

16. Reduce to its lowest terms

$$\frac{x^4-2x^3-25x^2+26x+120}{x^4-4x^3-19x^2+46x+120}.$$

17. Simplify the fractions :

(i.) $\dfrac{1}{x-\dfrac{x}{x+1}}+\dfrac{1}{x+\dfrac{x}{x-1}}+\dfrac{1-\dfrac{3}{x}}{1-\dfrac{x}{3}}$.

(ii.) $\dfrac{a+b}{a-b}-\dfrac{2ab}{a^2-b^2}-\dfrac{2a^2b^2}{a^4-b^4}$.

18. Solve the equations :

(i.) $5x - [3x + 6 - (2x - 11)] = 6x + 11 - \{2x - 3 - (15 - 2x)\}$

(ii.) $(ax + b)(bx - a) = a(bx^2 - a)$

(iii.) $\dfrac{7x + 3}{5} = \dfrac{5y - 7}{4} = \dfrac{2x + 3y - 5}{3}$.

19. If one part of £400 is put out at 4 per cent. per annum, and the other part at 5 per cent. per annum ; and if the yearly income obtained is £18. 15s., what are the parts ?

20. Given $2x = 3$, $4y = 3$, and $z = -2$; find the numerical values of the following expressions :

(i.) $\sqrt{(8y + 2z + 7)} + \sqrt{(6x - 8y + z)}$;

(ii.) $\dfrac{x}{y} + \dfrac{4x + 2z}{2x - z}$;　　(iii.) $xy + yz - zx$.

21. Simplify the expression

$$\frac{2(x^2 + 2xa + a^2)(x^2 - 2xa + a^2) - 2(x^2 - a^2)^2 + 5ax(a + x)^2 - 20a^2x^2}{(x - a)(x + a)}.$$

22. Reduce $\dfrac{8x^3 - 10x^2 - 16x - 3}{6x^4 - 22x^3 + 31x^2 - 23x - 7}$ to its lowest terms ; and find its value when $x = -\dfrac{2}{3}$.

23. If $\dfrac{x}{a} = \dfrac{y}{b}$, shew that either fraction is equal to $\dfrac{x + y}{a + b}$ or $\dfrac{x - y}{a - b}$.

24. Show that the product of

$$1 - \frac{a}{x + 2a},\ \ 1 + \frac{b}{x - 2b},\ \ \text{and}\ \ x + 2(a - b) - \frac{4ab}{x}\ \ \text{is}\ \ \frac{(x + a)(x - b)}{x} ;$$

write down the value of each factor and of the product, when $x = 3a = 3b$.

25. Solve the equations :

(i.) $ax + b = 3ax + c$;　　(ii.) $\dfrac{2}{x - a} - \dfrac{1}{x - b} = \dfrac{x}{x^2 - a^2}$;

(iii.) $\begin{cases} \dfrac{x}{2} - y = \dfrac{7}{2}, \\[2mm] \dfrac{x}{25} + \dfrac{y}{2} = -\dfrac{3}{10}. \end{cases}$

26. Of two squares of carpet one measures 44 feet more round than the other, and 187 square feet more in area. What are their sizes ?

27.   When $a = 7$, and $x = -16$, find the numerical value of

$$\frac{a+x}{a-x} + \sqrt{\frac{a+x}{a+2x}}.$$

28.   Find the value of

$$\frac{x^4 - 4x^3y + 6x^2y^2 - 5xy^3 + 2y^4}{2x^4 - 5x^3y + 6x^2y^2 - 4xy^3 + y^4},$$

when $x = 5$ and $y = 3$.

29.   Shew that

$$\frac{x^2}{a^2} + \left(\frac{z-x}{b}\right)^2 \quad \text{and} \quad \frac{z^2}{a^2+b^2} + \frac{a^2+b^2}{a^2b^2}\left(x - \frac{za^2}{a^2+b^2}\right)^2$$

are identical expressions; that is to say, that one may be deduced from the other.   Find their values when $x = 3$, $z = 4$, and $a = b = 5$.

30.   Simplify the following expressions :

(i.) $\left[\dfrac{1+x}{1-x} - \dfrac{1-y}{1+y}\right] \div \left[1 + \dfrac{(1+x)(1-y)}{(1-x)(1+y)}\right]$;

(ii.) $\left[1 - \dfrac{1}{1+x} - \dfrac{x}{1-x^2} + \dfrac{x^2}{1+x^3}\right] \times \dfrac{x-1}{x}.$

31.   Solve the equations :

(i.) $x + \dfrac{9b^2}{a} = \dfrac{3bx}{a} + a$;   (ii.) $\dfrac{x+a}{a} + \dfrac{x}{x-a} = \dfrac{x-a}{a}$;

32.   A herd of 125 cattle is sold for £2575.   There were half as many oxen again as there were cows; and the oxen fetched altogether £25 less than the cows.   What was the price of each ox, and of each cow?

33.   Find the numerical value of $\sqrt{\dfrac{a+b}{3a-b}} - \dfrac{a+10b}{a-b}$, when $a = 5$ and $b = -1$.

34.   The rent of a shop is two-sevenths of the rent of the whole house of which it is a part.   Being separately rated, its occupier pays £10. 15s. a year less in rates than the occupier of the rest of the house.   The rates are 3s. 7d. in the pound.   What is the rent of the whole house?

ANSWERS TO ALGEBRA.

I. a. Page 3.　　**1.** 70.　　**2.** 125.　　**3.** 105.
4. 343.　　**5.** 30.　　**6.** 32.　　**7.** 144.　　**8.** 48.
9. 189.　　**10.** 200.　　**11.** 27.　　**12.** 1000.　　**13.** 3.
14. 1.　　**15.** 567.　　**16.** 4.　　**17.** 125.　　**18.** 81.　　**19.** 1.
20. 243.　　**21.** 24.　　**22.** 81.　　**23.** 8.　　**24.** 56.　　**25.** 16.

I. b. Page 4.　　**1.** 700.　　**2.** 686.　　**3.** 96.
4. 135.　　**5.** 15.　　**6.** 60.　　**7.** 162.　　**8.** 0.
9. 0.　　**10.** 3000.　　**11.** 3.　　**12.** 1.　　**13.** $5\frac{1}{3}$.　　**14.** 2.
15. 36.　　**16.** $1\frac{1}{8}$.　　**17.** 40.　　**18.** 0.　　**19.** $11\frac{1}{4}$.　　**20.** $13\frac{1}{2}$.

I. c. Page 5.　　**1.** 8.　　**2.** 29.　　**3.** 4.
4. 6.　　**5.** 9.　　**6.** 0.　　**7.** 31.　　**8.** 12.
9. 31.　　**10.** 49.　　**11.** 1.　　**12.** $1\frac{1}{3}$.　　**13.** $2\frac{1}{6}$.
14. $3\frac{1}{3}$.　　**15.** 10.　　**16.** $5\frac{1}{3}$.　　**17.** 6.　　**18.** $4\frac{1}{3}$.　　**19.** $5\frac{1}{6}$.

II. a. Page 8.　　**1.** $16a$.　　**2.** $24x$.　　**3.** $32p$.
4. $40d$.　　**5.** $-26x$.　　**6.** $-40b$.　　**7.** $-47y$.　　**8.** $-93m$.
9. $3xy$.　　**10.** pq.　　**11.** $-5abc$.　　**12.** $3xyz$.　　**13.** a^2.
14. $-b^3$.　　**15.** $11a^2b^2$.　　**16.** $-9a^2x$.　　**17.** $-43abcd$.
18. $-11pqx$.　　**19.** $\frac{11}{6}x$.　　**20.** $\frac{8}{5}a$.　　**21.** $-3b$.　　**22.** $-x^2$.

II. b. Page 10.　　**1.** $3a-6c$.　　**2.** $6x$.　　**3.** $l+m+n$.
4. $5a-2d$.　　**5.** $16x-7y$.　　**6.** $4a$.　　**7.** $8+2x-2z$.
8. $11ab-5kl+5xy$.　　**9.** $4ax$.　　**10.** $2ab+8cd$.

II. c. Page 12.　　**1.** xy.　　**2.** x^2+x+1.　　**3.** x^2+4x.
4. a^2b+b^3.　　**5.** $7m^3+3m^2-1$.　　**6.** $3ax^3-cx+2d$.
7. $2py^2+qy-3r$.　　**8.** $-2a+8a^3$.　　**9.** $4+y+2y^2$.
10. $8a^3x^2+x$.　　**11.** x^3+1.　　**12.** z^2+3xz.
13. $-\frac{1}{4}x^3+\frac{3}{8}ax^2+\frac{5}{8}a^2x$.　　**14.** $-\frac{1}{4}x^2-xy+\frac{3}{5}y^2$.
15. $-a^3-\frac{1}{2}a^2b+\frac{1}{4}ab^2+b^3$.

III. Page 14.

1. $a + b + 2c$.   2. $a - 4b - 2c$.

3. $9x - 15y - 14z$.   4. $-m + 4n - 4p$.   5. $p - 5q + 2r$.

6. $-x + 8y + 3z$.   7. $3x + 8y - 2z$.   8. $4n + p$.

9. $-cd - ac - bd$.   10. $2pq + 4qr - 8rs$.   11. $mn - 22np + 3pm$.

12. $2x^3 - 6x^2 + 2x$.   13. $-3x^3 + x$.   14. $a^3 - abc$.

15. $-12 + 9bc + 6b^2c^2$.   16. $p^3 - q^3 - 6pqr$.   17. $-x^3$

18. $x^3 + 3x^2 + 5x + 7$.   19. $-7a^3 - 5a^2 + 12$.   20. $5x + 7x^2 - 7x^3$.

21. $2 - 2x + x^2 - x^3 - 2x^4 + x^5$.   22. $-m^3 + 22m^2n - 16mn^2 + 2n^3$.

23. $\frac{5}{8}x^2 + \frac{1}{6}ax - \frac{1}{3}$.   24. $\frac{3}{4}x^3 - \frac{1}{2}x^2y - \frac{1}{6}y^2$.

25. $-\frac{1}{8}a^3 - \frac{2}{3}a^2x - \frac{1}{2}ax^2$.

IV. a. Page 17.

1. $35x$.   2. $6b$.   3. x^5.

4. $30x^3$.   5. $42c^7$.   6. $45y^7$.   7. $15m^8$.

8. $24a^{10}$.   9. $30a^2x^2$.   10. $12q^2r^2$.   11. a^7x^4.

12. $12x^3y^7$.   13. $5a^6b^5$.   14. $54x^8y^4$.   15. $105a^2b^3c^4$.

16. $60a^4b^3c^3d^2$.   17. $a^3bc - a^3c^2$.   18. $x^5y^2z^2 - x^6yz^3 + 4x^3y^2z^7$.

19. $15a^3b^2c^4 - 9ab^4c^4$.   20. $3a^5b^2 - 15a^4b^2 + 18a^4b$.

21. $3xy^3z - 9x^2yz^2 - 6yz$.   22. $2a^5bx - 6a^4bx^2$.

IV. b. Page 18.

1. $-2a$.   2. $-12x$.   3. x^5.

4. $-15m^4$.   5. $-12q^3$.   6. $16y^6$.   7. $-9m^6$.

8. $-16x^8$.   9. $36a^2b^2c^2$.   10. $-36a^5b$.   11. $-24pqst$.

12. $-3a^2 + \frac{9}{2}ab - 6ac$.   13. $-\frac{5}{2}x^2 + \frac{5}{3}xy + \frac{10}{3}x$.

14. $\frac{1}{4}a^2x - \frac{1}{16}abx - \frac{3}{8}acx$.   15. $-2a^5x^3 + \frac{7}{2}a^4x^4$.

16. $\frac{5}{2}a^4x^2 - \frac{5}{3}a^3x^3 + a^2x^4$.   17. $\frac{21}{2}x^3y - x^2y^2$.

IV. c. Page 19.

1. $a^2 + 12a + 35$.   2. $x^2 + x - 12$.

3. $a^2 - 13a + 42$.   4. $y^2 - 16$.   5. $x^2 + x - 72$.

6. $c^2 - 64$.   7. $p^2 - 100$.   8. $d^2 + 14d + 49$.

9. $-x^2 + 8x - 16$.   10. $y^2 - 9$.   11. $a^2 - 9a + 20$.

12. $y^2 + 14y + 49$.   13. $6a^2 - 11a - 10$.   14. $2x^2 - 9x - 35$.

15. $6x^2 + x - 12$.   16. $2x^2 - 3ax - 9a^2$.   17. $6a^2 + 5ab - 6b^2$.

18. $25c^2 - 16d^2$.   19. $12x^2 - 17xy - 5y^2$.

20. $4y^2 - 9z^2$.   21. $x^2y^2 - 4b^2$.

IV. d. Page 21.

1. $2x^3 - 7x^2 - x + 2.$ **2.** $8a^3 + 10a^2 - 7a - 6.$
3. $6y^3 - 11y^2 + 6y - 1.$ **4.** $12x^3 + x^2 - 25.$ **5.** $6x^3 - 25x^2 + 28x - 49.$
6. $-10c^3 + 13c^2 - 10c + 3.$ **7.** $2a^4 - 5a^3 + a^2 - 12.$
8. $6k^4 - 11k^3 - 2k^2 + 4k + 1.$ **9.** $x^4 + x^2y^2 + y^4.$ **10.** $a^4 + 4x^4.$
11. $x^5 - 6x^4 + 9x^3 - x.$ **12.** $a^6 - 36a^2 + 60a - 25.$
13. $4y^8 - 16y^6 + 16y^4 - 1.$ **14.** $4x^5 - x^3 + 4x.$
15. $-a^6 + 2a^5b^2 - a^4b^4 + b^6.$ **16.** $a^6 - 3a^4x^2 + 3a^2x^4 - x^6.$
17. $a^7 - 7a^6 + 21a^5 - 35a^4 + 35a^3 - 21a^2 + 7a - 1.$
18. $-x^8 + 4x^6y^2 - 6x^4y^4 + 4x^2y^6 - y^8.$
19. $\frac{2}{9}x^3 - \frac{3}{4}y^3.$ **20.** $\frac{9}{8}x^4 - \frac{3}{2}ax^3 + \frac{1}{2}a^2x^2 - \frac{2}{9}a^4.$
21. $\frac{1}{4}x^4 - \frac{43}{36}x^2 + \frac{9}{16}.$ **22.** $\frac{1}{4}a^4 + x^4.$

V. a. Page 23.

1. $2x.$ **2.** $2a^4.$ **3.** $5a^3.$
4. $3b^4.$ **5.** $-2pq^2.$ **6.** $l^2m.$ **7.** $8x^6.$ **8.** $-5z^6.$
9. $a^3.$ **10.** $-5a^2c^5.$ **11.** $y^3.$ **12.** $3x - 2.$ **13.** $5a^2 - 7b^2.$
14. $-x + y + z.$ **15.** $-10a^2 + 5ab - 1.$ **16.** $-x^2 - 9ax + 4.$
17. $-a^2 + 3ab + 2b^2.$ **18.** $2a - 3b + 4c.$
19. $-\frac{1}{3}x^2 + 2y^2.$ **20.** $3x - 2y - 4.$ **21.** $-\frac{6}{7}a^2x^2 + \frac{3}{2}ax^3.$

V. b. Page 24.

1. $a + 1.$ **2.** $b + 1.$ **3.** $x + 6.$
4. $x + 4.$ **5.** $m + 13.$ **6.** $x - 5a.$ **7.** $a - 3b.$
8. $-x + 3.$ **9.** $x - 8.$ **10.** $4a + 3x.$ **11.** $5x - 6y.$
12. $3a + 7c.$ **13.** $2m - 7n.$ **14.** $5x + 7y.$
15. $2x^2 - 3x + 2.$ **16.** $3x^2y - 5xy^2 + 4y^3.$

V. c. Page 26.

1. $a - 2.$ **2.** $y + 4.$ **3.** $3a + 2.$
4. $2x + 3a.$ **5.** $2x - 5y.$ **6.** $x^2 - 3x + 2.$ **7.** $x - 4$; rem. $x - 1.$
8. $3m - 5.$ **9.** $3y - 1.$ **10.** $7x^2 + 5x - 3$; rem. 20.   **11.** $x - 2a.$
12. $y^2 + 3y + 9.$ **13.** $x^2 - 2xy + 2y^2.$ **14.** $3a^2 + 4a + 2.$
15. $a^4 + 4a^2 + 8.$ **16.** $4m^2 + 14m + 9.$ **17.** $3x^2 - 4x + 5$; rem. $2x + 7.$
18. $-a + b^3 - 3.$ **19.** $x^5 - x^4 + x - 1$; rem. 2.
20. $2a^3 - 4a^2 + 4a - 2$; rem. 4.   **21.** $a^2 + 2ab + b^2 + a + b + 1.$
22. $1 + a + a^2 + 2x - 2ax + 4x^2.$ **23.** $\frac{1}{4}a^2 - 3ax + 9x^2.$
24. $6x - \frac{1}{3}y - \frac{1}{2}.$ **25.** $\frac{4}{9}a^4 + \frac{1}{2}a^3x + \frac{9}{16}a^2x^2 + \frac{81}{128}ax^3.$

VI. a. Page 28.   1. $2x+y$.   2. $-x+y+z$.   3. $3b$.
4. $-5n+p$.   5. $-8x-17y$.   6. m^2.   7. $5a-b$.
8. $-7q^2$.   9. $-x$.   10. $2a$.   11. $4x$.   12. $-a$.

VI. b. Page 30.   1. $22-8x$.   2. $2x$.   3. $3x-9$.
4. $127x-315$.   5. $7x-3y-3$.   6. $2y+4z$.   7. $x-4y$.
8. $a-2b+3c-3d+3e$.   9. $p+21q$.   10. $3x-4y$.   11. $6x+2y$.
12. $5x+7$.   13. $-50x$.   14. $42x^2+216xy+30y^2$.
15. (1) $(a-1)x^4+(2-b)x^3+(2-c)x^2$.
     (2) $-(1-a)x^4-(b-2)x^3-(c-2)x^2$.
16. (1) $(a^2-c)x^3+(a-b-5)x^2$.   (2) $-(c-a^2)x^3-(5-a+b)x^2$.

VI. c. Page 32.   1. 36.   2. -48.   3. 5.
4. 24.   5. 16.   6. -12.   7. -16.   8. 375.
9. 500.   10. 140.   11. -2000.   12. -224.   13. 40.
14. 3.   15. 1.   16. 0.   17. 29.   18. -13.
19. $60\frac{2}{3}\frac{3}{7}$.   20. $-\dfrac{18}{65}$.   21. $1\frac{7}{2}\frac{1}{2}\frac{1}{1}$.   22. $-13\frac{3}{4}$.   23. $-\dfrac{3}{4}$.
24. $1\frac{2}{3}$.   25. -3.   26. 6.   27. $\dfrac{5}{3}$.   28. (1) -12; (2) 1.

VII. a. Page 35.   1. 2.   2. 7.   3. 3.   4. 2.
5. $4\frac{1}{2}$.   6. 2.   7. 5.   8. $1\frac{1}{2}$.   9. 13.   10. 32.
11. 7.   12. 10.   13. $1\frac{2}{3}$.   14. 5.   15. 3.   16. 18.
17. -6.   18. 7.   19. -8.   20. 5.   21. $-\dfrac{3}{2}$.   22. $\dfrac{2}{3}$.

VII. b. Page 37.   1. 3.   2. 1.   3. 2.   4. $5\frac{1}{2}$.   5. 11.
6. $\dfrac{1}{3}$.   7. $\dfrac{2}{5}$.   8. 10.   9. -6.   10. 1.   11. 7.
12. -10.   13. 2.   14. 11.   15. 8.   16. -4.   17. $3\frac{1}{7}$.
18. $2\frac{1}{2}$.   19. -16.   20. 9.   21. 60.   22. 12.   23. $-\dfrac{11}{13}$.

VIII. a. Page 39.   1. $x-5$.   2. $15-y$.   3. $b-6$.
4. $\dfrac{3}{a}$.   5. $4x$.   6. $x-10$.   7. $75-x$.   8. x.
9. $\dfrac{2y}{x}$.   10. $a-8$.   11. $4x$.   12. $2ax$.   13. p^x.
14. ny shillings.   15. $\dfrac{y}{2}$.   16. $\dfrac{3n}{10}$.   17. $100-n$.   18. $\dfrac{20}{p}$.
19. $\dfrac{20y}{x}$.   20. $\dfrac{x}{10}$.   21. pq miles.   22. $\dfrac{m}{n}$ miles.   23. $\dfrac{y}{x}$.

VIII. b. Page 40. 1. $a, a+1, a+2$. 2. $b-3, b-2, b-1, b$.
3. $2n+1$. 4. $2n-2$. 5. $x-15$ years. 6. $n-x$ years.
7. $y-2x$ years. 8. $40+x$ years. 9. $2x-10$ years.
10. $20m+3=n-3$. 11. $13+x=4(25-x)$. 12. $\dfrac{30}{x}$.
13. $\dfrac{pq}{15}$ hours. 14. $\dfrac{120x}{7}$ miles. 15. $\dfrac{5x^2}{y}$. 16. $\dfrac{n}{3}$.
17. $20x-25+\dfrac{y}{12}$. 18. $52x-\dfrac{3}{5}y$. 19. 6 shillings. 20. a^2 days.
21. $\dfrac{pm}{192k}$. 22. $\dfrac{100b}{a}$. 23. $\dfrac{100b}{a+b}$. 24. $\pounds\left(a+\dfrac{ab}{100}\right)$.

IX. a. Page 43. 1. 9. 2. 3. 3. 4. 4. 12, 15.
5. 11, 19. 6. 5, 15. 7. 6, 13. 8. 9, 17. 9. A £35, B £65.
10. A £12, B £20, C £34. 11. A £21, B £17, C £34.
12. A 31, B 23, C 19 rupees. 13. 30, 45. 14. 112, 10.
15. 6. 16. 4. 17. 6. 18. 56. 19. 18, 30.
20. A £48, B £24, C £4. 21. A £219, B £73, C £219.
22. A 48, B 32 years. 23. A 42, B 50, C 18 years.
24. 7, 8. 25. 15, 16.

IX. b. Page 45. 1. £2. 2. £7. 3. A £27, B £1.
4. A £27, B £9. 5. Father 32 years, Son 8 years.
6. A 45 years, B 25 years. 7. 25 years.
8. A 20 years, B 10 years. 9. Father 33 years, Son 9 years.
10. 20 men, 30 women. 11. 1s. 6d. 12. 84 lbs., 28 lbs.
13. 27 days. 14. 19 florins, 13 shillings.
15. 20 half-crowns, 5 florins, 2 shillings.
16. Coffee $10\frac{1}{4}$ lbs., Tea $13\frac{3}{4}$ lbs. 17. £55. 18. 168, 72.
19. 48 years, 9 years, 3 years. 20. A £25, B £17, C 28.
21. 36. 22. 360. 23. 128. 24. Tea 3s. 4d., Sugar 3d.
25. Port 50s., Sherry 36s.

X. a. Page 50. 1. $x=13, y=6$. 2. $x=3, y=-3$.
3. $x=19, y=-6$. 4. $x=5, y=7$. 5. $x=13, y=1$.
6. $x=1, y=6$. 7. $x=11, y=-8$. 8. $x=108, y=144$.
9. $x=-4, y=10$. 10. $x=7, y=8$. 11. $x=9, y=15$.
12. $x=-2, y=-3$. 13. $x=5, y=6$. 14. $x=-5, y=-1$.
15. $x=-9, y=5$. 16. $x=3, y=-2$. 17. $x=10, y=7$.
18. $x=-5, y=6$.

X. b. Page 52.   **1.** $x=6,$ $y=8.$ **2.** $x=2,$ $y=7.$

3. $x=3,$ $y=5.$ **4.** $x=-10,$ $y=4.$ **5.** $x=-2,$ $y=-3.$

6. $x=5,$ $y=-4.$ **7.** $x=-1,$ $y=-1.$ **8.** $x=-3,$ $y=-5.$

9. $x=20,$ $y=-4.$ **10.** $x=4,$ $y=5.$ **11.** $x=4,$ $y=-5.$

12. $x=7,$ $y=8.$ **13.** $x=-3,$ $y=-6.$ **14.** $x=-4,$ $y=4.$

15. $x=20,$ $y=-12.$ **16.** $x=2,$ $y=3.$ **17.** $x=\frac{1}{4},$ $y=\frac{1}{5}.$

18. $x=5,$ $y=-\frac{1}{2}.$

X. c. Page 53.   **1.** $x=1,$ $y=2,$ $z=5.$

2. $x=6,$ $y=1,$ $z=1.$ **3.** $x=y=z=4.$

4. $x=5,$ $y=-6,$ $z=-7.$ **5.** $x=-2,$ $y=2,$ $z=2.$

6. $x=4,$ $y=-5,$ $z=8.$ **7.** $x=20,$ $y=-10,$ $z=1.$

8. $x=7,$ $y=-4,$ $z=3.$ **9.** $x=2,$ $y=-1,$ $z=5.$

10. $x=0,$ $y=3,$ $z=-4.$ **11.** $x=-6,$ $y=-4,$ $z=8.$

XI. Page 56.   **1.** 33, 21.   **2.** 74, 23.   **3.** 33, 18.   **4.** 55, 29.

5. Cow £17, Sheep £3.   **6.** Horse £24, Cow £18.

7. A 37 years, B 24 years.   **8.** A 28 years, B 52 years.

9. C $3\frac{3}{4}$ miles, D $4\frac{1}{2}$ miles.   **10.** Train 31 miles, Coach 7 miles.

11. 18 half-crowns, 23 shillings.   **12.** 22 shillings, 32 sixpences.

13. Waggon $2\frac{1}{2}$ tons, Cart $1\frac{1}{2}$ tons.   **14.** Boy $5s.$, Girl $4s.$

15. 84.   **16.** 36.   **17.** 54.   **18.** 88.

19. 44 sovereigns, 208 half-crowns, 600 shillings.

20. 30 half-crowns, 20 shillings, 20 threepenny pieces.

21. Man £10, Woman £8, Boy £6, Girl £2.

Miscellaneous Examples I. Page 57.   **1.** $-\frac{1}{2};\ \frac{1}{4};\ -\frac{1}{4};\ 3\frac{1}{4}.$

2. 4.   **3.** 1.   **4** $x^6-14x^4+49x^2-36$; $7x^2-2xy+y^2.$

5. $-29\frac{2}{3}$; $-3\frac{1}{12}$, $3\frac{19}{77}.$ **6.** (i) $x=5,\ y=5$; (ii) $x=3\frac{3}{8},\ y=\frac{3}{4}.$

7. $-x^2+3ax-5a^2.$ **8.** $12\frac{1}{4}$; $7\frac{7}{8}$; $-\dfrac{27}{1024}.$ **9.** $20x-21y-9.$

10. x^6-64 ; $x^3+2x^2y+3xy^2-y^3.$ **11.** $\dfrac{11}{168}$; $60\frac{23}{27}$; $-\dfrac{18}{65}.$

12. $-30.$ **13.** $4x^5+7x^3y^2-10x^2y^3+xy^4-20y^5$; $-4\cdot0803.$

14. $c^4-(a+d)bc^2+(a-d)adc+b^2d^2.$ **15.** $-1.$

16. $9x^2-\dfrac{y^2}{4}$; $9x-\dfrac{3y}{2}.$ **17.** $-5x^2+47xy-\dfrac{31}{2}x-18y^2+\dfrac{21}{2}y-\dfrac{3}{2}.$

18. (i) $x=9,\ y=12$; (ii) $x=12,\ y=3.$

XII. a. Page 59.

1. $x(x+a)$.
2. $a(2a-3)$.
3. $a^2(a-1)$.
4. $a^2(a-b)$.
5. $3m(m-2n)$.
6. $p^2(1+2q)$.
7. $x^2(x^3-5)$.
8. $y(y+x)$.
9. $5a^2(1-5b)$.
10. $12x(1+4xy)$.
11. $5c^3(2-5cd)$.
12. $27(1-6x)$.
13. $xy(xyz^2+3)$.
14. $17x(x-3)$.
15. $a(2a^2-a+1)$.
16. $3x(x^2+2a^2x-a^3)$.
17. $7p^2(1-p+2p^2)$.
18. $2b^2(2b^3+3a^2b-1)$.
19. $xy(x^2y^2-xy+2)$.
20. $13a^3b^2(2b^3+3a)$.

XII. b. Page 60.

1. $(x+y)(x+z)$.
2. $(x-z)(x+y)$.
3. $(a+2)(a+b)$.
4. $(a+c)(a+4)$.
5. $(a+x)(2+x)$.
6. $(3+p)(q-p)$.
7. $(a-b)(m-n)$.
8. $(a-y)(b-y)$.
9. $(p+r)(q-r)$.
10. $(2m+n)(x+y)$.
11. $(x-2y)(a-b)$.
12. $(2a+3b)(a-c)$.
13. $(a+b)(c^2+1)$.
14. $(c^2-2)(a-b)$.
15. $(a-1)(a^2+1)$.
16. $(2x+3)(x^2+1)$.
17. $(ax-by)(a+2)$.
18. $(a+bc)(xy-z)$.
19. $(7x-4)(x^2-3)$.
20. $(x-4a)(3x^2-2a^2)$.

XII. c. Page 61.

1. $(x+1)(x+2)$.
2. $(y+2)(y+3)$.
3. $(y+3)(y+4)$.
4. $(a-1)(a-2)$.
5. $(a-2)(a-4)$.
6. $(b-3)(b-2)$.
7. $(b+6)(b+7)$.
8. $(b-5)(b-8)$.
9. $(z-9)(z-4)$.
10. $(x-8)(x-7)$.
11. $(x-9)(x-6)$.
12. $(z+11)(z+4)$.
13. $(b-6)(b-6)$.
14. $(a+7)(a+8)$.
15. $(a-9)(a-3)$.
16. $(x+5)(x+4)$.
17. $(x-9)(x-1)$.
18. $(x-8)(x-8)$.
19. $(y-17)(y-6)$.
20. $(y-19)(y-5)$.
21. $(y+27)(y+27)$.
22. $(a+3b)(a+7b)$.
23. $(a+b)(a+11b)$.
24. $(a-11b)(a-12b)$.
25. $(m^2+7)(m^2+1)$.
26. $(m^2+2n^2)(m^2+7n^2)$.
27. $(xy-2)(xy-3)$.
28. $(ab-9)(ab-6)$.
29. $(13+y)(1+y)$.
30. $(27-a)(8-a)$.

XII. d. Page 62.

1. $(x+2)(x-1)$.
2. $(x-3)(x+2)$.
3. $(x-5)(x+4)$.
4. $(y-2)(y+6)$.
5. $(y+7)(y-3)$.
6. $(y-9)(y+4)$.
7. $(a+11)(a-3)$.
8. $(a-15)(a+2)$.
9. $(a+12)(a-11)$.
10. $(b-15)(b+3)$.
11. $(b+17)(b-3)$.
12. $(b+13)(b-3)$.
13. $(m-8)(m+7)$.
14. $(m+7)(m-12)$.
15. $(m-7)(m+8)$.
16. $(p+5)(p-13)$.
17. $(p-9)(p+12)$.
18. $(p-10)(p+11)$.
19. $(x-6)(x+8)$.
20. $(x-15)(x+8)$.
21. $(x-12)(x+11)$.
22. $(y^2-3)(y^2+16)$.
23. $(y-8x)(y+12x)$.
24. $(y+14x)(y-7x)$.
25. $(a^2-8b^2)(a^2-9b^2)$.
26. $(a+16b)(a-15b)$.
27. $(ab-7)(ab+2)$.
28. $(ab-7c)(ab+5c)$.
29. $(8-b)(12+b)$.
30. $(9-b)(8+b)$.

XII. e. Page 64.

1. $(2a+1)(a+1)$.
2. $(3a+1)(a+1)$.
3. $(4a+1)(a+1)$.
4. $(a+2)(2a+1)$.
5. $(a+3)(3a+1)$.
6. $(2a+1)(a+3)$.
7. $(5a+2)(a+1)$.
8. $(2a+5)(a+2)$.

9. $(2a+3)(a+2)$.   **10.** $(x+4)(2x+1)$.   **11.** $(x+3)(2x-1)$.

12. $(x+2)(3x-1)$.   **13.** $(y+1)(3y-2)$.   **14.** $(y-3)(3y+2)$.

15. $(y+5)(2y-1)$.   **16.** $(2b+1)(b-3)$.   **17.** $(2b+3)(3b-1)$.

18. $(b+3)(2b-5)$.   **19.** $(4m-3)(m+2)$.   **20.** $(2m-3)(2m+1)$.

21. $(3m+1)(2m-3)$.   **22.** $(2x-5y)(2x+y)$.   **23.** $(3x-2y)(2x-y)$.

24. $(6x-y)(x-2y)$.   **25.** $(3a-2b)(4a-3b)$.   **26.** $(3a+2b)(2a-3b)$.

27. $(6a-b)(a+6b)$.   **28.** $(2+y)(1-2y)$.   **29.** $(3-y)(1+8y)$.

30. $(4-y)(2+5y)$.   **31.** $(4-3x)(1+5x)$.   **32.** $(2-3a)(3-2a)$.

33. $(7+b)(4-5b)$.

XII. f.   Page 65.
1. $(a+3)(a-3)$.   **2.** $(a+7)(a-7)$.

3. $(a+9)(a-9)$.   **4.** $(x+5)(x-5)$.   **5.** $(8+x)(8-x)$.

6. $(9+2x)(9-2x)$.   **7.** $(2y+1)(2y-1)$.   **8.** $(y+3a)(y-3a)$.

9. $(2y+5)(2y-5)$.   **10.** $(3y+7x)(3y-7x)$.   **11.** $(2m+9)(2m-9)$.

12. $(6a+1)(6a-1)$.   **13.** $(3a+5b)(3a-5b)$.   **14.** $(11+4y)(11-4y)$.

15. $(5+c^2)(5-c^2)$.   **16.** $(7a^2+10b)(7a^2-10b)$.

17. $(2pq+9)(2pq-9)$.   **18.** $(a^2b^2c+3)(a^2b^2c-3)$.

19. $(x^3+2a^2)(x^3-2a^2)$.   **20.** $(x^2+5z^2)(x^2-5z^2)$.

21. $(a^5+pq^2)(a^5-pq^2)$.   **22.** $(4a^8+3b^3)(4a^8-3b^3)$.

23. $(5x^6+2)(5x^6-2)$.   **24.** $(a^6b^4c^2+3x)(a^3b^4c^2-3x)$.

25. $70 \times 8 = 560$.   **26.** $100 \times 2 = 200$.   **27.** $1002 \times 1000 = 1002000$.

28. $100 \times 64 = 6400$.   **29.** $500 \times 50 = 25000$.   **30.** $1000 \times 872 = 872000$.

XII. g.   Page 66.
1. $(a-b)(a^2+ab+b^2)$.   **2.** $(a+b)(a^2-ab+b^2)$.

3. $(1+x)(1-x+x^2)$.   **4.** $(1-y)(1+y+y^2)$.

5. $(2x+1)(4x^2-2x+1)$.   **6.** $(x-2z)(x^2+2xz+4z^2)$.

7. $(a+3b)(a^2-3ab+9b^2)$.   **8.** $(xy-1)(x^2y^2+xy+1)$.

9. $(1-2a)(1+2a+4a^2)$.   **10.** $(b-2)(b^2+2b+4)$.

11. $(3+x)(9-3x+x^2)$.   **12.** $(4-p)(16+4p+p^2)$.

13. $(5a+1)(25a^2-5a+1)$.   **14.** $(6-b)(36+6b+b^2)$.

15. $(xy+7)(x^2y^2-7xy+49)$.   **16.** $(10x+1)(100x^2-10x+1)$.

17. $(8a-1)(64a^2+8a+1)$.   **18.** $(abc-3)(a^2b^2c^2+3abc+9)$.

19. $(2x-7)(4x^2+14x+49)$.   **20.** $(x+6y)(x^2-6xy+36y^2)$.

21. $(x^2-3z)(x^4+3x^2z+9z^2)$.   **22.** $(m-10n^2)(m^2+10mn^2+100n^4)$.

23. $(a-9b)(a^2+9ab+81b^2)$.   **24.** $(5a^2+8b)(25a^4-40a^2b+64b^2)$.

XII. h.   Page 68.
1. $(x+y+z)(x+y-z)$.   **2.** $(x-y+z)(x-y-z)$.

3. $(a+2b+c)(a+2b-c)$.   **4.** $(a+3c+1)(a+3c-1)$.

5. $(2x-1+a)(2x-1-a)$.   **6.** $(a+b+c)(a-b-c)$.

7. $(2a+b-1)(2a-b+1)$.   **8.** $(3+a+x)(3-a-x)$.

9. $(2a - 3b + c)(2a - 3b - c)$.   10. $35x(x + 2y)$.

11. $(11a - 1)(a + 7)$.   12. $3b(4a - 3b)$.

13. $(x + 2b - 3c)(x - 2b + 3c)$.   14. $(x + y + m - n)(x + y - m + n)$.

15. $(5x - y)(x + 5y)$.   16. $(a - x + 2b)(a - x - 2b)$.

17. $(x + a + z)(x + a - z)$.   18. $(1 + a + b)(1 - a - b)$.

19. $(5 + 2x - 3y)(5 - 2x + 3y)$.   20. $(c + a - b)(c - a + b)$.

21. $(x - 1 + m + 2n)(x - 1 - m - 2n)$.

22. $(x^2 + y^2 + z^2 + a^2)(x^2 + y^2 - z^2 - a^2)$.   23. $4n(m + p)$.

24. $(a^2 + a + 1)(a^2 - a + 1)$.   25. $(a^2 b^2 + 4)(ab + 2)(ab - 2)$.

26. $(16x^2 + 9y^2)(4x + 3y)(4x - 3y)$.   27. $b^2(4a^2 + b^2)(2a + b)(2a - b)$.

28. $m(2m + n)(4m^2 - 2mn + n^2)(2m - n)(4m^2 + 2mn + n^2)$.

29. $x^4(1 + y^2)(1 + y)(1 - y)$.   30. $a^2 b(b^2 + 9)(b + 3)(b - 3)$.

31. $x(20a + x)(20a - x)$.

32. $(1 + 3y)(1 - 3y + 9y^2)(1 - 3y)(1 + 3y + 9y^2)$.

33. $b^3(6b + a)(36b^2 - 6ab + a^2)$.   34. $2(5z + 1)(25z^2 - 5z + 1)$.

35. $3(7 - x)(49 + 7x + x^2)$.   36. $3xy^2(2x - 3)(x + 4)$.

37. $n^4(2m^4 + n^2)(m^8 + 2n)(m^2 - 2n)$.   38. $(7x^2 - y^2)(14x^2 + y^2)$.

39. $(a + 1)(a - 1)(b + 1)(b - 1)$.   40. $(x + 1)(x - 1)(x - 2)$.

41. $(a + b + 1)\{(a + b)^2 - (a + b) + 1\}$.

42. $(x - 2y)(x^2 + 2xy + 4y^2)(a + 2b)(a - 2b)$.

43. $(2p - 3q)(1 + 2p + 3q)$.   44. $(17 - m)(7 + m)$.

45. $6b^2(4a + 3b)(a - 2b)$.   46. $x^2(4 + x^2 y^2)(2 + xy)(2 - xy)(15 + x^4 y^4)$.

47. $(x^2 + 2x + 4)(x^2 - 2x + 4)$.   48. $(x^2 + 3xy + y^2)(x^2 - 3xy + y^2)$.

49. $(a^2 + 4ab - b^2)(a^2 - 4ab - b^2)$.

50. $(x^2 + x + 1)(x^2 - x + 1)(x^4 - x^2 + 1)$.

51. $(a^2 + 2ab + b^2 + c^2)(a + b + c)(a + b - c)$.   52. $2c(c^2 + 3d^2)$.

53. $(a - b)(a + b + c)$.   54. $(a + b)(a + b - c)$.   55. $2a(a - 3b)$.

56. $(x - y)(2x - 2y + 5)(2x - 2y - 5)$.

57. $(a + b + 2c)(a^2 + b^2 + 4c^2 - ab - 2bc - 2ca)$.

58. $(a - 3b + c)(a^2 + 9b^2 + c^2 + 3ab + 3bc - ca)$.

59. $(a + 2c + 1)(a^2 + 4c^2 + 1 - 2ac - a - 2c)$.

60. $(2a + 3b + c)(4a^2 + 9b^2 + c^2 - 6ab - 3bc - 2ca)$.

XII. k. Page 69.

1. $a^2 - 2ab + b^2 - c^2$.   2. $4x^2 - y^2 + 4xz + z^2$.

3. $1 - 6x^2 + x^4$.   4. $c^4 - 9c^2 - 12c - 4$.

5. $a^2 + 2ab + b^2 - c^2 + 2cd - d^2$.   6. $p^2 - 2pq + q^2 - x^2 + 2xy - y^2$.

7. $a^8 - 2a^4 b^4 + b^8$.   8. $1 - 3x^4 + 3x^8 - x^{12}$.   9. $a^6 - 14a^4 + 49a^2 - 36$.

10. $729 - y^6$.   11. $1 + c^4 + c^8$.   12. $a^2(a + 2)(a - 3)$.

13. $3x^2(x + 3)$.   14. $(2a - 3)(3a + 1)$.   15. $(x - 2)(x^2 - x + 1)$.

128   ALGEBRA.

XIII. a. Page 71. **1.** xy. **2.** $6abc^2$.
3. $5x^2$. **4.** $17xyz$. **5.** ab^2c^2. **6.** $x-y$.
7. $(a-b)^2$. **8.** $a(3a-2b)$. **9.** $3a+2b$. **10.** $x^2(x+y)$.
11. $a^2x^2(a-x)^2$. **12.** $(x-2)^2$. **13.** y. **14.** $y^3(x-y)$.
15. $(a-x)^2$. **16.** $c^2(a-c)^2$. **17.** $x(x-7)$. **18.** $x^3(x-5)$.
19. $x-4$. **20.** $c^2(3c-4)$. **21.** m. **22.** $a^2x^2(3a-2)$.

XIII. b. Page 75. **1.** $2x^2-x+3$. **2.** y^2-y+1.
3. $2x^2-5$. **4.** a^2-3a+2. **5.** $2x^2-x-2$. **6.** q^2-2q+1.
7. a^2+a-3. **8.** $y(y^2+2y+1)$. **9.** $5x(3x^2+2)$. **10.** $2m+7$.
11. $3(x-2)$. **12.** $a(a+1)$. **13.** $x+3$. **14.** $3a^2-2ax+x^2$.
15. $2-a$. **16.** $x(3+4x)$. **17.** $1+a$. **18.** $x(3+4x)$.
19. x^2-2x+1. **20.** $2x^2-7$.

XIV. a. Page 77. **1.** $6x^2y^2z^2$. **2.** $162a^3b^5$.
3. $210a^2c^5x^6yz$. **4.** $60a^2b^3x^2y$. **5.** $3630a^2b^5cx^4y^3z^3$.
6. $a^2(a-1)$. **7.** $4m^2(3m-4)$. **8.** $b(b^2-1)$.
9. $(x+2)(x-2)(x^2-2x+4)$. **10.** $2ab(3a+1)(3a-1)$.
11. $(m-2)(m-3)(m+7)$. **12.** $y^3(1+3y)(1-3y)$.
13. $(x+3y)(x^2-3xy+9y^2)(x-2y)$. **14.** $(c+3x)(c-6x)(c-2x)$.
15. $a(a+1)(a-3)(a-5)$. **16.** $6(x+2y)(x-2y)(x-4y)$
17. $12x(x+2a)^2(x-2a)$. **18.** $a^3c^2(a+c)^2(a-c)$.
19. $4a^2x^2(a-2x)^2$. **20.** $a^3(2-a)^3$.
21. $(2x-3)^2(x+1)(2x+3)$. **22.** $x(x-4)^2(2x+1)(3x+5)$.
23. $60x^3y^4(x-y)^3(x+y)(x^2+xy+y^2)$. **24.** $x^2(x+2)(2x-3)(7x-3)^2$.
25. $ax(3a+x)(2a-3x)(5a+2x)(2a-5x)$.

XIV. b. Page 77. **1.** $(x+1)(x+2)(x-4)(x-5)$.
2. $(y+3)(y^2-3)(y^2-8)$. **3.** $(m-1)(m+1)(m+2)(m+3)$.
4. $(2x^2-3)(x^2-x+2)(2x^2-x+3)$.
5. H.C.F. $x^3(1-x)$, L.C.M. $x^4(1-x)^3$. **6.** $a^4x^2(a+x)^3(a-x)^2$.
7. H.C.F. $2x+3$, L.C.M. $(2x+3)(3x-2)(3x-4)$.
8. H.C.F. $a+2b$, L.C.M. $(a+2b)(a-2b)(a+3b)(a-b)^2$.
9. $(1+x)(1-x)(1+x-x^4)$. **10.** $a^2(a+2b)^2$.
11. H.C.F. $a(3a-2x)$ L.C.M. $2a^2x(3a-2x)^2(3a+2x)(2a-3x)$.
12. $xy(x-y)(x^2+xy+y^2)(x^2-xy+y^2)$.

Miscellaneous Examples II. Page 78. **1.** $(x+18)(x-1)$.
2. $(x-2)(x+1)$. **3.** $(x+2)(x-1)$. **4.** $(2x-3)(x+4)$.
5. $(x-2)(x^2+2x+4)$. **6.** $(x+y)^2(x-y)$.

7. $(x+y+z)(x+y-z)$.   8. $(x+a)(x-a)(2x+3a)$.

9. $(x-3y+z)(x-3y-z)$.   10. $(x-9)(y+4)$.

11. $(1+x)(1+y)(1-x)(1-y)$.   12. $(a-d+b-c)(a-d-b+c)$.

13. $(x+2)(x+3)(x-3)$.   14. $(x-y)(x+y-6)$.

15. $(x^2+2xy+4y^2)(x^2-2xy+4y^2)$.   16. L.C.M. $30(x^3-1)(x^2-9)$.

17. L.C.M. $36(x^2-1)(x^2-4)(x^2-9)$.   18. L.C.M. $(x^2-4)(x-6)(3x-2)$.

19. L.C.M. $12(x^2-4)(2x-1)$.   20. $3y(3x+2y)$.   21. $3x^2-2x-7$.

22. x^2-5x+1 ; $(x^2-5x+1)(x^2-7)$, $(x^2-5x+1)(3x-8)$.

23. x^2-4x+3 ; $(x^2-4x+3)(x+7)$, $(x^2-4x+3)(2x-1)$.

24. $2x^2-x+1$; $(2x^2-x+1)(x-1)(2x+1)$, $(2x^2-x+1)(x+3)$.

25. x^2+3 ; $(x^2+3)(4x-9)$, $(x^2+3)(2x+1)(x-2)$.

26. $a^3+2a^2x+2ax^2+x^3$.   27. $x^4+\dfrac{625}{4}$.   28. $4x^2(7x-13)$.

29. $12x(x^3+15)$; $18\frac{3}{4}$.   30. $(a+b)(2a+b)(3a+b)$.

XV. a. Page 81.

1. $\dfrac{2y^3}{3xz^3}$.   2. $\dfrac{3p^2m^2}{5k}$.

3. $\dfrac{3a}{5bx^2}$.   4. $\dfrac{z}{5x}$.   5. $\dfrac{x}{2x-y}$.   6. $\dfrac{1}{2ab}$.

7. $\dfrac{yz^2}{y+2z}$.   8. $\dfrac{2(xy-2)}{3x}$.   9. $\dfrac{x}{x-3}$.   10. $\dfrac{7a^2}{5c(x-c)}$.

11. $\dfrac{x-5}{5x}$.   12. $\dfrac{2a+b}{a(2a-b)}$.   13. $\dfrac{a^2-ab+b^2}{a-2b}$.   14. $\dfrac{2c-d}{c+3d}$.

15. $\dfrac{x-7}{3x+1}$.   16. $\dfrac{x-1}{x+4}$.   17. $\dfrac{3(a^2+2a+4)}{4(a+3)}$.   18. $\dfrac{2a}{3a-x}$.

In each of the following examples the H.C.F. is given in [ ].

XV. b. Page 82.

1. $\dfrac{x-1}{3x^2+1}[x^2+2]$.   2. $\dfrac{a+1}{a-3}[a^2-a+2]$.

3. $\dfrac{y-3}{3y+1}[y^2+y+1]$.   4. $\dfrac{m(m+1)}{m^2+m+1}[m-2]$.

5. $\dfrac{a^2-3ab+7b^2}{a(a-7b)}[a+3b]$.   6. $\dfrac{3x-2a}{x-4a}[3x^2+2ax+a^2]$.

7. $\dfrac{5x^2+x+1}{2x^2-x-1}[x-1]$.   8. $\dfrac{c^2+5cd+3d^2}{2(c^2+6cd+4d^2)}[c-3d]$.

9. $\dfrac{x^2+3x+8}{8x^2+3x+1}[x^2-3x+1]$.   10. $\dfrac{y(y^2+7y+9)}{y^2+8y+11}[y(y-1)]$.

11. $\dfrac{1+2x}{2+3x}[1-2x+3x^2]$.   12. $\dfrac{1-3x}{2+x+5x^2}[2+x-x^2]$.

XV. c. Page 83.
1. $\dfrac{c}{a}$.
2. $\dfrac{3acd^2}{2b^3}$.
3. $\dfrac{pqxy}{6b}$.
4. $3a^2$.

5. $2(x-1)$.
6. $\dfrac{b}{4(ab-2)}$.
7. $\dfrac{cd}{c+d}$.
8. $\dfrac{5}{6}$.

9. $\dfrac{b(3b+4a)}{b+5}$.
10. 1.
11. $\dfrac{y-6}{y+3}$.
12. $\dfrac{a^2-3a+9}{a+2}$.

13. $\dfrac{2a+1}{a+2}$.
14. $\dfrac{5b+1}{b}$.
15. 2.
16. $8ab+1$.

17. $\dfrac{1}{x}$.
18. $\dfrac{a}{a+b}$.
19. $\dfrac{x+3a}{x-6a}$.
20. 1.

XV. d. Page 86.
1. $\dfrac{5a}{24}$.
2. $\dfrac{5x}{4}$.
3. $\dfrac{az+2ax-3ay}{xyz}$.

4. $\dfrac{y}{30}$.
5. $\dfrac{2ab+a^2-b^2}{ab}$.
6. $\dfrac{ap^2q^2+bq-cp^2}{p^2q^2}$.
7. $\dfrac{3a-1}{3}$.

8. $\dfrac{b-10}{5}$.
9. $\dfrac{4}{35}$.
10. $\dfrac{2x+1}{3x}$.
11. $\dfrac{2}{3}$.
12. 1.

13. $\dfrac{x^2+3y^2}{3xy}$.
14. $\dfrac{ab-6ac}{6bc}$.
15. $\dfrac{ay-3xy+4x}{2xy}$.
16. $\dfrac{2a-3b}{ab}$.

XV. e. Page 88.
1. $\dfrac{2a-5}{(a-2)(a-3)}$.
2. $\dfrac{2}{(x-4)(x-2)}$.

3. $\dfrac{(a-b)x}{(x-a)(x-b)}$.
4. $\dfrac{2(a^2+x^2)}{a^2-x^2}$.
5. $\dfrac{x}{x^2-1}$.

6. $\dfrac{2(a+1)}{a^2-4}$.
7. $\dfrac{2(x^2-2xy+2y^2)}{x^2-4y^2}$.
8. $\dfrac{a(5x+13a)}{6x(x^2-a^2)}$.

9. $\dfrac{x+5}{(x-2)(x+1)}$.
10. $\dfrac{3y+4}{(y-3)(y+1)}$.
11. $\dfrac{1}{(1-a)^2}$.

12. $\dfrac{x+3y}{(x+y)^2}$.
13. $\dfrac{6xy}{(x+y)^2(x-y)}$.
14. $\dfrac{5bc-c^2}{(b+c)(b-c)^2}$.

15. $\dfrac{x+y}{xy}$.
16. $\dfrac{4a^2+b^2}{b(2a+b)}$.
17. $\dfrac{x-1}{x^3+1}$.
18. $\dfrac{b^2}{b^3+8}$.

19. $\dfrac{x(x^2-2y^2)}{x^3-y^3}$.
20. $\dfrac{2(a+2)}{(a-2)(a-1)(a+5)}$.
21. $\dfrac{x^2}{x-1}$.

22. $\dfrac{2+a-2a^2}{2+a}$.
23. $\dfrac{x^4-x^2+1}{x^2(x+1)}$.
24. $\dfrac{8}{x(x-2)(x-4)}$.

25. $\dfrac{9x+7}{4x^2-1}$.
26. $\dfrac{8-11a^2}{6(1-a^2)}$.
27. $\dfrac{15}{2(9-4x^2)}$.
28. $\dfrac{x^2}{(x-a)^3}$.

29. $\dfrac{2a^2+5a+7}{(a+1)^4}$.
30. $\dfrac{2}{(2y-1)(y+1)(2y-3)}$.
31. $\dfrac{7}{(4-x)(3+x)}$.

32. 0.
33. $\dfrac{4x^2}{(x-2)^2(x^2+4)}$.
34. $\dfrac{2}{(y-2)(y-3)(y-4)}$.

35. $\dfrac{1}{3+x}$.
36. $\dfrac{7y}{(x-3y)(x-2y)(x+2y)}$.
37. $\dfrac{4a-5}{6(a^2-1)}$.

38. $\dfrac{a^2b}{a^3-b^3}.$ **39.** 0.   **40.** 0,   **41.** $\dfrac{x^3-2ax^2-a^3}{(x+a)(x-a)^3}.$

42. $\dfrac{4+7x^2}{16-x^4}.$ **43.** $\dfrac{3-4x^2-x^4}{2(1-x^4)}.$ **44.** $\dfrac{16(m^2+5)}{3(m^4-16)}.$

45. $\dfrac{a^4+a^2b^2+2b^4}{a^4-b^4}.$ **46.** $\dfrac{x}{(x-2)^2(x-4)}.$ **47.** $\dfrac{18}{x(x-6)}.$

XV. f. Page 91.   **1.** $\dfrac{1-6x^2}{1-4x^2}.$ **2.** $\dfrac{1+a}{9-a^2}.$ **3.** $\dfrac{4a-5}{6(a^2-1)}.$

4. $\dfrac{12y^2-4y+7}{3(4y^2-9)}.$ **5.** 0.   **6.** $\dfrac{a}{a+b}.$ **7.** 0.   **8.** $\dfrac{2x^7y^3}{x^6-y^6}.$

9. 0.   **10.** $\dfrac{2x}{(x-a)(x-b)}.$ **11.** $\dfrac{3x^2+2x-2}{(x-1)^2(2x-1)}.$ **12.** $\dfrac{b^2}{(a-b)^3}.$

13. $\dfrac{x+c}{(x-a)(x-b)}.$ **14.** $\dfrac{a-z}{(a-x)(a-y)}.$ **15.** $\dfrac{b}{a+b}.$

16. $\dfrac{2a(2a^2-5x^2)}{(a^2-x^2)(a^2-4x^2)}.$ **17.** $\dfrac{48x^3}{(a^2-x^2)(a^2-9x^2)}.$ **18.** 0.

19. $\dfrac{a^2+b^2+c^2-bc-ca-ab}{(b-c)(c-a)(a-b)}.$ **20.** $\dfrac{2(yz+zx+xy-x^2-y^2-z^2)}{(y-z)(z-x)(x-y)}.$

21. 0.   **22.** $\dfrac{a^2x}{x^4-a^4}.$ **23.** $\dfrac{2x^4}{a^8-x^8}.$ **24.** 0.   **25.** $\dfrac{8x^2}{(x+2)^2(x-2)^2}.$

XV. g. Page 95.   **1.** $\dfrac{z}{xz+y}.$ **2.** $\dfrac{ad}{bd-c}.$ **3.** $\dfrac{a^2}{1+a}.$

4. $b(1-a).$ **5.** $a+x.$ **6.** $\dfrac{1}{xy}.$ **7.** $b.$ **8.** $p^2.$

9. $\dfrac{a(a-3)}{a-4}.$ **10.** $\dfrac{y}{3}.$ **11.** $\dfrac{n+1}{n^2(n+4)}.$ **12.** $\dfrac{x+1}{x-1}.$

13. $\dfrac{b+4}{b+2}.$ **14.** $a-b.$ **15.** $\dfrac{2cd}{c^2+d^2}.$ **16.** $b.$

17. $\dfrac{4(x-1)}{7(x+4)}.$ **18.** $\dfrac{2a+1}{a+1}.$ **19.** $\dfrac{x^3}{x^2-1}.$ **20.** $\dfrac{5d-2c}{4d-c}.$

21. $\dfrac{x(x-1)}{x^2+x-1}.$ **22.** $\dfrac{x^2+2}{x(x^2+3)}.$ **23.** $1-y.$ **24.** $1+x.$

25. $\dfrac{2+y}{2}.$ **26.** $\dfrac{a(df-e)}{bdf-be-cf}.$ **27.** $x-1.$ **28.** $\dfrac{c}{c-3a}.$

XV. h. Page 96.   **1.** $\dfrac{1-x}{1+x}.$ **2.** $\dfrac{4x+1}{3x+1}.$ **3.** $\dfrac{a-b}{a+b}.$

4. $-\dfrac{1}{a+2b}.$ **5.** $2x.$ **6.** $\dfrac{2y(3x^2+y^2)}{x^2-y^2}.$ **7.** $\dfrac{bx-c}{x-1}.$

8. $\dfrac{1}{a-x}.$ **9.** $\dfrac{10}{(x-2)(x-7)}.$ **10.** $\dfrac{1}{3}.$ **11.** $\dfrac{2}{x(x+a)}.$

12. 2.　　　13. $\dfrac{2x+3}{3(x+6)}$.　　14. $\dfrac{ac+bd}{ac-bd}$.　　15. $\dfrac{x-1}{x^2(1+x^2)}$.

16. $\dfrac{2ax}{(x^2-a^2)(x-2a)^2(x+2a)}$.　　17. $-\dfrac{a}{2}$.　　18. $\dfrac{15a-2}{9a}$.　　19. 1.

20. 0.　　21. 0.　　22. 1.　　23. 1.　　24. 0.　　25. $\dfrac{4a^5b^2}{a^8-256b^8}$.

26. $\dfrac{2(3b+2)}{2b-1}$.　　27. 1.　　28. $\dfrac{x}{(x-y)^4}$.　　29. $\dfrac{2m^2}{m^2-1}$.

Miscellaneous Examples III.　Page 97.　　1. $\dfrac{x-1}{x+1}$.

2. $\dfrac{2}{a+2}$.　　3. $m+n-a$.　　4. $\dfrac{8xy}{x^2-4y^2}$.　　5. $\dfrac{1}{1-4x^2}$.

6. $\dfrac{2x+3}{3x-4}$.　　7. $\dfrac{2x+1}{x^3-1}$.　　8. $\dfrac{y}{y^2-9x^2}$.　　9. $\dfrac{2xy}{x^2-y^2}$.

10. $\dfrac{x+y}{y}$.　　11. $-abx$.　　12. $\dfrac{1}{6(3a-2)}$.　　13. $-\dfrac{1}{ab}$.

14. $\dfrac{1}{a^2b(b-a)}$.　　15. $\dfrac{x^2+x-6}{x^2-1}$.　　16. $\dfrac{x^4-x^2y^2+y^4}{x^2-y^2}$.　　17. $\dfrac{c+a}{a-b}$.

18. $\dfrac{x-4}{x-1}$.　　19. $x-y$.　　20. $\dfrac{xy}{x+y}$.　　21. $\dfrac{x-2}{x-1}$.

XVI. a.　Page 101.　　1. 3.　　2. 4.　　3. -1.　　4. 9.　　5. 0.

6. 7.　　7. 19.　　8. $\dfrac{12}{5}$.　　9. $\dfrac{9}{4}$.　　10. 0.

11. -3.　　12. $3\frac{1}{3}$.　　13. $\dfrac{2}{15}$.　　14. $-\dfrac{5}{2}$.　　15. -10.

16. -2.　　17. 5.　　18. 13.　　19. 2.　　20. 3.

21. 10.　　22. 0.　　23. $\dfrac{7}{6}$.　　24. $\dfrac{15}{4}$.　　25. $1\frac{1}{2}$.

26. $\dfrac{1}{6}$.　　27. 3.　　28. $2\frac{1}{4}$.　　29. 14.　　30. $\dfrac{1}{4}$.

XVI. b.　Page 104.　　1. $a-b$.　　2. $\dfrac{5a}{4}$.

3. $a+b$.　　4. $-ab$.　　5. $a-b$.　　6. $\dfrac{c^2-cd+d^2}{c-d}$.

7. $a+b+c$.　　8. $a-b$.　　9. $\dfrac{2a+b}{2ab}$.　　10. $a+b$.

11. $2a+3b$.　　12. 0.　　13. $\dfrac{ab^2+bc^2+ca^2}{a^2b+b^2c+c^2a}$.　　14. $\dfrac{2b^2-a^2-c^2}{2(a-b)}$.

15. q.　　16. a^2-c^2.　　17. $\dfrac{m+n}{2}$.

18. $x = b$, $y = -a$.

19. $x = c + d$, $y = c - d$.

20. $x = \dfrac{bc}{b-a}$, $y = \dfrac{ac}{b-a}$.

21. $x = 2a$, $y = 3b$.

22. $x = a$, $y = b$.

23. $x = (a+b)^2$, $y = a^2 - b^2$.

24. $x = \dfrac{b^2}{a}$, $y = -\dfrac{a^2}{b}$.

25. $x = pq(p+q)$, $y = pq(p-q)$.

26. $x = \dfrac{a+b}{2}$, $y = \dfrac{b-a}{2}$.

27. $x = a - 2b$, $y = 2a - b$.

Miscellaneous Examples IV. Page 105.

1. $1\frac{6}{35}$. **2.** 7.

3. $\dfrac{2}{3}$. **4.** $4\frac{1}{2}$. **5.** $1\frac{2}{9}$. **6.** $\dfrac{3}{25}$. **7.** $13\frac{6}{7}$. **8.** $-\dfrac{2}{5}$.

9. $\dfrac{(a-b)^2}{a+b}$. **10.** $\dfrac{bc}{ad}$. **11.** $\dfrac{a}{a+b}$. **12.** $\dfrac{2a}{3}$. **13.** $x = 2$, $y = -1$.

14. $x = \dfrac{1}{50}$, $y = \dfrac{29}{10}$. **15.** $x = 7\frac{1}{10}$, $y = 2$.

16. $x = \dfrac{33}{10}$, $y = \dfrac{9}{20}$. **17.** $x = 11$, $y = 8$.

18. $x = \dfrac{c(a+b)}{a^2+b^2}$, $y = \dfrac{c(b-a)}{a^2+b^2}$. **19.** $a = 17$, $b = -31$.

XVII. Page 108.

1. $\dfrac{11}{24}$. **2.** $\dfrac{15}{32}$. **3.** $\dfrac{17}{35}$. **4.** $\dfrac{16}{29}$.

5. $\dfrac{14}{27}$. **6.** $49\frac{1}{11}'$ past 9. **7.** $18\frac{6}{11}'$ and $36'$ past 5.

8. $48'$ past 10. **9.** $38\frac{2}{11}'$ past 1. **10.** $10\frac{10}{11}'$ and $43\frac{7}{11}'$ past 5.

11. $16\frac{4}{11}'$ and $49\frac{1}{11}'$ past 12. **12.** Cloth 15d., canvas 6d.

13. 15 horses, 16 cows. **14.** Males 15,500, females 17,500.

15. $\dfrac{(p+1)a}{2p+1}$ miles. **16.** $\dfrac{c(n+a)}{a}$ miles.

17. $\dfrac{na}{a+2b}$ hours, $\dfrac{na(a+b)}{b}$ miles. **18.** 210 miles.

19. £840. **20.** $A\ 4\frac{1}{2}$ miles, $B\ 4$ miles, $C\ 3\frac{3}{4}$ miles.

Miscellaneous Examples V. Page 110.

1. 400. **2.** 350.

3. A £7. 10s., B £9, C £7. 4s. **4.** 20, 24, 17.

5. 100, 120. **6.** 15 ft. 9 in., 4 ft. 6 in. **7.** $23\frac{2}{5}$, $15\frac{3}{5}$.

8. $76\frac{1}{2}$, $202\frac{1}{2}$. **9.** 55. **10.** 72, 110.

11. $\dfrac{29}{91}$. **12.** 27. **13.** 5s. 6d., 7s. 4d.

14. 240. **15.** 17, 23. **16.** $\dfrac{53}{112}$. **17.** £4. 16s., £7. 4s.

18. 59. **19.** 107. **20.** 14s., 16s.

Examples XVIII.   Page 114.   **1.** a^{15}.   **2.** a^8.   **3.** $2x^8$.

4. x^3.   **5.** a.   **6.** 1.   **7.** $\dfrac{1}{9}$.   **8.** $2a^6b^6$.

9. 6.   **10.** $1\frac{1}{2}$.   **11.** 4.   **12.** -3.   **13.** a^{9p}.

14. 1.   **15.** $\dfrac{1}{a^{3mn}}$.   **16.** 1.   **17.** $\dfrac{1}{2}$.   **18.** $(a+b)^4$.

19. $(a-b)^2$.   **20.** $\dfrac{1}{(x+y)^m}$.   **21.** $\dfrac{a^p}{3b^q}$.   **22.** $b^m c^n$; 18.

Miscellaneous Examples VI.   Page 115.   **1.** $\dfrac{1}{8}$.

2. $x^5+(a-1)x^4y+x^3y^2+x^2y^3+(a-1)xy^4+y^5$; mx^3+3x+7.

3. $2x^2+2x+3$; $(2x^2+2x+3)(x^2-x+4)(x^2+x+4)$.

4. (i) $(x-1)^2$; (ii) $\dfrac{x^2+y^2}{2(x^4+x^3y^2+y^4)}$.

5. (i) $9\frac{2}{5}$; (ii) $\dfrac{m+n}{2}$; (iii) $x=4,\ y=8$.   **6.** 6 ft., 3 ft.   **7.** 4.

8. $(a+1)x^2+(a^2+1)x+a^3$; $-\dfrac{5}{8}$.   **9.** $\dfrac{x^2+3x-10}{x^2-x-12}$.

10. (i) $3y-2x$; (ii) $\dfrac{a(2a+b)}{(2a-b)^2}$.

11. (i) 4; (ii) $x=3\frac{1}{2},\ y=4$; (iii) $5a=b$.

12. 4 hrs. $5\frac{5}{11}$ min.; 4 hrs. $38\frac{2}{11}$ min.   **13.** (i) $\dfrac{7}{24}$; (ii) $9\frac{11}{42}$.

14. $x^2+4y^2+9z^2+6yz+3xz-2xy$.   **15.** 250.

16. $\dfrac{x^2+x-12}{x^2-x-12}$.   **17.** (i) $-\dfrac{1}{x}$; (ii) $\dfrac{a^4+b^4}{a^4-b^4}$.

18. (i) 23; (ii) $\dfrac{a}{a+b}$; (iii) $x=1,\ y=3$.   **19.** £125, £275.

20. (i) 4; (ii) $2\frac{2}{5}$; (iii) $2\frac{5}{8}$.   **21.** $\dfrac{5ax(x-a)}{x+a}$.

22. $\dfrac{4x+3}{3x^2-5x+7}$; $\dfrac{1}{35}$.   **24.** $\dfrac{4}{5},\ 2,\ \dfrac{5a}{3},\ \dfrac{8a}{3}$.

25. (i) $\dfrac{b-c}{2a}$; (ii) $\dfrac{a(a-2b)}{b-2a}$; (iii) $x=5,\ y=-1$.

26. 14 feet, 3 feet.   **27.** $\dfrac{24}{115}$.   **28.** $-\dfrac{1}{7}$.   **29.** $\dfrac{2}{5}$.

30. (i) $\dfrac{x+y}{1-xy}$; (ii) $\dfrac{x^3}{1+x^3}$.   **31.** (i) $a+3b$; (ii) $\dfrac{2a}{3}$.

32. £17, £26.   **33.** $1\frac{1}{3}$.   **34.** £140.

EUCLID'S ELEMENTS.

BOOK I.

DEFINITIONS.

1.   A **point** is that which has position, but no magnitude.

2.   A **line** is that which has length without breadth.

3.   The extremities of a line are points, and the intersection of two lines is a point.

4.   A **straight line** is that which lies evenly between its extreme points.

Any portion cut off from a straight line is called a **segment** of it.

5.   A **surface** (or superficies) is that which has length and breadth, but no thickness.

6.   The boundaries of a surface are lines.

7.   A **plane surface** is one in which any two points being taken, the straight line between them lies wholly in that surface.

A plane surface is frequently referred to simply as a *plane*.

NOTE.   Euclid regards a point merely as a *mark of position*, and he therefore attaches to it no idea of size and shape.

Similarly he considers that the properties of a line arise only from its *length* and *position*, without reference to that minute breadth which every line must really have *if actually drawn*, even though the most perfect instruments are used.

The definition of a surface is to be understood in a similar way.

8. A **plane angle** is the inclination of two lines to one another, which meet together, but are not in the same direction.

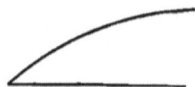

[Definition 8 is not required in Euclid's Geometry, the only angles employed by him being those formed by *straight* lines.   See Def. 9.]

9. A **plane rectilineal angle** is the inclination of two *straight* lines to one another, which meet together, but are not in the same straight line.

The point at which the straight lines meet is called the **vertex** of the angle, and the straight lines themselves the **arms** of the angle.

NOTE.   When there are several angles at one point, each is expressed by three letters, of which the letter that refers to the vertex is put between the other two.   Thus the angle contained by the straight lines OA, OB is named the angle AOB or BOA; and the angle contained by OA, OC is named the angle AOC or COA.   But if there is only one angle at a point, it may be expressed by a single letter, as *the angle at* O.

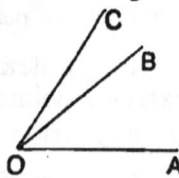

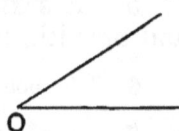

Of the two straight lines OB, OC shewn in the adjoining diagram, we recognize that OC is *more inclined* than OB to the straight line OA: this we express by saying that the angle AOC is greater than the angle AOB.   Thus an angle must be regarded as having *magnitude*.

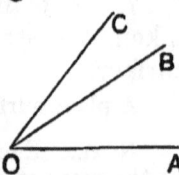

It must be carefully observed that the size of an angle in no way depends on the length of its arms, but only on their *inclination* to one another.

The angle AOC is the *sum* of the angles AOB and BOC; and AOB is the *difference* of the angles AOC and BOC.

[Another view of an angle is recognized in many branches of mathematics; and though not employed by Euclid, it is here given because it furnishes more clearly than any other a conception of what is meant by the *magnitude* of an angle.

Suppose that the straight line OP in the diagram is capable of revolution about the point O, like the hand of a watch, but in the opposite direction; and suppose that in this way it has passed successively from the position OA to the positions occupied by OB and OC.   Such a line must have undergone *more turning* in passing from OA to

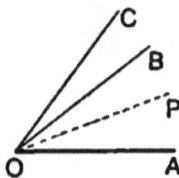

OC, than in passing from OA to OB ; and consequently the angle AOC is said to be greater than the angle AOB.]

Angles which lie on either side of a common arm are called **adjacent** angles.

For example, when one straight line OC is drawn from a point in another straight line AB, the angles COA, COB are *adjacent.*

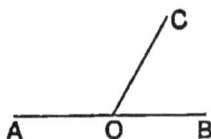

When two straight lines, such as AB, CD, cross one another at E, the two angles CEA, BED are said to be **vertically opposite.** The two angles CEB, AED are also vertically opposite to one another.

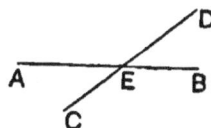

10. When a straight line standing on another straight line makes the adjacent angles equal to one another, each of the angles is called a **right angle** ; and the straight line which stands on the other is called a **perpendicular** to it.

11. An **obtuse angle** is an angle which is greater than a right angle.

12. An **acute angle** is an angle which is less than a right angle.

[In the adjoining figure the straight line OB may be supposed to have arrived at its present position, from the position occupied by OA, by revolution about the point O in *either* of the two directions indicated by the arrows : thus two straight lines drawn from a point may be considered as forming *two* angles (marked (i) and (ii) in the figure), of which the greater (ii) is said to be **reflex.**

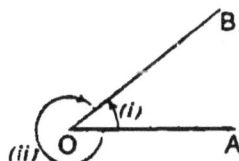

If the arms OA, OB are in the same straight line, the angle formed by them on either side is called a **straight angle.**]

13. A term or boundary is the extremity of anything.

14. Any portion of a plane surface bounded by one or more lines is called a plane figure.

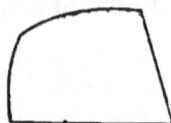

> The sum of the bounding lines is called the **perimeter** of the figure.
> Two figures are said to be equal in **area** when they enclose equal portions of a plane surface.

15. A **circle** is a plane figure contained by one line, which is called the **circumference**, and is such that all straight lines drawn from a certain point within the figure to the circumference are equal to one another; this point is called the **centre** of the circle.

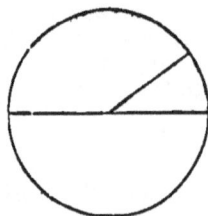

16. A **radius** of a circle is a straight line drawn from the centre to the circumference.

17. A **diameter** of a circle is a straight line drawn through the centre, and terminated both ways by the circumference.

18. A **semicircle** is the figure bounded by a diameter of a circle and the part of the circumference cut off by the diameter.

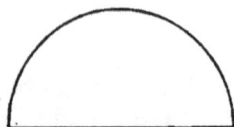

19. A **segment of a circle** is the figure bounded by a straight line and the part of the circumference which it cuts off.

20. **Rectilineal figures** are those which are bounded by straight lines.

21. A **triangle** is a plane figure bounded by *three* straight lines.

> Any one of the angular points of a triangle may be regarded as its **vertex**; and the opposite side is then called the **base**.

22. A **quadrilateral** is a plane figure bounded by *four* straight lines.

The straight line which joins opposite angular points in a quadrilateral is called a **diagonal**.

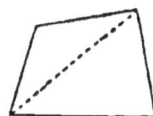

23. A **polygon** is a plane figure bounded by more than four straight lines.

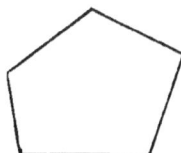

TRIANGLES.

24. An **equilateral triangle** is a triangle whose three sides are equal.

25. An **isosceles triangle** is a triangle two of whose sides are equal.

26. A **scalene triangle** is a triangle which has three unequal sides.

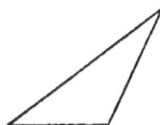

27. A **right-angled triangle** is a triangle which has a right angle.

The side opposite to the right angle in a right-angled triangle is called the **hypotenuse**.

28. An **obtuse-angled triangle** is a triangle which has an obtuse angle.

29. An **acute-angled triangle** is a triangle which has *three* acute angles.

[It will be seen hereafter (Book I. Proposition 17) *that every triangle must have at least two acute angles.*]

QUADRILATERALS.

30. A **square** is a four-sided figure which
has all its sides equal and all its angles right
angles.

[It may be shewn that if a quadrilateral has all its
sides equal and *one* angle a right angle, then *all* its
angles will be right angles.]

31. An **oblong** is a four-sided figure which has all its angles
right angles, but not all its sides equal.

32. A **rhombus** is a four-sided figure
which has all its sides equal, but its
angles are not right angles.

33. A **rhomboid** is a four-sided figure which has its opposite
sides equal to one another, but all its sides are not equal nor its
angles right angles.

34.   All other four-sided figures are called **trapeziums.**

It is usual now to restrict the term *trapezium* to
a quadrilateral which has two of its sides *parallel.*
[See Def. 35.]

35.   **Parallel straight lines** are such as,
being in the same plane, do not meet, however far they are produced in either direction.

36.   A **Parallelogram** is a four-sided
figure which has its opposite sides
parallel.

37. A **rectangle** is a parallelogram
which has one of its angles a right angle.

THE POSTULATES.

Let it be granted,

1. *That a straight line may be drawn from any one point to any other point.*

2. *That a* finite, *that is to say a terminated, straight line may be* produced *to any length in that straight line.*

3. *That a circle may be described from any centre, at any distance from that centre, that is, with a radius equal to any finite straight line drawn from the centre.*

NOTES ON THE POSTULATES.

1. In order to draw the diagrams required in Euclid's Geometry certain instruments are necessary. These are

    (i) A *ruler* with which to draw straight lines.

    (ii) A *pair of compasses* with which to draw circles.

In the *Postulates*, or requests, Euclid claims the use of these instruments, and assumes that they suffice for the purposes mentioned above.

2. It is important to notice that the Postulates include no means of *direct measurement*: hence the straight ruler is not supposed to be *graduated*; and the compasses are not to be employed for *transferring distances* from one part of a diagram to another.

3. When we draw a straight line from the point A to the point B, we are said to *join* AB.

To *produce* a straight line means to *prolong* or *lengthen* it.

The expression *to describe* is used in Geometry in the sense of *to draw*.

ON THE AXIOMS.

The science of Geometry is based upon certain simple statements, the truth of which is so evident that they are accepted without proof.

These self-evident truths, called by Euclid *Common Notions*, are known as the **Axioms**.

GENERAL AXIOMS.

1. *Things which are equal to the same thing are equal to one another.*

2. *If equals be added to equals, the wholes are equal.*

3. *If equals be taken from equals, the remainders are equal.*

4. *If equals be added to unequals, the wholes are unequal, the greater sum being that which includes the greater of the unequals.*

5. *If equals be taken from unequals, the remainders are unequal, the greater remainder being that which is left from the greater of the unequals.*

6. *Things which are double of the same thing, or of equal things, are equal to one another.*

7. *Things which are halves of the same thing, or of equal things, are equal to one another.*

9.* *The whole is greater than its part.*

* To preserve the classification of general and geometrical axioms, we have placed Euclid's *ninth* axiom before the *eighth*.

GEOMETRICAL AXIOMS.

8. *Magnitudes which can be made to coincide with one another, are equal.*

10. *Two straight lines cannot enclose a space.*

11. *All right angles are equal.*

12. *If a straight line meet two straight lines so as to make the interior angles on one side of it together less than two right angles, these straight lines will meet if continually produced on the side on which are the angles which are together less than two right angles.*

That is to say, if the two straight lines AB and CD are met by the straight line EH at F and G, in such a way that the angles BFG, DGF are together less than two right angles, it is asserted that AB and CD will meet if continually produced in the direction of B and D.

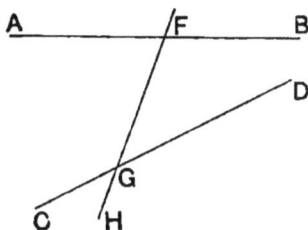

NOTES ON THE AXIOMS.

1. The necessary characteristics of an Axiom are

(i) That it should be *self-evident*; that is, that its truth should be immediately accepted without proof.

(ii) That it should be *fundamental*; that is, that its truth should not be derivable from any other truth more simple than itself.

(iii) That it should supply a basis for the establishment of further truths.

These characteristics may be summed up in the following definition.

DEFINITION. An **Axiom** is a self-evident truth, which neither requires nor is capable of proof, but which serves as a foundation for future reasoning.

2. Euclid's Axioms may be classified as *general* and *geometrical*.

General Axioms apply to *magnitudes of all kinds*. Geometrical Axioms refer specially to *geometrical magnitudes*, as lines, angles, and figures.

3. Axiom 8 is Euclid's test of the equality of two geometrical magnitudes. It implies that any line, angle, or figure, may be taken up from its position, and without change in size or form, laid down upon a second line, angle, or figure, for the purpose of comparison, and it states that two such magnitudes are equal when one can be exactly placed over the other without overlapping.

This process is called **superposition**, and the first magnitude is said to be **applied** to the other.

4. Axiom 12 has been objected to on the double ground that it cannot be considered self-evident, and that its truth may be deduced from simpler principles. It is employed for the first time in the 29th Proposition of Book I., where a short discussion of the difficulty will be found.

E.C.   Q

INTRODUCTORY.

1. Little is known of Euclid beyond the fact that he lived about three centuries before Christ (325-285) at Alexandria, where he became famous as a writer and teacher of Mathematics.

Among the works ascribed to him, the best known and most important is *The Elements*, written in Greek, and consisting of Thirteen Books. Of these it is now usual to read Books I.-IV. and VI. (which deal with Plane Geometry), together with parts of Books XI. and XII. (on the Geometry of Solids). The remaining Books deal with subjects which belong to the theory of Arithmetic.

2. Plane Geometry deals with the properties of all lines and figures that may be drawn upon a plane surface.

Euclid in his first Six Books confines himself to the properties of straight lines, rectilineal figures, and circles.

3. The subject is divided into a number of separate discussions, called **propositions**.

Propositions are of two kinds, **Problems** and **Theorems**.

A **Problem** proposes to perform some geometrical construction, such as to draw some particular line, or to construct some required figure.

A **Theorem** proposes to prove the truth of some geometrical statement.

4. A Proposition consists of the following parts :

The *General Enunciation*, the *Particular Enunciation*, the *Construction*, and the *Proof*.

(i) The **General Enunciation** is a preliminary statement, describing in general terms the purpose of the proposition.

(ii) The **Particular Enunciation** repeats in special terms the statement already made, and refers it to a diagram, which enables the reader to follow the reasoning more easily.

(iii) The **Construction** then directs the drawing of such straight lines and circles as may be required to effect the purpose of a problem, or to prove the truth of a theorem.

(iv) The **Proof** shews that the object proposed in a problem has been accomplished, or that the property stated in a theorem is true.

5. Euclid's reasoning is said to be **Deductive**, because by a connected chain of argument it *deduces* new truths from truths already proved or admitted. Thus each proposition, though in one sense complete in itself, is derived from the Postulates, Axioms, or former propositions, and itself leads up to subsequent propositions.

6. The initial letters Q.E.F., placed at the end of a problem, stand for **Quod erat Faciendum**, *which was to be done.*

The letters Q.E.D. are appended to a theorem, and stand for **Quod erat Demonstrandum**, *which was to be proved.*

7. A **Corollary** is a statement the truth of which follows readily from an established proposition; it is therefore appended to the proposition as an inference or deduction, which usually requires no further proof.

8. The attention of the beginner is drawn to the special use of the *future tense* in the Particular Enunciations of Euclid's propositions.

The future is only used in a statement of which the truth is *about to be proved.* Thus: " *The triangle* ABC SHALL BE *equilateral*" means that the triangle *has yet to be proved* equilateral. While, " *The triangle* ABC IS *equilateral*" means that the triangle *has already been proved* (or given) equilateral.

9. The following symbols and abbreviations may be employed in writing out the propositions of Book I., though their use is not recommended to beginners.

∴	*for* therefore,	par^l (or ‖)	*for*	parallel,
=	,, is, or are, equal to,	par^m	,,	parallelogram,
∠	,, angle,	sq.	,,	square,
rt. ∠	,, right angle,	rectil.	,,	rectilineal,
· △	,, triangle,	st. line	,,	straight line,
perp.	,, perpendicular,	pt.	,,	point ;

and all obvious contractions of words, such as opp., adj., diag., etc., for opposite, adjacent, diagonal, etc.

SECTION I.

PROPOSITION 1.   PROBLEM.

To describe an equilateral triangle on a given finite straight line.

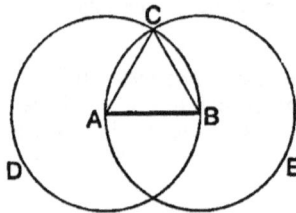

Let AB be the given straight line.
It is required to describe an equilateral triangle on AB.

Construction.   With centre A, and radius AB, describe
   the circle BCD.   *Post.* 3.
With centre B, and radius BA, describe the circle ACE.
                                                        Post. 3.
From the point C at which the circles cut one another,
   draw the straight lines CA and CB to the points A and B.
                                                        Post. 1.
   Then shall the triangle ABC be equilateral.

Proof.   Because A is the centre of the circle BCD,
               therefore AC is equal to AB.   *Def.* 15.
   And because B is the centre of the circle ACE,
               therefore BC is equal to AB.   *Def.* 15.
      Therefore AC and BC are each equal to AB.
   But things which are equal to the same thing are equal
to one another.   *Ax.* 1.
            Therefore AC is equal to BC.
   Therefore AC, AB, BC are equal to one another.
      Therefore the triangle ABC is equilateral ; .
and it is described on the given straight line AB.   Q.E.F.

PROPOSITION 2. PROBLEM.

From a given point to draw a straight line equal to a given straight line.

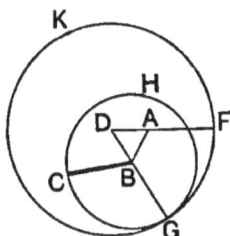

Let A be the given point, and BC the given straight line.
It is required to draw from A a straight line equal to BC.

Construction. Join AB ; *Post.* 1.
and on AB describe an equilateral triangle DAB. I. 1.
With centre B, and radius BC, describe the circle CGH.
Post. 3.
Produce DB to meet the circle CGH at G. *Post.* 2.
With centre D, and radius DG, describe the circle GKF.
Produce DA to meet the circle GKF at F. *Post.* 2.
Then AF *shall be equal to* BC.

Proof. Because B is the centre of the circle CGH,
therefore BC is equal to BG. *Def.* 15.
And because D is the centre of the circle GKF,
therefore DF is equal to DG. *Def.* 15.
And DA, a part of DF, is equal to DB, a part of DG ; *Def.* 24.
therefore the remainder AF is equal to the remainder BG.
Ax. 3.
But BC has been proved equal to BG ;
therefore AF and BC are each equal to BG.
And things which are equal to the same thing are equal
to one another. *Ax.* 1.
Therefore AF is equal to BC ;
and it has been drawn from the given point A. Q.E.F.

PROPOSITION 3.   PROBLEM.

From the greater of two given straight lines to cut off a part equal to the less.

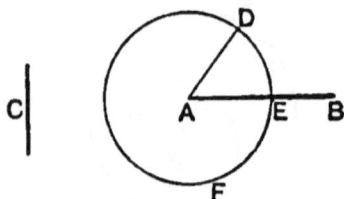

Let AB and C be the two given straight lines, of which AB is the greater.

It is required to cut off from AB a part equal to C.

Construction.   From the point A draw the straight line AD equal to C ;   I. 2.
and with centre A and radius AD, describe the circle DEF, cutting AB at E.   *Post* 3.
        Then AE *shall be equal to* C.

Proof.   Because A is the centre of the circle DEF,
            therefore AE is equal to AD.   *Def.* 15.
            But C is equal to AD.   *Constr.*
    Therefore AE and C are each equal to AD.
            Therefore AE is equal to C ;   *Ax.* 1.
and it has been cut off from the given straight line AB.
                                                    Q.E.F.

EXERCISES ON PROPOSITIONS 1 TO 3.

1.   If the two circles in Proposition 1 cut one another again at F, prove that AFB is an equilateral triangle.

2.   If the two circles in Proposition 1 cut one another at C and F, prove that the figure ACBF is a rhombus.

3.   AB is a straight line of given length : shew how to draw from A a line double the length of AB.

4. Two circles are drawn with the same centre O, and two radii OA, OB are drawn in the smaller circle. If OA, OB are produced to cut the outer circle at D and E, prove that AD=BE.

5. AB is a straight line, and P, Q are two points, one on each side of AB. Shew how to find points in AB, whose distance from P is equal to PQ. How many such points will there be?

6. In the figure of Proposition 2, if AB is equal to BC, shew that D, the vertex of the equilateral triangle, will fall on the circumference of the circle CGH.

7. In Proposition 2 the point A may be joined to either extremity of BC. Draw the figure, and prove the proposition in the case when A is joined to C.

8. On a given straight line AB describe an isosceles triangle having each of its equal sides equal to a given straight line PQ.

9. On a given base describe an isosceles triangle having each of its equal sides double of the base.

10. In a given straight line the points A, M, N, B are taken in order. On AB describe a triangle ABC, such that the side AC may be equal to AN, and the side BC to BM.

NOTE ON PROPOSITIONS 2 AND 3.

Propositions 2 and 3 are rendered necessary by the restriction tacitly imposed by Euclid, that compasses shall not be used to *transfer distances*. [See Notes on the Postulates.]

In carrying out the construction of Prop. 2 the point A may be joined to *either* extremity of the line BC; the equilateral triangle may be described on *either* side of the line so drawn; and the sides of the equilateral triangle may be produced in *either* direction. Thus there are in general $2 \times 2 \times 2$, or *eight*, possible constructions. The student should exercise himself in drawing the various figures that may arise.

PROPOSITION 4. THEOREM.

If two triangles have two sides of the one equal to two sides of the other, each to each, and have also the angles contained by those sides equal, then the triangles shall be equal in all respects; that is to say, their bases or third sides shall be equal, and their remaining angles shall be equal, each to each, namely those to which the equal sides are opposite; and the triangles shall be equal in area.

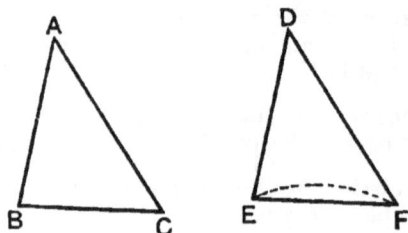

Let ABC, DEF be two triangles, in which
the side AB is equal to the side DE,
the side AC is equal to the side DF, and
the contained angle BAC is equal to the contained angle EDF.

Then   (i)   *the base* BC *shall be equal to the base* EF;
         (ii)   *the angle* ABC *shall be equal to the angle* DEF;
        (iii)   *the angle* ACB *shall be equal to the angle* DFE;
       (iv)   *the triangle* ABC *shall be equal to the triangle* DEF *in area.*

Proof. If the triangle ABC be applied to the triangle DEF,
so that the point A may lie on the point D,
and the straight line AB along the straight line DE;
then because AB is equal to DE,   *Hyp.*
therefore the point B must coincide with the point E.
And because AB falls along DE,
and the angle BAC is equal to the angle EDF,   *Hyp.*
therefore AC must fall along DF.
And because AC is equal to DF,   *Hyp.*
therefore the point C must coincide with the point F.
Then since B coincides with E, and C with F,
therefore the base BC must coincide with the base EF;

for if not, two straight lines would enclose a space; which
is impossible. *Ax.* 10.

Thus the base BC coincides with the base EF, and is
therefore equal to it. *Ax.* 8.

And the remaining angles of the triangle ABC coincide
with the remaining angles of the triangle DEF, and are
therefore equal to them;

namely, the angle ABC is equal to the angle DEF,

and the angle ACB is equal to the angle DFE.

And the triangle ABC coincides with the triangle DEF,
and is therefore equal to it in area. *Ax.* 8.

That is, the triangles are equal in all respects. Q.E.D.

NOTE. The sides and angles of a triangle are known as its *six
parts*. A triangle may also be considered in regard to its *area*.

Two triangles are said to be **equal in all respects**, or **identically
equal**, when the sides and angles of one are respectively equal to
the sides and angles of the other. We have seen that such triangles
may be made to *coincide* with one another by *superposition*, so that
they are also equal in *area*. [See Note on Axiom 8.]

[It will be shewn later that triangles can be equal in *area* without
being equal in their several parts; that is to say, triangles can have
the same *area* without having the same *shape*.]

EXERCISES ON PROPOSITION 4.

1. ABCD is a square : prove that the diagonals AC, BD are equal
to one another.

2. ABCD is a square, and L, M, and N are the middle points of
AB, BC, and CD : prove that

    (i) LM = MN.   (ii) AM = DM.

    (iii) AN = AM.   (iv) BN = DM.

    [Draw a separate figure in each case.]

3. ABC is an isosceles triangle : from the equal sides AB, AC
two equal parts AX, AY are cut off, and BY and CX are joined.
Prove that BY = CX.

4. ABCD is a quadrilateral having the opposite sides BC, AD
equal, and also the angle BCD equal to the angle ADC : prove that
BD is equal to AC.

PROPOSITION 5. THEOREM

The angles at the base of an isosceles triangle are equal to one another ; and if the equal sides be produced, the angles on the other side of the base shall also be equal to one another.

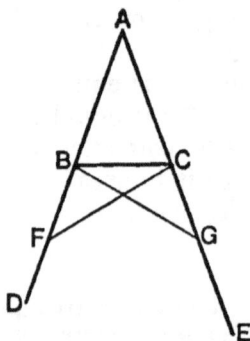

Let ABC be an isosceles triangle, in which
the side AB is equal to the side AC,
and let the straight lines AB, AC be produced to D and E.

Then (i) *the angle* ABC *shall be equal to the angle* ACB ;
(ii) *the angle* CBD *shall be equal to the angle* BCE.

Construction. In BD take any point F ;
and from AE cut off a part AG equal to AF.   I. 3.
Join FC, GB.

Proof. Then in the triangles FAC, GAB,
Because { FA is equal to GA,   *Constr.*
and AC is equal to AB,   *Hyp.*
also the contained angle at A is common to the two triangles :
therefore the triangle FAC is equal to the triangle GAB in all respects ;   I. 4.
that is, the base FC is equal to the base GB,
and the angle ACF is equal to the angle ABG,
also the angle AFC is equal to the angle AGB.

Again, because AF is equal to AG,
and AB, a part of AF, is equal to AC, a part of AG ; *Hyp.*
therefore the remainder BF is equal to the remainder CG.

Then in the two triangles BFC, CGB,

Because
{
BF is equal to CG,   *Proved.*
and FC is equal to GB,   *Proved.*
also the contained angle BFC is equal to the
contained angle CGB,   *Proved.*
}

therefore the triangle BFC is equal to the triangle CGB in
all respects ;   I. 4.

so that the angle FBC is equal to the angle GCB,
and the angle BCF to the angle CBG.

Now it has been shewn that the angle ABG is equal to the
angle ACF,

and that the angle CBG, a part of ABG, is equal to the angle
BCF, a part of ACF ;

therefore the remaining angle ABC is equal to the remaining angle ACB ;   *Ax.* 3.

and these are the angles at the base of the triangle ABC.

Also it has been shewn that the angle FBC is equal to the
angle GCB ;

and these are the angles on the other side of the base.   Q.E.D.

COROLLARY. *Hence if a triangle is equilateral it is also
equiangular.*

NOTE. The difficulty which beginners find with this proposition
arises from the fact that the triangles
to be compared overlap one another
in the diagram. This difficulty may
be diminished by detaching each
pair of triangles from the rest of the
figure, as shewn in the margin.

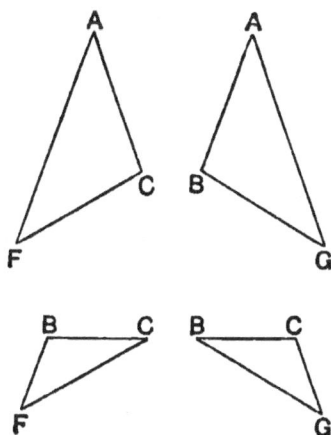

·

PROPOSITION 6. THEOREM.

If two angles of a triangle be equal to one another, then the sides also which subtend, or are opposite to, the equal angles, shall be equal to one another.

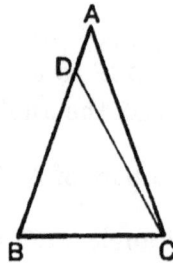

Let ABC be a triangle, in which
the angle ABC is equal to the angle ACB.
Then shall the side AC be equal to the side AB.

Construction. For if AC be not equal to AB,
one of them must be greater than the other.
If possible, let AB be the greater ;
.and from it cut off BD equal to AC.   I. 3.
Join DC.

Proof. Then in the triangles DBC, ACB,

Because
{
DB is equal to AC,   *Constr.*
and BC is common to both,
also the contained angle DBC is equal to the
contained angle ACB ;   *Hyp*
}

therefore the triangle DBC is equal to the triangle ACB
in area,   I. 4.
the part equal to the whole ; which is absurd. *Ax.* 9.

Therefore AB is not unequal to AC ;
that is, AB is equal to AC.   Q.E.D.

COROLLARY. *Hence if a triangle is equiangular it is also equilateral.*

NOTE ON PROPOSITIONS 5 AND 6.

The enunciation of a theorem consists of two clauses. The first clause tells us what we are to *assume*, and is called the **hypothesis** ; the second tells us what *it is required to prove*, and is called the **conclusion.**

For example, the enunciation of Proposition 5 assumes that in a certain triangle ABC *the side* AB = *the side* AC : this is the *hypothesis*. From this it is required to prove that *the angle* ABC = *the angle* ACB : this is the *conclusion*.

If we interchange the hypothesis and conclusion of a theorem, we enunciate a new theorem which is called the **converse** of the first.

For example, in Prop. 5

it is *assumed* that $\qquad\qquad$ AB = AC ;
it is *required to prove* that the angle ABC = the angle ACB. ∫

Now in Prop. 6

it is *assumed* that the angle ABC = the angle ACB ;⎫
it is *required to prove* that $\quad$ AB = AC. $\qquad$⎭

Thus we see that Prop. 6 is the converse of Prop. 5 ; for *the hypothesis of each is the conclusion of the other*.

In Proposition 6 Euclid employs for the first time an *indirect method of proof* frequently used in geometry. It consists in shewing that the theorem *cannot be untrue* ; since, if it were, we should be led to some *impossible conclusion*. This form of proof is known as **Reductio ad Absurdum,** and is most commonly used in demonstrating the converse of some foregoing theorem.

The converse of *all* true theorems are not themselves necessarily true. [See Note on Prop 8.]

EXERCISES ON PROPOSITION 5.

1. ABCD is a rhombus, in which the diagonal BD is drawn : shew that

(i) the angle ABD = the angle ADB ;

(ii) the angle CBD = the angle CDB ;

(iii) the angle ABC = the angle ADC.

2. ABC, DBC are two isosceles triangles drawn on the same base BC, but on *opposite* sides of it : prove (by means of I. 5) that the angle ABD = the angle ACD.

3. ABC, DBC are two isosceles triangles drawn on the same base BC and on *the same* side of it : employ I. 5 to prove that the angle ABD = the angle ACD.

PROPOSITION 7. THEOREM

On the same base, and on the same side of it, there cannot be two triangles having their sides which are terminated at one extremity of the base equal to one another, and likewise those which are terminated at the other extremity equal to one another.

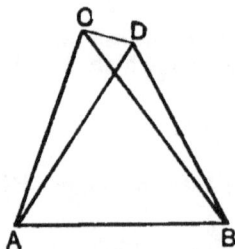

If it be possible, on the same base AB, and on the same side of it, let there be two triangles ACB, ADB in which
the side AC is equal to the side AD,
and also the side BC is equal to the side BD.

CASE I.   When the vertex of each triangle is without the other triangle.

Construction.   Join CD.

Proof.   Then in the triangle ACD,
because AC is equal to AD,   *Hyp.*
therefore the angle ACD is equal to the angle ADC. I. 5.

But the whole angle ACD is greater than its part, the angle BCD ;
therefore also the angle ADC is greater than the angle BCD;
still more then is the angle BDC greater than the angle BCD.

Again, in the triangle BCD,
because BC is equal to BD,   *Hyp*
therefore the angle BDC is equal to the angle BCD : I. 5.
but it was shewn to be greater ; which is impossible.

CASE II. When one of the vertices, as D, is within the other triangle ACB.

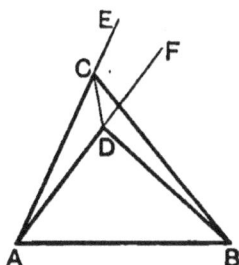

Construction.   As before, join CD ;
and produce AC, AD to E and F.

Proof.   Then in the triangle ACD,
because AC is equal to AD,   *Hyp.*
therefore the angle ECD is equal to the angle FDC,
these being the angles on the other side of the base. I. 5.
But the angle ECD is greater than its part, the angle BCD ;
therefore the angle FDC is also greater than the angle
BCD :
still more then is the angle BDC greater than the angle
BCD.
            Again, in the triangle BCD,
        because BC is equal to BD,   *Hyp.*
therefore the angle BDC is equal to the angle BCD : I. 5.
but it has been shewn to be greater ; which is impossible.

The case in which the vertex of one triangle is on a
side of the other needs no demonstration.
Therefore AC cannot be equal to AD, and *at the same
time,* BC equal to BD.   Q.E.D.

NOTE.   The sides AC, AD are called **conterminous** sides ; similarly
the sides BC, BD are conterminous.

PROPOSITION 8.   THEOREM.

If two triangles have two sides of the one equal to two sides of the other, each to each, and have likewise their bases equal, then the angle which is contained by the two sides of the one shall be equal to the angle which is contained by the two sides of the other.

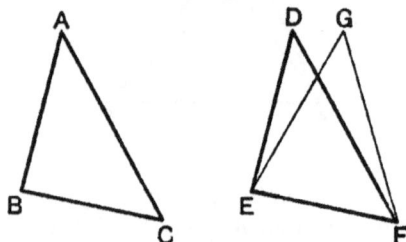

Let ABC, DEF be two triangles, in which
the side AB is equal to the side DE,
the side AC is equal to the side DF,
and the base BC is equal to the base EF.

Then shall the angle BAC be equal to the angle EDF.

Proof. If the triangle ABC be applied to the triangle DEF,
so that the point B falls on the point E,
and the base BC along the base EF ;
then because BC is equal to EF,   *Hyp.*
therefore the point C must coincide with the point F.

Then since BC coincides with EF,
it follows that BA and AC must coincide with ED and DF :
for if they did not, but took some other position, as EG, GF,
then on the same base EF, and on the same side of it, there
would be two triangles EDF, EGF, having their *conterminous*
sides equal : namely ED equal to EG, and FD equal to FG.
But this is impossible.   I. 7.

Therefore the sides BA, AC coincide with the sides ED, DF.
That is, the angle BAC coincides with the angle EDF, and is
therefore equal to it.   *Ax.* 8.

Q.E.D.

NOTE 1.   In this Proposition the three sides of one triangle are given equal respectively to the three sides of the other ; and from this it is shewn that the two triangles may be made *to coincide with one another.*

Hence we are led to the following important Corollary.

COROLLARY.   *If in two triangles the three sides of the one are equal to the three sides of the other, each to each, then the triangles are equal in all respects.*

[An alternative proof, which is independent of Prop. 7, will be found on page 26.]

NOTE 2.   Proposition 8 furnishes an instance of a true theorem of which the *converse* is not necessarily true.

It is proved above that *if the sides of one triangle are severally equal to the sides of another, then the angles of the first triangle are severally equal to the angles of the second.*

The *converse* of this enunciation would be as follows : *If the angles of one triangle are severally equal to the angles of another, then the sides of the first triangle are equal to the sides of the second.*

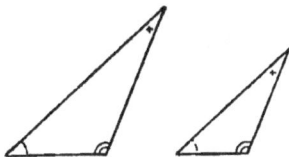

But this, as the diagram in the margin shews, is by no means necessarily true.

EXERCISES ON PROPOSITION 8.

1.   Shew (by drawing a diagonal) that the opposite angles of a rhombus are equal.

2.   If ABCD is a quadrilateral, in which AB = CD and AD = CB, prove that the angle ADC = the angle ABC.

3.   If ABC and DBC are two isosceles triangles drawn on the same base BC, prove (by means of I. 8) that the angle ABD = the angle ACD, taking (i) the case where the triangles are on the *same* side of BC, (ii) the case where they are on *opposite* sides of BC.

4.   If ABC, DBC are two isosceles triangles drawn on opposite sides of the same base BC, and if AD be joined, prove that each of the angles BAC, BDC will be divided into two equal parts.

5.   If in the figure of Ex. 4 the line AD meets BC in E, prove that BE = EC.

PROPOSITION 8.   ALTERNATIVE PROOF.

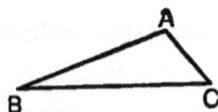

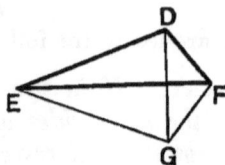

Let **ABC** and **DEF** be two triangles, which have the sides **BA, AC** equal respectively to the sides **ED, DF,** and the base **BC** equal to the base **EF.**
Then shall the angle **BAC** *be equal to the angle* **EDF.**

For apply the triangle **ABC** to the triangle **DEF**, so that **B** may fall on **E**, and **BC** along **EF**, and so that the point **A** may be on the side of **EF** remote from **D** ;
then **C** must fall on **F**, since **BC** is equal to **EF**.
Let **GEF** be the new position of the triangle **ABC**.
Join **DG**.

CASE I.   When **DG** intersects **EF**.
Then because **ED = EG**,
∴   the angle **EDG** = the angle **EGD**.   I. 5.
Again because **FD = FG**,
∴   the angle **FDG** = the angle **FGD**.   I. 5.
Hence the whole angle **EDF** = the whole angle **EGF** ; *Ax.* 2.
. that is, the angle **EDF** = the angle **BAC**.

Two cases remain which may be dealt with in a similar manner : namely,

CASE II.   When **DG** meets **EF** produced.

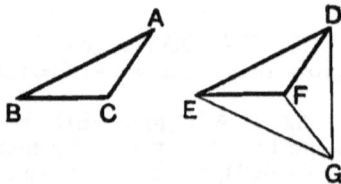

CASE III.   When one pair of sides, as **DF, FG** are in one straight line,

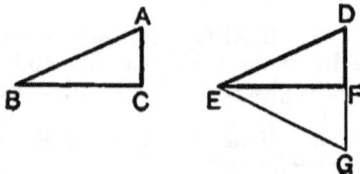

QUESTIONS AND EXERCISES FOR REVISION.

1. Define *adjacent angles, a right angle, vertically opposite angles.*

2. Explain the words *enunciation, hypothesis, conclusion.*

3. Distinguish between the meanings of the following statements:
   (i)   then AB *is* equal to PQ ;
   (ii) then AB *shall be* equal to PQ.

4. When are two theorems said to be *converse* to one another. Give an example.

5. Shew by an example that the converse of a true theorem is not itself necessarily true.

6. What is a *corollary*? Quote the corollary to Proposition 5 ; and shew how its truth follows from that proposition.

7. Name the six *parts* of a triangle. When are triangles said to be *equal in all respects*?

8. What do you understand by the expression *geometrical magnitudes*? Give examples?

9. What is meant by *superposition*? Explain the test by which Euclid determines if two geometrical magnitudes are equal to one another. Illustrate by an example.

10. Quote and explain the *third postulate*. What restrictions does Euclid impose on the use of compasses, and what problems are thereby made necessary?

11. Define an *axiom*. Quote the axioms referred to in (i) in Proposition 2 ; (ii) in Proposition 7.

12. Prove by the method of *superposition* that two squares are equal in area, if a side of one is equal to a side of the other.

13. Two quadrilaterals ABCD, EFGH have the sides AB, BC, CD, DA equal respectively to the sides EF, FG, GH, HE, and have also the angle BAD equal to the angle FEH. Shew that the figures may be made to coincide with one another.

14. AB, AC are the equal sides of an isosceles triangle ABC ; and L, M, N are the middle points of AB, BC, and CA respectively : prove that
   (i)   LM = MN.   (ii)   BN = CL.
   (iii) the angle ALM = the angle ANM.

PROPOSITION 9. PROBLEM.

To bisect a given rectilineal angle, that is, to divide it into two equal parts.

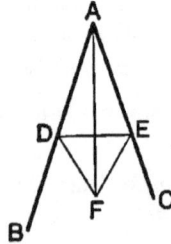

Let BAC be the given angle.
It is required to bisect the angle BAC.

Construction. In AB take any point D ;
and from AC cut off AE equal to AD.   I. 3.
Join DE ;
and on DE, on the side remote from A, describe an equilateral triangle DEF.   I. 1.
Join AF.
Then shall the straight line AF *bisect the angle* BAC.

Proof. For in the two triangles DAF, EAF,

Because $\begin{cases} \text{DA is equal to EA,} & \textit{Constr.} \\ \text{and AF is common to both ;} \\ \text{and the third side DF is equal to the third side} \\ \text{EF ;} & \textit{Def. 24.} \end{cases}$

therefore the angle DAF is equal to the angle EAF. I. 8.
Therefore the given angle BAC is bisected by the straight
line AF.   Q.E.F.

EXERCISES.

1. If in the above figure the equilateral triangle DFE were described on the same side of DE as A, what different cases would
arise ? And under what circumstances would the construction fail ?

2. In the same figure, shew that AF also bisects the angle DFE.

3. Divide an angle into four equal parts.

PROPOSITION 10. PROBLEM.

To bisect a given finite straight line, that is, to divide it into two equal parts.

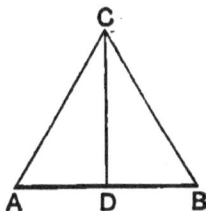

Let AB be the given straight line.
It is required to divide AB into two equal parts.

Constr. On AB describe an equilateral triangle ABC; I. 1. and bisect the angle ACB by the straight line CD, meeting AB at D. I. 9.
Then shall AB be bisected at the point D.

Proof. For in the triangles ACD, BCD,

Because { AC is equal to BC, *Def.* 24. and CD is common to both; also the contained angle ACD is equal to the contained angle BCD; *Constr.*

therefore the triangle ACD is equal to the triangle BCD in all respects : I. 4.
so that the base AD is equal to the base BD.

Therefore the straight line AB is bisected at the point D.
Q.E.F.

EXERCISES.

1. Shew that the straight line which bisects the vertical angle of an isosceles triangle, also bisects the base.

2. On a given base describe an isosceles triangle such that the sum of its equal sides may be equal to a given straight line.

PROPOSITION 11. PROBLEM.

To draw a straight line at right angles to a given straight line, from a given point in the same.

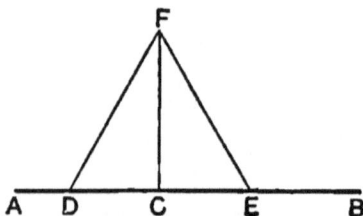

Let AB be the given straight line, and C the given point in it.

It is required to draw from C a straight line at right angles to AB.

Construction. In AC take any point D,
and from CB cut off CE equal to CD.   I. 3.
On DE describe the equilateral triangle DFE.   I. 1.
Join CF.
Then shall CF be at right angles to AB.

Proof.   For in the triangles DCF, ECF,
DC is equal to EC,   *Constr.*
Because { and CF is common to both;
and the third side DF is equal to the third side EF :   *Def. 24.*
Therefore the angle DCF is equal to the angle ECF : I. 8.
and these are adjacent angles.

But when one straight line, standing on another, makes the adjacent angles equal, each of these angles is called a right angle ;   *Def. 10.*
therefore each of the angles DCF, ECF is a right angle.
Therefore CF is at right angles to AB,
and has been drawn from a point C in it.   Q.E.F.

EXERCISE.

In the figure of the above proposition, shew that *any* point in FC, or FC produced, is equidistant from D and E.

PROPOSITION 12. PROBLEM.

To draw a straight line perpendicular to a given straight line of unlimited length, from a given point without it.

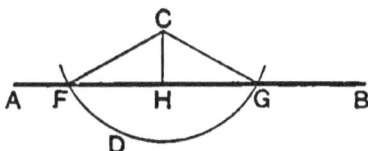

Let AB be the given straight line of unlimited length, and let C be the given point without it.

It is required to draw from C a straight line perpendicular to AB.

Construction. On the side of AB remote from C take any point D;
and with centre C, and radius CD, describe the circle FDG, cutting AB at F and G

Bisect FG at H; 1. 10.
and join CH.
Then shall CH be perpendicular to AB.
Join CF and CG.

Proof. Then in the triangles FHC, GHC,

Because
{
FH is equal to GH, *Constr.*
and HC is common to both;
and the third side CF is equal to the third side CG, being radii of the circle FDG; *Def.* 15.
}

therefore the angle CHF is equal to the angle CHG; I. 8.
and these are adjacent angles.

But when one straight line, standing on another, makes the adjacent angles equal, each of these angles is called a right angle, and the straight line which stands on the other is called a perpendicular to it. *Def.* 10.

Therefore CH is perpendicular to AB,
and has been drawn from the point C without it. Q.E.F.

NOTE. The line AB must be of unlimited length, that is, capable of production to an indefinite length in either direction, to ensure its being intersected in two points by the circle FDG.

QUESTIONS AND EXERCISES FOR REVISION.

1. Distinguish between a *problem* and a *theorem*.

2. When are two figures said to be *identically equal*? Under what conditions has it so far been proved that two *triangles* are identically equal?

3. Explain the method of proof known as *Reductio ad Absurdum*. Quote the enunciations of the propositions in which this method has so far been used.

4. Quote the corollaries of Propositions 5 and 6, and shew that each is the converse of the other.

5. What is meant by saying that Euclid's reasoning is *deductive*? Shew, for instance, that the proof of Proposition 5 is a deductive argument.

6. Two forts defend the mouth of a river, one on each side; the forts are 4000 yards apart, and their guns have a range of 3000 yards. Taking *one inch* to represent a length of 1000 yards, draw a diagram shewing what part of the river is exposed to the fire of both forts.

7. Define *the perimeter* of a rectilineal figure. A square and an equilateral triangle each have a perimeter of 3 feet: compare the lengths of their sides.

8. Shew how to draw a rhombus each of whose sides is equal to a given straight line PQ, which is also to be one diagonal of the figure.

9. A and B are two given points. Shew how to draw a rhombus having A and B as opposite vertices, and having each side equal to a given line PQ. Is this always possible?

10. Two circles are described with the same centre O; and two radii OA, OB are drawn to the inner circle, and produced to cut the outer circle at D and E: prove that

     (i)   DB = EA;
     (ii)   the angle BAD = the angle ABE;
     (iii) the angle ODB = the angle OEA.

EXERCISES ON PROPOSITIONS 1 TO 12.

1.   Shew that the straight line which joins the vertex of an isosceles triangle to the middle point of the base is perpendicular to the base.

2.   Shew that the straight lines which join the extremities of the base of an isosceles triangle to the middle points of the opposite sides, are equal to one another.

3.   Two given points in the base of an isosceles triangle are equi-distant from the extremities of the base : shew that they are also equidistant from the vertex.

4.   If the opposite sides of a quadrilateral are equal, shew that the opposite angles are also equal.

5.   Any two isosceles triangles XAB, YAB stand on the same base AB : shew that the angle XAY is equal to the angle XBY ; and if XY be joined, that the angle AXY is equal to the angle BXY.

6.   Shew that the opposite angles of a rhombus are bisected by the diagonal which joins them.

7.   Shew that the straight lines which bisect the base angles of an isosceles triangle form with the base a triangle which is also isosceles.

8.   ABC is an isosceles triangle having AB equal to AC ; and the angles at B and C are bisected by straight lines which meet at O : shew that OA bisects the angle BAC.

9.   Shew that the triangle formed by joining the middle points of the sides of an equilateral triangle is also equilateral.

10.   The equal sides BA, CA of an isosceles triangle BAC are pro-duced beyond the vertex A to the points E and F, so that AE is equal to AF ; and FB, EC are joined : shew that FB is equal to EC.

11.   Shew that the diagonals of a rhombus bisect one another at right angles.

12.   In the equal sides AB, AC of an isosceles triangle ABC two points X and Y are taken, so that AX is equal to AY ; and CX and BY are drawn intersecting in O : shew that

     (i)   the triangle BOC is isosceles ;
     (ii)   AO bisects the vertical angle BAC ;
     (iii)   AO, if produced, bisects BC at right angles.

13.   Describe an isosceles triangle, having given the base and the length of the perpendicular drawn from the vertex to the base.

14.   In a given straight line find a point that is equidistant from two given points.   In what case is this impossible ?

PROPOSITION 13. THEOREM.

The adjacent angles which one straight line makes with another straight line, on one side of it, are either two right angles or are together equal to two right angles.

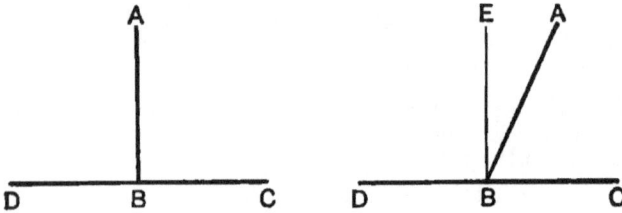

Let the straight line AB meet the straight line DC.

Then the adjacent angles DBA, ABC shall be either two right angles, or together equal to two right angles.

CASE I. For if the angle DBA is equal to the angle ABC, each of them is a right angle. *Def.* 10.

CASE II. But if the angle DBA is not equal to the angle ABC,

from B draw BE at right angles to CD. I. 11.

Proof. Now the angle DBA is made up of the two angles DBE, EBA;

to each of these equals add the angle ABC;
then the two angles DBA, ABC are together equal to the
three angles DBE, EBA, ABC. *Ax.* 2.

Again, the angle EBC is made up of the two angles EBA, ABC;

to each of these equals add the angle DBE;
then the two angles DBE, EBC are together equal to the
three angles DBE, EBA, ABC. *Ax.* 2.

But the two angles DBA, ABC have been shewn to be equal
to the same three angles;
therefore the angles DBA, ABC are together equal to the
angles DBE, EBC. *Ax.* 1.

But the angles DBE, EBC are two right angles; *Constr.*
therefore the angles DBA, ABC are together equal to two
right angles. Q. E. D.

DEFINITIONS.

(i) The **complement** of an acute angle is its *defect from* a right angle, that is, the angle by which it falls short of a right angle.

Thus two angles are **complementary,** when their sum is a right angle.

(ii) The **supplement** of an angle is its *defect from* two right angles, that is, the angle by which it falls short of two right angles.

Thus two angles are **supplementary,** when their sum is two right angles.

COROLLARY. *Angles which are complementary or supplementary to the same angle are equal to one another.*

EXERCISES.

1. If the two exterior angles formed by producing a side of a triangle both ways are equal, shew that the triangle is isosceles.

2. *The bisectors of the adjacent angles which one straight line makes with another contain a right angle.*

NOTE In the adjoining diagram AOB is a given angle; and one of its arms AO is produced to C : the adjacent angles AOB, BOC are bisected by OX, OY.

Then OX and OY are called respectively the **internal** and **external bisectors** of the angle AOB.

Hence Exercise 2 may be thus enunciated :

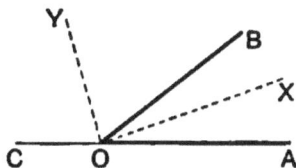

The internal and external bisectors of an angle are at right angles to one another.

3. Shew that the angles AOX and COY are complementary.

4. Show that the angles BOX and COX are supplementary ; and also that the angles AOY and BOY are supplementary.

PROPOSITION 14. THEOREM.

If, at a point in a straight line, two other straight lines, on opposite sides of it, make the adjacent angles together equal to two right angles, then these two straight lines shall be in one and the same straight line.

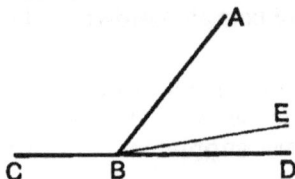

At the point B in the straight line AB, let the two straight lines BC, BD, on the opposite sides of AB, make the adjacent angles ABC, ABD together equal to two right angles.

Then BD shall be in the same straight line with BC.

Proof. For if BD be not in the same straight line with BC, if possible, let BE be in the same straight line with BC.

Then because AB meets the straight line CBE, therefore the adjacent angles CBA, ABE are together equal to two right angles. I. 13.
But the angles CBA, ABD are also together equal to two right angles. *Hyp.*
Therefore the angles CBA, ABE are together equal to the angles CBA, ABD. *Ax.* 11.
From each of these equals take the common angle CBA; then the remaining angle ABE is equal to the remaining angle ABD; the part equal to the whole; which is impossible.
Therefore BE is not in the same straight line with BC.
And in the same way it may be shewn that no other line but BD can be in the same straight line with BC.
Therefore BD is in the same straight line with BC. Q.E.D.

EXERCISE.

ABCD is a rhombus; and the diagonal AC is bisected at O. If O is joined to the angular points B and D; shew that OB and OD are in one straight line.

PROPOSITION 15. THEOREM.

If two straight lines intersect one another, then the vertically opposite angles shall be equal.

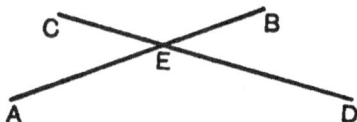

Let the two straight lines AB, CD cut one another at the point E.

Then (i) *the angle* AEC *shall be equal to the angle* DEB;
(ii) *the angle* CEB *shall be equal to the angle* AED.

Proof. Because AE meets the straight line CD,
therefore the adjacent angles CEA, AED are together equal
to two right angles.   I. 13.

Again, because DE meets the straight line AB,
therefore the adjacent angles AED, DEB are together equal
to two right angles.   I. 13.
Therefore the angles CEA, AED are together equal to the
angles AED, DEB.

From each of these equals take the common angle AED;
then the remaining angle CEA is equal to the remaining
angle DEB.   *Ax.* 3.

In the same way it may be proved that the angle CEB
is equal to the angle AED.   Q.E.D.

COROLLARY 1. *From this it follows that, if two straight lines cut one another, the four angles so formed are together equal to four right angles.*

COROLLARY 2. *Consequently, when any number of straight lines meet at a point, the sum of the angles made by consecutive lines is equal to four right angles.*

PROPOSITION 16. THEOREM.

If one side of a triangle be produced, then the exterior angle shall be greater than either of the interior opposite angles.

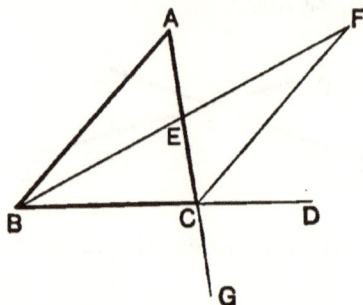

Let ABC be a triangle, and let BC be produced to D.

Then shall the exterior angle ACD *be greater than either of the interior opposite angles* ABC, BAC.

Construction.   Bisect AC at E;   I. 10.
Join BE; and produce it to F, making EF equal to BE. I. 3.
Join FC.

Proof.   Then in the triangles AEB, CEF,

Because
{
          AE is equal to CE,   *Constr.*
      and EB is equal to EF;   *Constr.*
also the angle AEB is equal to the vertically
opposite angle CEF;   I. 15.
}

therefore the triangle AEB is equal to the triangle CEF in all respects: I. 4.

    so that the angle BAE is equal to the angle ECF.

But the angle ECD is greater than its part, the angle ECF;

therefore the angle ECD is greater than the angle BAE;

that is, the angle ACD is greater than the angle BAC.

In the same way, if BC be bisected, and the side AC produced to G, it may be proved that the angle BCG is greater than the angle ABC.

    But the angle BCG is equal to the angle ACD: I. 15
therefore also the angle ACD is greater than the angle ABC.

                                          Q.E.D.

PROPOSITION 17. THEOREM.

Any two angles of a triangle are together less than two right angles.

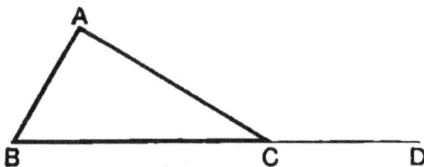

Let ABC be a triangle.

Then shall any two of the angles of the triangle ABC be together less·than two right angles.

Construction. Produce the side BC to D.

Proof. Then because BC, a side of the triangle ABC, is produced to D;
therefore the exterior angle ACD is greater than the interior opposite angle ABC.   I. 16.
To each of these add the angle ACB:
then the angles ACD, ACB are together greater than the angles ABC, ACB.   *Ax.* 4.
But the adjacent angles ACD, ACB are together equal to two right angles.   I. 13.
Therefore the angles ABC, ACB are together less than two right angles.
Similarly it may be shewn that the angles BAC, ACB, as also the angles CAB, ABC, are together less than two right angles.   Q.E.D.

NOTE. It follows from this Proposition that *every triangle must have at least two acute angles :* for if one angle is obtuse, or a right angle; each of the other angles must be less than a right angle.

EXERCISES.

1. Enunciate this Proposition so as to shew that it is the converse of Axiom 12.

2. If any side of a triangle is produced both ways, the exterior angles so formed are together greater than two right angles.

3. Shew how a proof of Proposition 17 may be obtained by joining each vertex in turn to any point in the opposite side.

PROPOSITION 18.   THEOREM.

If one side of a triangle be greater than another, then the angle opposite to the greater side shall be greater than the angle opposite to the less.

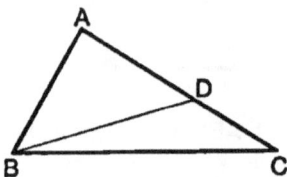

Let ABC be a triangle, in which the side AC is greater than the side AB.

Then shall the angle ABC be greater than the angle ACB.

Construction.   From AC cut off a part AD equal to AB.   I. 3.
Join BD.

Proof.   Then in the triangle ABD,
                because AB is equal to AD,
therefore the angle ABD is equal to the angle ADB.   I. 5.

But the exterior angle ADB of the triangle DCB is greater than the interior opposite angle DCB, that is, greater than the angle ACB.   I. 16.

Therefore also the angle ABD is greater than the angle ACB; still more then is the angle ABC greater than the angle ACB.   Q.E.D.

Euclid enunciated Proposition 18 as follows :

The greater side of every triangle has the greater angle opposite to it.

[This form of enunciation is found to be a common source of difficulty with beginners, who fail to distinguish what is *assumed* in it and what is *to be proved*. If Euclid's enunciations of Props. 18 and 19 are adopted, it is important to remember that in each case the part of the triangle *first named* points out the hypothesis.]

PROPOSITION 19. THEOREM.

If one angle of a triangle be greater than another, then the side opposite to the greater angle shall be greater than the side opposite to the less.

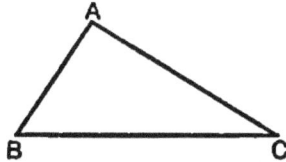

Let ABC be a triangle in which the angle ABC is greater than the angle ACB.

Then shall the side AC be greater than the side AB.

Proof.   For if AC be not greater than AB,
        it must be either equal to, or less than AB.

But AC is not equal to AB,
for then the angle ABC would be equal to the angle ACB; I. 5.
        but it is not.   *Hyp.*

Neither is AC less than AB;
for then the angle ABC would be less than the angle ACB; I. 18.
        but it is not.   *Hyp.*

That is, AC is neither equal to, nor less than AB.
        Therefore AC is greater than AB.   Q.E.D.

NOTE. The mode of demonstration used in this Proposition is known as the **Proof by Exhaustion**. It is applicable to cases in which *one* of certain suppositions must necessarily be true ; and it consists in shewing that each of these suppositions is false *with one exception* : hence the truth of the remaining supposition is inferred.

Euclid enunciated Proposition 19 as follows :

The greater angle of every triangle is subtended by the greater side, or, has the greater side opposite to it.

[For Exercises on Props. 18 and 19 see page 44.]
E.C.   s

PROPOSITION 20. THEOREM.

Any two sides of a triangle are together greater than the third side.

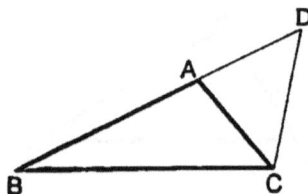

Let ABC be a triangle.

Then shall any two of its sides be together greater than the third side :

namely, BA, AC, *shall be greater than* CB ;

AC, CB *shall be greater than* BA ;

and CB, BA *shall be greater than* AC.

Construction.   Produce BA to D, making AD equal to AC. I. 3.
Join DC.

Proof.   Then in the triangle ADC,
because AD is equal to AC,   *Constr.*
therefore the angle ACD is equal to the angle ADC. I. 5.
But the angle BCD is greater than its part the angle ACD ;
therefore also the angle BCD is greater than the angle ADC,
that is, than the angle BDC.

And in the triangle BCD,
because the angle BCD is greater than the angle BDC,
therefore the side BD is greater than the side CB.   I. 19.

But BA and AC are together equal to BD ;
therefore BA and AC are together greater than CB.

Similarly it may be shewn
that AC, CB are together greater than BA ;
and CB, BA are together greater than AC.   Q.E.D.

[For Exercises see page 44.]

PROPOSITION 21. THEOREM.

If from the ends of a side of a triangle, there be drawn two straight lines to a point within the triangle, then these straight lines shall be less than the other two sides of the triangle, but shall contain a greater angle.

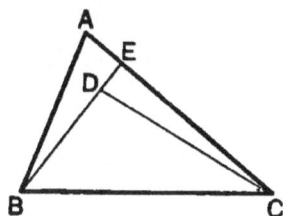

Let ABC be a triangle, and from B, C, the ends of the side BC, let the straight lines BD, CD be drawn to a point D within the triangle

Then (i) BD *and* DC *shall be together less than* BA *and* AC ;

(ii) *the angle* BDC *shall be greater than the angle* BAC.

Construction. Produce BD to meet AC in E.

Proof. (i) In the triangle BAE, the two sides BA, AE are together greater than the third side BE ;   I. 20.

to each of these add EC ;

then BA, AC are together greater than BE, EC. *Ax.* 4.

Again, in the triangle DEC, the two sides DE, EC are together greater than DC ;   I. 20.

to each of these add BD ;

then BE, EC are together greater than BD, DC.

But it has been shewn that BA, AC are together greater than BE, EC :

still more then are BA, AC greater than BD, DC.

(ii) Again, the exterior angle BDC of the triangle DEC is greater than the interior opposite angle DEC ;   I. 16.

and the exterior angle DEC of the triangle BAE is greater than the interior opposite angle BAE, that is, than the angle BAC ;   I. 16.

still more then is the angle BDC greater than the angle BAC.

Q.E.D.

ON PROPOSITIONS 18 AND 19.

1.   The hypotenuse is the greatest side of a right-angled triangle.

2.   If two angles of a triangle are equal to one another, the sides also, which subtend the equal angles, are equal to one another. Prove this [*i.e.* Prop. 6] indirectly by using the result of Prop. 18.

3.   BC, the base of an isosceles triangle ABC, is produced to any point D ; shew that AD is greater than either of the equal sides.

4.   If in a quadrilateral the greatest and least sides are opposite to one another, then each of the angles adjacent to the least side is greater than its opposite angle.

5.   In a triangle ABC, if AC is not greater than AB, shew that any straight line drawn through the vertex A and terminated by the base BC, is less than AB.

6.   ABC is a triangle, in which OB, OC bisect the angles ABC, ACB respectively : shew that, if AB is greater than AC, then OB is greater than OC.

ON PROPOSITION 20.

7.   The difference of any two sides of a triangle is less than the third side.

8.   In a quadrilateral, if two opposite sides which are not parallel are produced to meet one another ; shew that the perimeter of the greater of the two triangles so formed is greater than the perimeter of the quadrilateral.

9.   The sum of the distances of any point from the three angular points of a triangle is greater than half its perimeter.

10.   The perimeter of a quadrilateral is greater than the sum of its diagonals.

11.   Obtain a proof of Proposition 20 by bisecting an angle by a straight line which meets the opposite side.

ON PROPOSITION 21.

12.   In Proposition 21 shew that the angle BDC is greater than the angle BAC by joining AD, and producing it towards the base.

13.   The sum of the distances of any point within a triangle from its angular points is less than the perimeter of the triangle.

QUESTIONS FOR REVISION.

1. Define the *complement* of an angle. When are two angles said to be *supplementary*? Shew that two angles which are supplementary to the same angle are equal to one another.

2. What is meant by an angle being *bisected internally and externally*?

Prove that the internal and external bisectors of an angle are at right angles to one another.

3. Prove that the sum of the angles formed by any number of straight lines drawn from a point is equal to four right angles.

4. Why must every triangle have *at least two acute angles*? Quote the enunciation of the proposition from which this inference is drawn.

5. In the enunciation *The greater side of a triangle has the greater angle opposite to it*, point out what is assumed and what is to be proved.

6. What is meant by the *Proof by Exhaustion*? Illustrate the use of this method by naming the steps in the proof of Proposition 19.

7. What inference may be drawn respecting the triangles whose sides measure

(i)   4 inches,   5 inches,   4 inches ;

(ii)   8 inches,   9 inches,   10 inches ;

(iii)   6 inches,   10 inches,   4 inches ?

8. Quote the enunciations of propositions which, from a hypothesis relating to the *sides* of triangle, establish a conclusion relating to the *angles*.

9. Quote the enunciations of propositions which, from a hypothesis relating to the *angles* of a triangle, establish a conclusion relating to the sides.

10. Explain why parallel straight lines must be *in the same plane*.

11. Prove by means of Prop. 7 that on a given base and on the same side of it only one equilateral triangle can be drawn.

12. In an isosceles triangle, if the equal sides are produced, shew that the angles on the other side of the base must be obtuse.

PROPOSITION 22. PROBLEM.

To describe a triangle having its sides equal to three given straight lines, any two of which are together greater than the third.

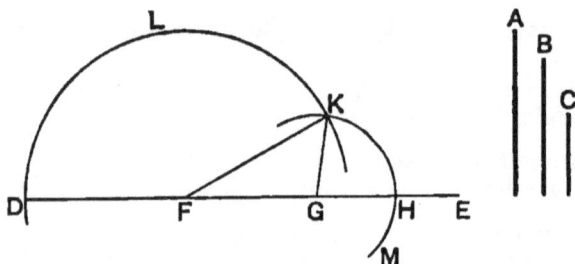

Let A, B, C be the three given straight lines, of which any two are together greater than the third.

It is required to describe a triangle of which the sides shall be equal to A, B, C.

Construction. Take a straight line DE terminated at the point D, but unlimited towards E.

Make DF equal to A, FG equal to B, and GH equal to C. I. 3.

With centre F and radius FD, describe the circle DLK.

With centre G and radius GH, describe the circle MHK cutting the former circle at K.

Join FK, GK.

Then shall the triangle KFG have its sides equal to the three straight lines A, B, C.

Proof. Because F is the centre of the circle DLK,

therefore FK is equal to FD : *Def.* 15.

but FD is equal to A ; *Constr.*

therefore also FK is equal to A. *Ax.* 1.

Again, because G is the centre of the circle MHK,

therefore GK is equal to GH : *Def.* 15.

but GH is equal to C ; *Constr.*

therefore also GK is equal to C. *Ax.* 1.

And FG is equal to B. *Constr.*

Therefore the triangle KFG has its sides KF, FG, GK equal respectively to the three given lines A, B, C. Q.E.F.

PROPOSITION 23. PROBLEM.

At a given point in a given straight line, to make an angle equal to a given rectilineal angle.

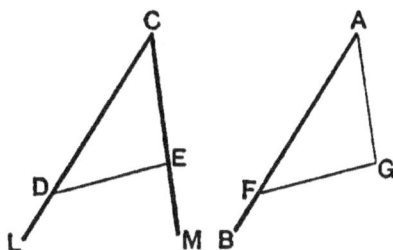

Let AB be the given straight line, and A the given point in it, and let LCM be the given angle.

It is required to draw from A a straight line making with AB an angle equal to the given angle DCE.

Construction. In CL, CM take any points D and E; and join DE.
From AB cut off AF equal to CD.   I. 3.
On AF describe the triangle FAG, having the remaining sides AG, GF equal respectively to CE, ED.   I. 22.
Then shall the angle FAG be equal to the angle DCE.

Proof.   For in the triangles FAG, DCE,
Because {   FA is equal to DC,   *Constr.*
              and AG is equal to CE ;   *Constr.*
              and the base FG is equal to the base DE: *Constr.*
therefore the angle FAG is equal to the angle DCE.   I. 8.

That is, AG makes with AB, at the given point A. an angle equal to the given angle DCE.   Q.E.F.

EXERCISE.

On a given base describe a triangle, whose remaining sides shall be equal to two given straight lines. Point out how the construction fails, if any one of the three given lines is greater than the sum of the other two.

Proposition 24. Theorem.

If two triangles have two sides of the one equal to two sides of the other, each to each, but the angle contained by the two sides of one greater than the angle contained by the corresponding sides of the other ; then the base of that which has the greater angle shall be greater than the base of the other.

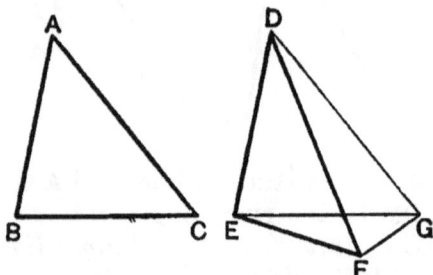

Let ABC, DEF be two triangles, in which
the side BA is equal to the side ED,
and the side AC is equal to the side DF,
but the angle BAC is greater than the angle EDF.
Then shall the base BC be greater than the base EF.

Of the two sides DE, DF, let DE be that which is not greater than the other.*

Construction. At D in the straight line ED, and on the same side of it as DF, make the angle EDG equal to the angle BAC.　　　　　　　　　　　　　　　I. 23.
Make DG equal to DF or AC ;　　　I. 3.
and join EG, GF.

Proof. Then in the triangles BAC, EDG,
Because ⎰ BA is equal to ED,　　　　　　　　　*Hyp.*
and AC is equal to DG,　　　　　*Constr.*
also the contained angle BAC is equal to the contained angle EDG ;　　　　*Constr.*
therefore the triangle BAC is equal to the triangle EDG in all respects :　　　　　　　　　　　　　I. 4.
so that the base BC is equal to the base EG.

Again, in the triangle FDG,
because DG is equal to DF,
therefore the angle DFG is equal to the angle DGF. I. 5.

But the angle DGF is greater than its part the angle EGF;
therefore also the angle DFG is greater than the angle EGF;
still more then is the angle EFG greater than the angle EGF.

And in the triangle EFG,
because the angle EFG is greater than the angle EGF,
therefore the side EG is greater than the side EF; I. 19.
but EG was shewn to be equal to BC;
therefore BC is greater than EF. Q.E.D.

---

*The object of this step is to make the point F fall *below* EG. Otherwise F might fall *above*, *upon*, or *below* EG; and each case would require separate treatment. But as it is not *proved* that this condition fulfils its object, this demonstration of Prop. 24 must be considered defective.

An alternative construction and proof are given below.

Construction. At D in ED make the angle EDG equal to the angle BAC; and make DG equal to DF. Join EG.

Then, as before, it may be shewn that the triangle EDG = the triangle BAC in all respects.

Now if EG passes through F, then EG is greater than EF; that is, BC is greater than EF.

But if not, bisect the angle FDG by DK, meeting EG at K. Join FK.

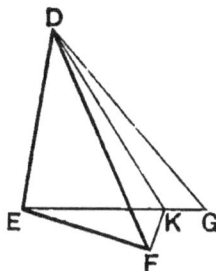

Proof. Then in the triangles FDK, GDK,

Because { FD = GD,
and DK is common to both,
and the angle FDK = the angle GDK; *Constr.*
∴ FK = GK. I. 4.

But in the triangle EKF, the two sides EK, KF are greater than EF;
that is, EK, KG are greater than EF.
Hence EG (or BC) is greater than EF.

PROPOSITION 25. THEOREM.

If two triangles have two sides of the one equal to two sides of the other, each to each, but the base of one greater than the base of the other ; then the angle contained by the sides of that which has the greater base, shall be greater than the angle contained by the corresponding sides of the other.

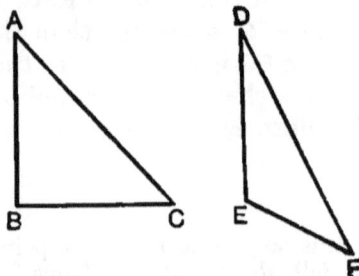

Let ABC, DEF be two triangles in which
the side BA is equal to the side ED,
and the side AC is equal to the side DF,
but the base BC is greater than the base EF.
Then shall the angle BAC be greater than the angle EDF.

Proof. For if the angle BAC be not greater than the angle EDF, it must be either equal to, or less than the angle EDF.

But the angle BAC is not equal to the angle EDF,
for then the base BC would be equal to the base EF; I. 4.
but it is not.   *Hyp.*
Neither is the angle BAC less than the angle EDF,
for then the base BC would be less than the base EF; I. 24.
but it is not.   *Hyp.*
Therefore the angle BAC is neither equal to, nor less than the angle EDF ;
that is, the angle BAC is greater than the angle EDF. Q.E.D.

EXERCISE.

In a triangle ABC, the vertex A is joined to X, the middle point of the base BC ; shew that the angle AXB is obtuse or acute, according as AB is greater or less than AC.

Proposition 26. Theorem.

If two triangles have two angles of the one equal to two angles of the other, each to each, and a side of one equal to a side of the other, these sides being either adjacent to the equal angles, or opposite to equal angles in each; then shall the triangles be equal in all respects.

Case I. When the equal sides are *adjacent* to the equal angles in the two triangles.

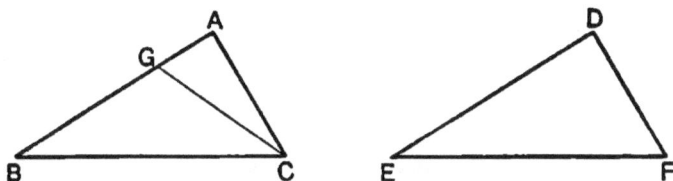

Let ABC, DEF be two triangles, in which
the angle ABC is equal to the angle DEF,
and the angle ACB is equal to the angle DFE,
and the side BC is equal to the side EF.
Then shall the triangle ABC be equal to the triangle DEF in all respects; that is, AB shall be equal to DE, and AC to DF, and the angle BAC shall be equal to the angle EDF.

For if AB be not equal to DE, one must be greater than the other. If possible, let AB be greater than DE.

Construction. From BA cut off BG equal to ED,   I. 3.
and join GC.

Proof. Then in the two triangles GBC, DEF,

Because
{
GB is equal to DE,   *Constr.*
and BC is equal to EF,   *Hyp.*
also the contained angle GBC is equal to the contained angle DEF;   *Hyp.*
}

therefore the triangle GBC is equal to the triangle DEF in all respects;   I. 4.
so that the angle GCB is equal to the angle DFE.
But the angle ACB is equal to the angle DFE;   *Hyp.*
therefore also the angle GCB is equal to the angle ACB; *Ax.* 1.
the part equal to the whole, which is impossible.

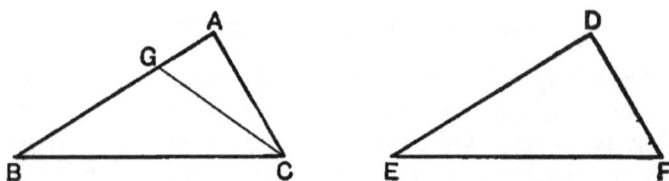

Therefore AB is not unequal to DE;
that is, AB is equal to DE.

Hence in the triangles ABC, DEF,

	AB is equal to DE,	*Proved.*
Because	and BC is equal to EF;	*Hyp.*
	also the contained angle ABC is equal to the contained angle DEF;	*Hyp.*

therefore the triangle ABC is equal to the triangle DEF in all respects :   I. 4.

so that the side AC is equal to the side DF;
and the angle BAC is equal to the angle EDF.

Q.E.D.

CASE II. When the equal sides are *opposite* to equal angles in the two triangles.

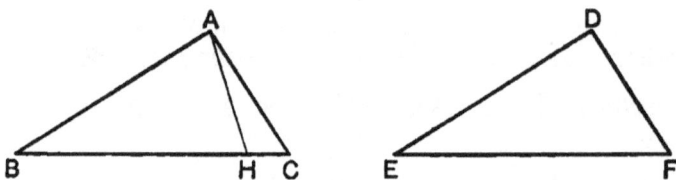

Let ABC, DEF be two triangles, in which
the angle ABC is equal to the angle DEF,
and the angle ACB is equal to the angle DFE,
and the side AB is equal to the side DE.

Then the triangle ABC shall be equal to the triangle DEF in all respects ;

namely, BC shall be equal to EF,
and AC shall be equal to DF,
and the angle BAC shall be equal to the angle EDF.

For if BC be not equal to EF, one must be greater than the other. If possible, let BC be greater than EF.

Construction. From BC cut off BH equal to EF,   I. 3.
and join AH.

Proof. Then in the triangles ABH, DEF,

Because
{
 AB is equal to DE,   *Hyp.*
 and BH is equal to EF,   *Constr.*
 also the contained angle ABH is equal to the contained angle DEF;   *Hyp.*
}

therefore the triangle ABH is equal to the triangle DEF in all respects;   I. 4.

so that the angle AHB is equal to the angle DFE.

But the angle DFE is equal to the angle ACB; *Hyp.*
therefore the angle AHB is equal to the angle ACB; *Ax.* 1.
that is, an exterior angle of the triangle ACH is equal to an interior opposite angle; which is impossible.   I. 16.

Therefore BC is not unequal to EF,
that is, BC is equal to EF.

Hence in the triangles ABC, DEF,

Because
{
 AB is equal to DE,   *Hyp.*
 and BC is equal to EF;   *Proved.*
 also the contained angle ABC is equal to the contained angle DEF;   *Hyp.*
}

therefore the triangle ABC is equal to the triangle DEF in all respects;   I. 4.

so that the side AC is equal to the side DF,
and the angle BAC is equal to the angle EDF.

Q.E.D.

COROLLARY. *In both cases of this Proposition it is seen that the triangles may be made to coincide with one another; and they are therefore equal in area.*

ON THE IDENTICAL EQUALITY OF TRIANGLES.

Three cases have been already dealt with in Propositions 4, 8, and 26, the results of which may be summarized as follows :

Two triangles are equal in all respects when the following three parts in each are severally equal :

1. Two sides, and the included angle.   *Prop.* 4.

2. The three sides.   *Prop.* 8, *Cor.*

3. (*a*) Two angles, and the adjacent side ;   } *Prop.* 26.
    (*b*) Two angles, and a side opposite one of them.

Two triangles are not, however, necessarily equal in all respects when *any three parts* of one are equal to the corresponding parts of the other.   For example

(i)   When the *three angles* of one are equal to the *three angles* of the other, each to each, the adjoining diagram shews that the triangles need not be equal in all respects.

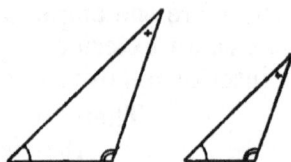

(ii)   When *two sides and one angle* in one are equal to *two sides and one angle* in the other, the given angles being *opposite* to equal sides, the diagram shews that the triangles need not be equal in all respects.
For it will be seen that if AB = DE, and AC = DF, and the angle ABC = the angle DEF, then the shorter of the given sides in the triangle DEF may lie in either of the positions DF or DF'.

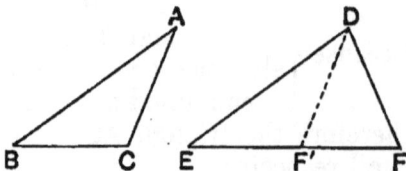

In cases (i) and (ii) a further condition must be given before we can prove that the two triangles are identically equal.
[See Theorem B, p. 96.]

We observe that in each of the three cases in which two triangles have been proved equal in all respects, namely in Propositions 4, 8, 26, it is shewn that the triangles may be made to *coincide with one another* ; so that they are equal in *area*.   Euclid however restricted himself to the use of Prop. 4, when he required to deduce the equality in *area* of two triangles from the equality of certain of their parts.   This restriction is now generally abandoned.

EXERCISES ON PROPOSITIONS 12-26.

1.   If **BX** and **CY**, the bisectors of the angles at the base **BC** of an isosceles triangle **ABC**, meet the opposite sides in **X** and **Y**, shew that the triangles **YBC, XCB** are equal in all respects.

2.   Shew that the perpendiculars drawn from the extremities of the base of an isosceles triangle to the opposite sides are equal.

3.   *Any point on the bisector of an angle is equidistant from the arms of the angle.*

4.   Through **O**, the middle point of a straight line **AB**, any straight line is drawn, and perpendiculars **AX** and **BY** are dropped upon it from **A** and **B** : shew that **AX** is equal to **BY**.

5.   If the bisector of the vertical angle of a triangle is at right angles to the base, the triangle is isosceles.

6.   *The perpendicular is the shortest straight line that can be drawn from a given point to a given straight line ; and of others, that which is nearer to the perpendicular is less than the more remote ; and two, and only two equal straight lines can be drawn from the given point to the given straight line, one on each side of the perpendicular.*

7.   *From two given points on the same side of a given straight line, draw two straight lines, which shall meet in the given straight line, and make equal angles with it.*

Let **AB** be the given straight line, and **P, Q** the given points.

It is required to draw from **P** and **Q** to a point in **AB**, two straight lines that shall be equally inclined to **AB**.

Construction.   From **P** draw **PH** perpendicular to **AB** : produce **PH** to **P′**, making **HP′** equal to **PH**.   Draw **QP′**, meeting **AB** in **K**.   Join **PK**.

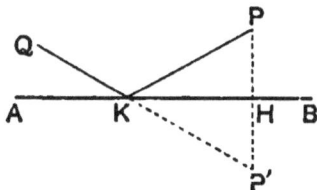

Then **PK, QK** shall be the required lines.   [Supply the proof.]

8.   In a given straight line find a point which is equidistant from two given intersecting straight lines.   In what case is this impossible?

9.   Through a given point draw a straight line such that the perpendiculars drawn to it from two given points may be equal.

In what case is this impossible?

SECTION II.

PARALLEL STRAIGHT LINES AND PARALLELOGRAMS.

DEFINITION.   Parallel straight lines are such as, being in the same plane, do not meet however far they are produced in both directions.

When two straight lines AB, CD are met by a third straight line EF, *eight* angles are formed, to which for the sake of distinction particular names are given.

Thus in the adjoining figure,

1, 2, 7, 8 are called **exterior** angles,

3, 4, 5, 6 are called **interior** angles,

4 and 6 are said to be **alternate** angles ; so also the angles 3 and 5 are alternate to one another.

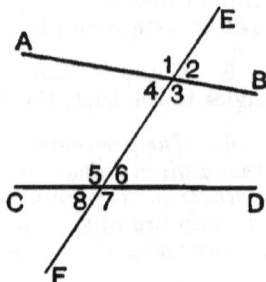

Of the angles 2 and 6, 2 is referred to as the exterior angle, and 6 as the **interior opposite** angle on the same side of EF.

2 and 6 are sometimes called **corresponding** angles.

So also, 1 and 5, 7 and 3, 8 and 4 are corresponding angles.

Euclid's treatment of parallel straight lines is based upon his twelfth Axiom, which we here repeat.

AXIOM 12.   If a straight line cut two straight lines so as to make the two interior angles on the same side of it together less than two right angles, these straight lines, being continually produced, will at length meet on that side on which are the angles which are together less than two right angles.

Thus in the figure given above, if the two angles 3 and 6 are together less than two right angles, it is asserted that AB and CD will meet towards B and D.

This Axiom is used to establish I. 29 : some remarks upon it will be found in a note on that Proposition.

PROPOSITION 27. THEOREM.

If a straight line, falling on two other straight lines, make the alternate angles equal to one another, then these two straight lines shall be parallel.

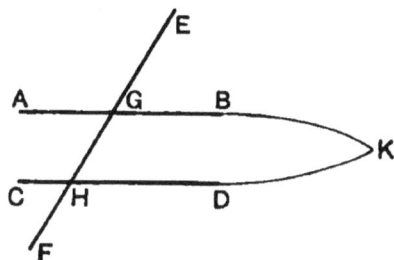

Let the straight line EF cut the two straight lines AB, CD at G and H, so as to make the alternate angles AGH, GHD equal to one another.

Then shall AB *and* CD *be parallel.*

Proof. For if AB and CD be not parallel,
they will meet, if produced, either towards B and D, or towards A and C.

If possible, let AB and CD, when produced, meet towards B and D, at the point K.

Then KGH is a triangle, of which one side KG is produced to A ;
therefore the exterior angle AGH is greater than the interior opposite angle GHK. I. 16.

But the angle AGH was given equal to the angle GHK : *Hyp.*
hence the angles AGH and GHK are both equal and unequal ;
which is impossible.

Therefore AB and CD cannot meet when produced towards B and D.

Similarly it may be shewn that they cannot meet towards A and C :

therefore AB and CD are parallel. Q.E.D.

E.C. T

PROPOSITION 28. THEOREM.

If a straight line, falling on two other straight lines, make an exterior angle equal to the interior opposite angle on the same side of the line; or if it make the interior angles on the same side together equal to two right angles, then the two straight lines shall be parallel.

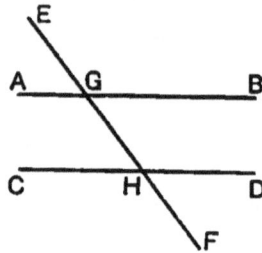

Let the straight line EF cut the two straight lines AB, CD in G and H : and

First, let the exterior angle EGB be equal to the interior opposite angle GHD.

Then shall AB *and* CD *be parallel.*

Proof. Because the angle EGB is equal to the angle GHD ; and because the angle EGB is also equal to the vertically opposite angle AGH ;   I. 15.
therefore the angle AGH is equal to the angle GHD ;
            but these are alternate angles ;
            therefore AB and CD are parallel.   I. 27.

Q.E.D.

Secondly, let the two interior angles BGH, GHD be together equal to two right angles.

Then shall AB *and* CD *be parallel.*

Proof. Because the angles BGH, GHD are together equal to two right angles ;   *Hyp.*
and because the adjacent angles BGH, AGH are also together equal to two right angles ;   I. 13.
therefore the angles BGH, AGH are together equal to the two angles BGH, GHD.

From these equals take the common angle BGH :
then the remaining angle AGH is equal to the remaining angle GHD : and these are alternate angles ;
            therefore AB and CD are parallel.   I. 27.

Q.E.D.

PROPOSITION 29. THEOREM.

If a straight line fall on two parallel straight lines, then it shall make the alternate angles equal to one another, and the exterior angle equal to the interior opposite angle on the same side ; and also the two interior angles on the same side equal to two right angles.

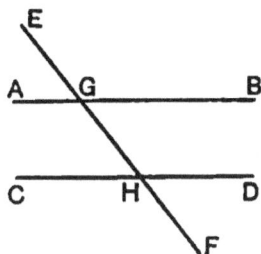

Let the straight line EF fall on the parallel straight lines AB, CD.

Then (i) *the angle* AGH *shall be equal to the alternate angle* GHD ;

  (ii) *the exterior angle* EGB *shall be equal to the interior opposite angle* GHD ;

  (iii) *the two interior angles* BGH, GHD *shall be together equal to two right angles.*

Proof. (i) For if the angle AGH be not equal to the angle GHD, one of them must be greater than the other.

If possible, let the angle AGH be greater than the angle GHD ;

    add to each the angle BGH :

then the angles AGH, BGH are together greater than the angles BGH, GHD.

But the adjacent angles AGH, BGH are together equal to two right angles ;   I. 13.

therefore the angles BGH, GHD are together less than two right angles ;

therefore, *by Axiom 12*, AB and CD meet towards B and D.

But they never meet, since they are parallel. *Hyp.*

Therefore the angle AGH is not unequal to the angle GHD : that is, the angle AGH is equal to the alternate angle GHD.

                 (*Over*)

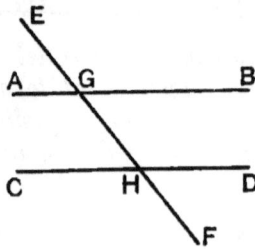

(ii) Again. because the angle AGH is equal to the vertically
   opposite angle EGB ;   I. 15.
   and because the angle AGH is equal to the angle GHD ;
                                                      Proved.

therefore the exterior angle EGB is equal to the interior
   opposite angle GHD.

(iii) Lastly, the angle EGB is equal to the angle GHD ;
                                                      Proved.

            add to each the angle BGH ;
then the angles EGB, BGH are together equal to the angles
   BGH, GHD.
But the adjacent angles EGB, BGH are together equal to
   two right angles :   I. 13.
therefore also the two interior angles BGH, GHD are together equal to two right angles.   Q.E.D.

EXERCISES ON PROPOSITIONS 27, 28, 29.

1. Two straight lines AB, CD bisect one another at O : shew
that the straight lines joining AC and BD are parallel.   [I. 27.]

2. *Straight lines which are perpendicular to the same straight line
are parallel to one another.*   [I. 27 or I. 28.]

3. *If a straight line meet two or more parallel straight lines, and is
perpendicular to one of them, it is also perpendicular to all the others.*
                                                           [I. 29.]

4. *If two straight lines are parallel to two other straight lines, each
to each, then the angles contained by the first pair are equal respectively
to the angles contained by the second pair.*   [I. 29.]

NOTE ON THE TWELFTH AXIOM.

Euclid's twelfth Axiom is unsatisfactory as the basis of a theory of parallel straight lines. It cannot be regarded as either simple or self-evident, and it therefore falls short of the essential characteristics of an axiom : nor is the difficulty entirely removed by considering it as a corollary to Proposition 17, of which it is the converse.

Of the many substitutes which have been proposed, we need only notice the following :

AXIOM. *Two intersecting straight lines cannot be both parallel to a third straight line.*

This statement is known as **Playfair's Axiom** ; and though it is not altogether free from objection, it is no doubt simpler and more fundamental than that employed by Euclid, and more readily admitted without proof.

Propositions 27 and 28 having been proved in the usual way, the first part of Proposition 29 is then given thus.

PROPOSITION 29. [ALTERNATIVE PROOF.]

If a straight line fall on two parallel straight lines, then it shall make the alternate angles equal.

Let the straight line EF meet the two parallel straight lines AB, CD at G and H.

Then shall the alternate angles AGH, GHD *be equal.*

For if the angle AGH is not equal to the angle GHD :

at G in the straight line HG make the angle HGP equal to the angle GHD, and alternate to it.   I. 23.

Then PG and CD are parallel.   I. 27.

But AB and CD are parallel :   *Hyp.*

therefore the two intersecting straight lines AG, PG are both parallel to CD :

which is impossible.   *Playfair's Axiom.*

Therefore the angle AGH is not unequal to the angle GHD ; that is, the alternate angles AGH, GHD are equal.   Q.E.D.

The second and third parts of the Proposition may then be deduced as in the text ; and Euclid's Axiom 12 follows as a Corollary.

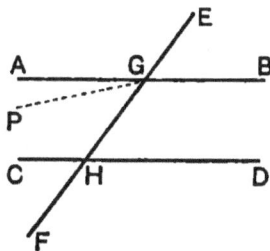

PROPOSITION 30.   THEOREM.

Straight lines which are parallel to the same straight line are parallel to one another.

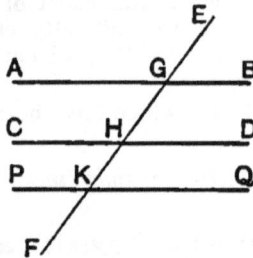

Let the straight lines AB, CD be each parallel to the straight line PQ.

Then shall AB *and* CD *be parallel to one another.*

Construction.   Draw any straight line EF cutting AB, CD, and PQ in the points G, H, and K.

Proof.   Then because AB and PQ are parallel, and EF meets them,
therefore the angle AGK is equal to the alternate angle GKQ
                                                                    I. 29.

And because CD and PQ are parallel, and EF meets them,
therefore the exterior angle GHD is equal to the interior opposite angle GKQ.   I. 29.

Therefore the angle AGH is equal to the angle GHD ;

and these are alternate angles ;

therefore AB and CD are parallel.   I. 27.

                                                            Q.E.D.

NOTE.   If PQ lies between AB and CD, the Proposition may be established in a similar manner, though in this case it scarcely needs proof ; for it is inconceivable that two straight lines, which do not meet an intermediate straight line, should meet one another.

The truth of this Proposition may be readily deduced from Playfair's Axiom, of which it is the converse.

For if AB and CD were not parallel, they would meet when produced.   Then there would be two intersecting straight lines both parallel to a third straight line : which is impossible.

Therefore AB and CD never meet ; that is, they are parallel.

PROPOSITION 31. PROBLEM.

To draw a straight line through a given point parallel to a given straight line.

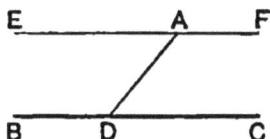

Let A be the given point, and BC the given straight line.
It is required to draw through A a straight line parallel to BC.

Construction. In BC take any point D ; and join AD.
At the point A in DA, make the angle DAE equal to the
angle ADC, and alternate to it, I. 23.
and produce EA to F.
Then shall EF be parallel to BC.

Proof. Because the straight line AD, meeting the two
straight lines EF, BC, makes the alternate angles EAD, ADC
equal ; *Constr.*
therefore EF is parallel to BC ; I. 27.
and it has been drawn through the given point A.
Q.E.F.

EXERCISES.

1. Any straight line drawn parallel to the base of an isosceles
triangle makes equal angles with the sides.

2. If from any point in the bisector of an angle a straight line is
drawn parallel to either arm of the angle, the triangle thus formed
is isosceles.

3. From a given point draw a straight line that shall make with
a given straight line an angle equal to a given angle.

4. From X, a point in the base BC of an isosceles triangle ABC,
a straight line is drawn at right angles to the base, cutting AB in Y,
and CA produced in Z : shew the triangle AYZ is isosceles.

5. If the straight line which bisects an exterior angle of a triangle
is parallel to the opposite side, shew that the triangle is isosceles.

PROPOSITION 32. THEOREM.

If a side of a triangle be produced, then the exterior angle shall be equal to the sum of the two interior opposite angles; also the three interior angles of a triangle are together equal to two right angles.

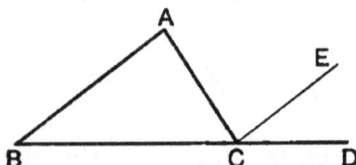

Let ABC be a triangle, and let one of its sides BC be produced to D.

Then (i) *the exterior angle* ACD *shall be equal to the sum of the two interior opposite angles* CAB, ABC ;

    (ii) *the three interior angles* ABC, BCA, CAB *shall be together equal to two right angles.*

Construction.   Through C draw CE parallel to BA. ı. 31.

Proof. (i) Then because BA and CE are parallel, and AC meets them,
therefore the angle ACE is equal to the alternate angle CAB.   ı. 29.

Again, because BA and CE are parallel, and BD meets them,
therefore the exterior angle ECD is equal to the interior opposite angle ABC.   ı. 29.
Therefore the whole exterior angle ACD is equal to the sum of the two interior opposite angles CAB, ABC.

    (ii) Again, since the angle ACD is equal to the sum of the angles CAB, ABC ;   *Proved.*
        to each of these equals add the angle BCA :
then the angles BCA, ACD are together equal to the three angles BCA, CAB, ABC.
But the adjacent angles BCA, ACD are together equal to two right angles.   ı. 13.
Therefore also the angles BCA, CAB, ABC are together equal to two right angles.   Q.E.D.

From this Proposition we draw the following important inferences.

1. *If two triangles have two angles of the one equal to two angles of the other, each to each, then the third angle of the one is equal to the third angle of the other.*

2. *In any right-angled triangle the two acute angles are complementary.*

3. *In a right-angled isosceles triangle each of the equal angles is half a right angle.*

4. *If one angle of a triangle is equal to the sum of the other two, the triangle is right-angled.*

5. *The sum of the angles of any quadrilateral figure is equal to four right angles.*

6. *Each angle of an equilateral triangle is two-thirds of a right angle.*

EXERCISES ON PROPOSITION 32.

1. Prove that the three angles of a triangle are together equal to two right angles,

> (i) by drawing through the vertex a straight line parallel to the base ;
> (ii) by joining the vertex to any point in the base.

2. If the base of any triangle is produced both ways, shew that the sum of the two exterior angles diminished by the vertical angle is equal to two right angles.

3. *If two straight lines are perpendicular to two other straight lines, each to each, the acute angle between the first pair is equal to the acute angle between the second pair.*

4. *Every right-angled triangle is divided into two isosceles triangles by a straight line drawn from the right angle to the middle point of the hypotenuse.*

Hence the joining line is equal to half the hypotenuse.

5. *Draw a straight line at right angles to a given finite straight line from one of its extremities, without producing the given straight line.*

[Let AB be the given straight line. On AB describe any isosceles triangle ACB. Produce BC to D, making CD equal to BC. Join AD. Then shall AD be perpendicular to AB.]

6. *Trisect a right angle.*

7. The angle contained by the bisectors of the angles at the base of an isosceles triangle is equal to an exterior angle formed by producing the base.

8. The angle contained by the bisectors of two adjacent angles of a quadrilateral is equal to half the sum of the remaining angles.

The following theorems were added as corollaries to Proposition 32 by Robert Simson, who edited Euclid's text in 1756.

COROLLARY 1.   *All the interior angles of any rectilineal figure, together with four right angles, are equal to twice as many right angles as the figure has sides.*

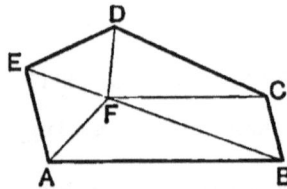

Let ABCDE be any rectilineal figure.
Take F, any point within it,
and join F to each of the angular points of the figure.

Then the figure is divided into as many triangles as it has sides.

And the three angles of each triangle are together equal to two right angles.   I. 32.

Hence *all* the angles of *all* the triangles are together equal to twice as many right angles as the figure has sides.

But all the angles of all the triangles make up all the interior angles of the figure, together with the angles at F, which are equal to four right angles.   I. 15, *Cor.*

Therefore all the interior angles of the figure, together with four right angles, are equal to twice as many right angles as the figure has sides.   Q.E.D.

COROLLARY 2. *If the sides of a rectilineal figure, which has no re-entrant angle, are produced in order, then all the exterior angles so formed are together equal to four right angles.*

For at each angular point of the figure, the interior angle and the exterior angle are together equal to two right angles. I. 13.

Therefore all the interior angles, with all the exterior angles, are together equal to twice as many right angles as the figure has sides.

But all the interior angles, with four right angles, are together equal to twice as many right angles as the figure has sides. I. 32, *Cor.* 1.

Therefore all the interior angles, with all the exterior angles, are together equal to all the interior angles, with four right angles.

Therefore the exterior angles are together equal to four right angles. Q.E.D.

EXERCISES ON SIMSON'S COROLLARIES.

[A polygon is said to be **regular** when it has all its sides and all its angles equal.]

1. Express in terms of a right angle the magnitude of each angle of (i) a regular hexagon, (ii) a regular octagon.

2. If one side of a regular hexagon is produced, shew that the exterior angle is equal to the angle of an equilateral triangle.

3. Prove Simson's first Corollary by joining one vertex of the rectilineal figure to each of the other vertices.

4. Find the magnitude of each angle of a regular polygon of n sides.

5. If the alternate sides of any polygon be produced to meet, the sum of the included angles, together with eight right angles, will be equal to twice as many right angles as the figure has sides.

PROPOSITION 33.   THEOREM.

The straight lines which join the extremities of two equal and parallel straight lines towards the same parts are themselves equal and parallel.

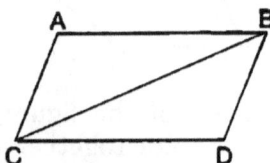

Let **AB** and **CD** be equal and parallel straight lines ; and let them be joined towards the same parts by the straight lines **AC** and **BD**.

Then shall **AC** *and* **BD** *be equal and parallel.*

Construction.   Join **BC**.

Proof.   Then because **AB** and **CD** are parallel, and **BC** meets them,
therefore the angle **ABC** is equal to the alternate angle **DCB**.   I. 29.

Now in the triangles **ABC**, **DCB**,

Because
{
          AB is equal to **DC**,   *Hyp.*
          and **BC** is common to both ;
also the angle **ABC** is equal to the angle **DCB** ;   *Proved.*
}

therefore the triangle **ABC** is equal to the triangle **DCB** in all respects ;   I. 4.

so that the base **AC** is equal to the base **DB**,
and the angle **ACB** equal to the angle **DBC**.

But these are alternate angles.
Therefore **AC** and **BD** are parallel :   I. 27.
and it has been shewn that they are also equal.

Q.E.D.

DEFINITION.   A **Parallelogram** is a four-sided figure whose opposite sides are parallel.

PROPOSITION 34. THEOREM.

The opposite sides and angles of a parallelogram are equal to one another, and each diagonal bisects the parallelogram.

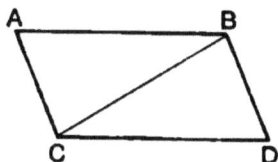

Let ACDB be a parallelogram, of which BC is a diagonal.

Then shall the opposite sides and angles of the figure be equal to one another ; and the diagonal BC shall bisect it,

Proof. Because AB and CD are parallel, and BC meets them,

therefore the angle ABC is equal to the alternate angle DCB ;   I. 29.

Again, because AC and BD are parallel, and BC meets them,

therefore the angle ACB is equal to the alternate angle DBC.   I. 29.

Hence in the triangles ABC, DCB,

Because ⎰ the angle ABC is equal to the angle DCB,
and the angle ACB is equal to the angle DBC ;
⎱ also the side BC is common to both ;

therefore the triangle ABC is equal to the triangle DCB in all respects ;   I. 26.

so that AB is equal to DC, and AC to DB ;
and the angle BAC is equal to the angle CDB.

Also, because the angle ABC is equal to the angle DCB,
and the angle CBD equal to the angle BCA,

therefore the whole angle ABD is equal to the whole angle DCA.

And the triangles ABC, DCB having been proved equal in all respects are equal in area.

Therefore the diagonal BC bisects the parallelogram ACDB.

Q.E.D.

EXERCISES ON PARALLELOGRAMS.

1. *If one angle of a parallelogram is a right angle, all its angles are right angles.*

2. *If the opposite sides of a quadrilateral are equal, the figure is a parallelogram.*

3. *If the opposite angles of a quadrilateral are equal, the figure is a parallelogram.*

4. *If a quadrilateral has all its sides equal and one angle a right angle, all its angles are right angles.*

5. *The diagonals of a parallelogram bisect each other.*

6. *If the diagonals of a quadrilateral bisect each other, the figure is a parallelogram.*

7. If two opposite angles of a parallelogram are bisected by the diagonal which joins them, the figure is equilateral.

8. If the diagonals of a parallelogram are equal, all its angles are right angles.

9. In a parallelogram which is not rectangular the diagonals are unequal.

10. Any straight line drawn through the middle point of a diagonal of a parallelogram and terminated by a pair of opposite sides, is bisected at that point.

11. *If two parallelograms have two adjacent sides of one equal to two adjacent sides of the other, each to each, and one angle of one equal to one angle of the other, the parallelograms are equal in all respects.*

12. *Two rectangles are equal if two adjacent sides of one are equal to two adjacent sides of the other, each to each.*

13. In a parallelogram the perpendiculars drawn from one pair of opposite angles to the diagonal which joins the other pair are equal.

14. If ABCD is a parallelogram, and X, Y respectively the middle points of the sides AD, BC; shew that the figure AYCX is a parallelogram.

MISCELLANEOUS EXERCISES ON SECTIONS I. AND II.

1. Shew that the construction in Proposition 2 may generally be performed in eight different ways. Point out the exceptional case.

2. The bisectors of two vertically opposite angles are in the same straight line.

3. In the figure of Proposition 16, if AF is joined, shew
  (i) that AF is equal to BC;
  (ii) that the triangle ABC is equal to the triangle CFA in all respects.

4. ABC is a triangle right-angled at B, and BC is produced to D: shew that the angle ACD is obtuse.

5. Shew that in any regular polygon of n sides each angle contains $\dfrac{2(n-2)}{n}$ right angles.

6. The angle contained by the bisectors of the angles at the base of any triangle is equal to the vertical angle together with half the sum of the base angles.

7. The angle contained by the bisectors of two exterior angles of any triangle is equal to half the sum of the two corresponding interior angles.

8. If perpendiculars are drawn to two intersecting straight lines from any point between them, shew that the bisector of the angle between the perpendiculars is parallel to (or coincident with) the bisector of the angle between the given straight lines.

9. If two points P, Q be taken in the equal sides of an isosceles triangle ABC, so that BP is equal to CQ, shew that PQ is parallel to BC.

10. ABC and DEF are two triangles, such that AB, BC are equal and parallel to DE, EF, each to each; shew that AC is equal and parallel to DF.

11. Prove the second Corollary to Prop. 32 by drawing through any angular point lines parallel to all the sides.

12. If two sides of a quadrilateral are parallel, and the remaining two sides equal but not parallel, shew that the opposite angles are supplementary; also that the diagonals are equal.

SECTION III.

THE AREAS OF PARALLELOGRAMS AND TRIANGLES.

Hitherto when two figures have been said to be *equal*, it has been implied that they are *identically* equal, that is, equal in all respects.

But figures may be equal *in area* without being equal in all respects, that is, without having the same shape.

The present section deals with parallelograms and triangles which are equal in area but not necessarily identically equal.

[The ultimate test of equality, as we have already seen, is afforded by Axiom 8, which asserts that magnitudes which *may be made to coincide with one another* are equal. Now figures which are not equal in all respects, cannot be made to coincide without first undergoing some change of form : hence the method of direct *superposition* is unsuited to the purposes of the present section.

We shall see however from Euclid's proof of Proposition 35, that two figures which are not identically equal, may nevertheless be so related to a third figure, that it is possible to infer the equality of their areas.]

DEFINITIONS.

1.   The **Altitude** of a parallelogram with reference to a given side as base, is the perpendicular distance between the base and the opposite side.

2.   The **Altitude** of a triangle with reference to a given side as base, is the perpendicular distance of the opposite vertex from the base.

[From this point the following symbols will be introduced into the text :

$$= \text{ for } \textit{is equal to} \text{ ; } \quad \therefore \text{ for } \textit{therefore.}$$

If it is thought desirable to shorten *written work* by the use of symbols and abbreviations, it is strongly recommended that only some well recognized system should be allowed, such, for example, as that given on page 11.]

PROPOSITION 35. THEOREM.

Parallelograms on the same base, and between the same parallels, are equal in area.

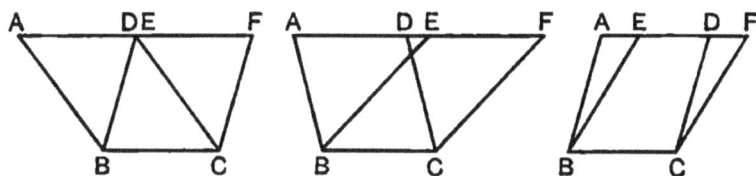

Let the parallelograms ABCD, EBCF be on the same base BC, and between the same parallels BC, AF.

Then shall the parallelogram ABCD *be equal in area to the parallelogram* EBCF.

CASE I. If the sides AD, EF, opposite to the base BC, are terminated at the same point D:
then each of the parallelograms ABCD, EBCF is double of the triangle BDC;   I. 34.
∴ they are equal to one another.   *Ax.* 6.

CASE II. But if the sides AD, EF are not terminated at the same point:
then because ABCD is a parallelogram,
∴ the side AD = the opposite side BC;   I. 34.
similarly EF = BC;
∴ AD = EF.   *Ax.* 1.
∴ the whole, or remainder, EA = the whole, or remainder, FD.
Then in the triangles FDC, EAB,

Because { FD = EA,   *Proved.*
and the side DC = the opposite side AB,   I. 34.
also the exterior angle FDC = the interior opposite angle EAB,
I. 29.
∴ the triangle FDC = the triangle EAB.   I. 4.

From the whole figure ABCF take the triangle FDC;
and from the same figure take the equal triangle EAB;
then the remainders are equal.   *Ax.* 3.
Therefore the parallelogram ABCD is equal to the parallelogram EBCF.
                                Q.E.D,

E.C.   U

PROPOSITION 36. THEOREM.

Parallelograms on equal bases, and between the same parallels, are equal in area.

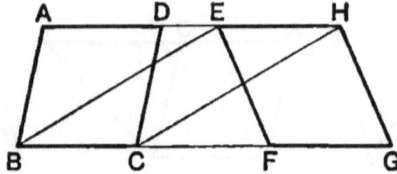

Let ABCD, EFGH be parallelograms on equal bases BC, FG, and between the same parallels AH, BG.

Then shall the parallelogram ABCD be equal to the parallelogram EFGH.

Construction. Join BE, CH.

Proof.   Then because BC = FG ;   *Hyp.*
          and the side FG = the opposite side EH ;   I. 34.
                    ∴ BC = EH :   *Ax.* 1.
          and BC is parallel to EH ;   *Hyp.*
     ∴ BE and CH are also equal and parallel.   I. 33.
          Therefore EBCH is a parallelogram.   *Def.* 36.

Now the parallelograms ABCD, EBCH are on the same base BC, and between the same parallels BC, AH ;
∴ the parallelogram ABCD = the parallelogram EBCH. I. 35.

Also the parallelograms EFGH, EBCH are on the same base EH, and between the same parallels EH, BG ;
∴ the parallelogram EFGH = the parallelogram EBCH. I. 35.

Therefore the parallelogram ABCD is equal to the parallelogram EFGH.   *Ax.* 1.

Q.E.D.

From the last two Propositions we infer that :

(i) *A parallelogram is equal in area to a rectangle of equal base and equal altitude.*

(ii) *Parallelograms on equal bases and of equal altitudes are equal in area.*

PROPOSITION 37. THEOREM.

Triangles on the same base, and between the same parallels, are equal in area.

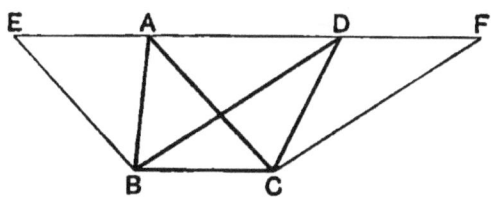

Let the triangles ABC, DBC be upon the same base BC, and between the same parallels BC, AD.

Then shall the triangle ABC be equal to the triangle DBC.

Construction. Through B draw BE parallel to CA, to meet DA produced in E ;   I. 31.
through C draw CF parallel to BD, to meet AD produced in F.

Proof. Then, by construction, each of the figures EBCA, DBCF is a parallelogram.   *Def.* 36.
And since they are on the same base BC, and between the same parallels BC, EF ;
∴ the parallelogram EBCA = the parallelogram DBCF. I. 35.

      Now the diagonal AB bisects EBCA ;   I. 34.
∴ the triangle ABC is half the parallelogram EBCA.
      And the diagonal DC bisects DBCF ;   I. 34.
∴ the triangle DBC is half the parallelogram DBCF.

      And the halves of equal things are equal.   *Ax.* 7.
Therefore the triangle ABC is equal to the triangle DBC.

                         Q.E.D.

[For Exercises see page 79.]

PROPOSITION 38.   THEOREM.

Triangles on equal bases, and between the same parallels, are equal in area.

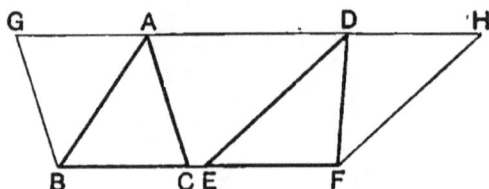

Let the triangles ABC, DEF be on equal bases BC, EF, and between the same parallels BF, AD.

Then shall the triangle ABC be equal to the triangle DEF.

Construction.   Through B draw BG parallel to CA, to meet DA produced in G;   I. 31.
through F draw FH parallel to ED, to meet AD produced in H.

Proof.   Then, by construction, each of the figures GBCA, DEFH is a parallelogram.   *Def.* 36.
And since they are on equal bases BC, EF, and between the same parallels BF, GH;
∴ the parallelogram GBCA = the parallelogram DEFH. I. 36.
                  Now the diagonal DF bisects GBCA;   I. 34.
∴ the triangle ABC is half the parallelogram GBCA.
                  And the diagonal DF bisects DEFH;   I. 34.
∴ the triangle DEF is half the parallelogram DEFH.
                  And the halves of equal things are equal.   *Ax.* 7.
Therefore the triangle ABC is equal to the triangle DEF.
                                                      Q.E.D.

From this Proposition we infer that:

(i)   *Triangles on equal bases and of equal altitude are equal in area.*

(ii) *Of two triangles of the same altitude, that is the greater which has the greater base; and of two triangles on the same base, or on equal bases, that is the greater which has the greater altitude.*

PROPOSITION 39. THEOREM.

Equal triangles on the same base, and on the same side of it, are between the same parallels.

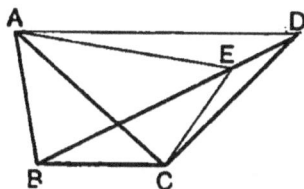

Let the triangles ABC, DBC which stand on the same base BC, and on the same side of it be equal in area.

Then shall the triangles ABC, DBC be between the same parallels; that is, if AD be joined, AD shall be parallel to BC.

Construction. For if AD be not parallel to BC,
if possible, through A draw AE parallel to BC, I. 31.
meeting BD, or BD produced, in E.
Join EC.

Proof. Now the triangles ABC, EBC are on the same base BC, and between the same parallels BC, AE ;
∴ the triangle ABC = the triangle EBC.   I. 37.
But the triangle ABC = the triangle DBC ;   *Hyp.*
∴ the triangle DBC = the triangle EBC ;
that is, the whole is equal to a part ; which is impossible.
∴ AE is not parallel to BC.

Similarly it can be shewn that no other straight line through A, except AD, is parallel to BC.

Therefore AD is parallel to BC.

Q.E.D.

From this Proposition it follows that :

Equal triangles on the same base have equal altitudes.

[For Exercises see page 79.]

PROPOSITION 40. THEOREM.

Equal triangles, on equal bases in the same straight line, and on the same side of it, are between the same parallels.

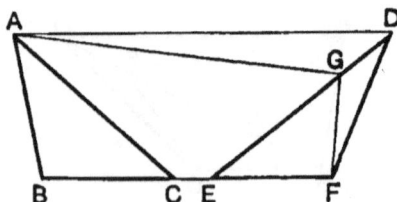

Let the triangles ABC, DEF which stand on equal bases BC, EF, in the same straight line BF, and on the same side of it, be equal in area.

Then shall the triangles ABC, DEF *be between the same parallels; that is, if* AD *be joined,* AD *shall be parallel to* BF.

Construction. For if AD be not parallel to BF,
    if possible, through A draw AG parallel to BF,   I. 31.
    meeting ED, or ED produced, in G.
        Join GF.

Proof. Now the triangles ABC, GEF are on equal bases BC, EF, and between the same parallels BF, AG;
    ∴ the triangle ABC = the triangle GEF.   I. 38.
    But the triangle ABC = the triangle DEF :   *Hyp.*
    ∴ the triangle DEF = the triangle GEF :
that is, the whole is equal to a part; which is impossible.
        ∴ AG is not parallel to BF.

Similarly it can be shewn that no other straight line through A, except AD, is parallel to BF.
        Therefore AD is parallel to BF.

                           Q.E.D.

From this Proposition it follows that :

(i)   *Equal triangles on equal bases have equal altitudes.*
(ii)   *Equal triangles of equal altitudes have equal bases.*

EXERCISES ON PROPOSITIONS 37–40.

DEFINITION.   Each of the three straight lines which join the angular points of a triangle to the middle points of the opposite sides is called a **Median** of the triangle.

ON PROP. 37.

1.   If, in the figure of Prop. 37, AC and BD intersect in K, shew that

     (i)   the triangles AKB, DKC are equal in area.
     (ii)   the quadrilaterals EBKA, FCKD are equal.

2.   In the figure of I. 16, shew that the triangles ABC, FBC are equal in area.

3.   On the base of a given triangle construct a second triangle, equal in area to the first, and having its vertex in a given straight line.

4.   Describe an isosceles triangle equal in area to a given triangle and standing on the same base.

ON PROP. 38.

5.   *A triangle is divided by each of its medians into two parts of equal area.*

6.   A parallelogram is divided by its diagonals into four triangles of equal area.

7.   ABC is a triangle, and its base BC is bisected at X ; if Y be any point in the median AX, shew that the triangles ABY, ACY are equal in area.

8.   In AC, a diagonal of the parallelogram ABCD, any point X is taken, and XB, XD are drawn : shew that the triangle BAX is equal to the triangle DAX.

9.   If two triangles have two sides of one respectively equal to two sides of the other, and the angles contained by those sides *supplementary*, the triangles are equal in area.

ON PROP. 39.

10.   *The straight line which joins the middle points of two sides of a triangle is parallel to the third side.*

11.   *If two straight lines AB, CD intersect in O, so that the triangle AOC is equal to the triangle DOB, shew that AD and CB are parallel.*

ON PROP. 40.

12.   Deduce Prop. 40 from Prop. 39 by joining AE, AF in the figure of page 78.

PROPOSITION 41. THEOREM.

If a parallelogram and a triangle be on the same base and between the same parallels, the parallelogram shall be double of the triangle.

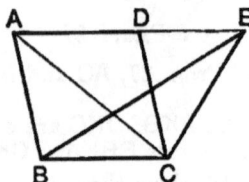

Let the parallelogram ABCD, and the triangle EBC be upon the same base BC, and between the same parallels BC, AE.

Then shall the parallelogram ABCD be double of the triangle EBC.

Construction.   Join AC.

Proof. Now the triangles ABC, EBC are on the same base BC, and between the same parallels BC, AE ;

∴ the triangle ABC = the triangle EBC.   I. 37.

And since the diagonal AC bisects ABCD ;   I. 34.
∴ the parallelogram ABCD is double of the triangle ABC.

Therefore the parallelogram ABCD is also double of the triangle EBC.   Q.E.D.

EXERCISES.

1. ABCD is a parallelogram, and X, Y are the middle points of the sides AD, BC ; if Z is any point in XY, or XY produced, shew that the triangle AZB is one quarter of the parallelogram ABCD.

2. Describe a right-angled isosceles triangle equal to a given square.

3. If ABCD is a parallelogram, and X, Y any points in DC and AD respectively : shew that the triangles AXB, BYC are equal in area.

4. ABCD is a parallelogram, and P is any point within it ; shew that the sum of the triangles PAB, PCD is equal to half the parallelogram.

PROPOSITION 42. PROBLEM.

To describe a parallelogram that shall be equal to a given triangle, and have one of its angles equal to a given angle.

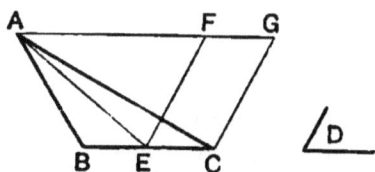

Let ABC be the given triangle, and D the given angle.

It is required to describe a parallelogram equal to ABC, and having one of its angles equal to D.

Construction. Bisect BC at E. I. 10.

At E in CE, make the angle CEF equal to D ; I. 23.

through A draw AFG parallel to EC ; I. 31.

and through C draw CG parallel to EF.

Then FECG shall be the parallelogram required.

Join AE.

Proof. Now the triangles ABE, AEC are on equal bases BE, EC, and between the same parallels ;

∴ the triangle ABE = the triangle AEC ; I. 38.

∴ the triangle ABC is double of the triangle AEC.

But FECG is a parallelogram by construction ; *Def.* 36.

and it is double of the triangle AEC,

being on the same base EC, and between the same parallels EC and AG. I. 41.

Therefore the parallelogram FECG is equal to the triangle ABC ;

and it has one of its angles CEF equal to the given angle D.

Q.E.F.

EXERCISES.

1. Describe a parallelogram equal to a given square standing on the same base, and having an angle equal to half a right angle.

2. Describe a rhombus equal to a given parallelogram and standing on the same base. When does the construction fail ?

DEFINITION.   If in the diagonal of a parallelogram any point is taken, and straight lines are drawn through it parallel to the sides of the parallelogram ; then of the four parallelograms into which the whole figure is divided, the two through which the diagonal passes are called **Parallelograms about that diagonal,** and the other two, which with these make up the whole figure, are called the **complements** of the parallelograms about the diagonal.

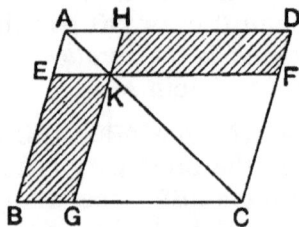

Thus in the figure given above, AEKH, KGCF are parallelograms about the diagonal AC ; and the shaded figures HKFD, EBGK are the complements of those parallelograms.

NOTE.   A parallelogram is often named by *two* letters only, these being placed at opposite angular points.

PROPOSITION 43. THEOREM.

The complements of the parallelograms about the diagonal of any parallelogram, are equal to one another.

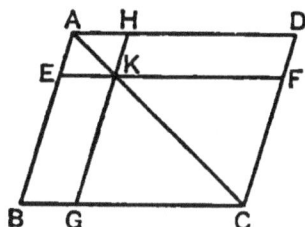

Let ABCD be a parallelogram, and KD, KB the complements of the parallelograms EH, GF about the diagonal AC.

Then shall the complement BK be equal to the complement KD.

Proof. Because EH is a parallelogram, and AK its diagonal,
∴ the triangle AEK = the triangle AHK. I. 34.

Similarly the triangle KGC = the triangle KFC.

Hence the triangles AEK, KGC are together equal to the triangles AHK, KFC.

But since the diagonal AC bisects the parallelogram ABCD;
∴ the whole triangle ABC = the whole triangle ADC. I. 34.

Therefore the remainder, the complement BK, is equal to the remainder, the complement KD. Q.E.D.

EXERCISES.

In the figure of Prop. 43, prove that

  (i) The parallelogram ED is equal to the parallelogram BH.

  (ii) If KB, KD are joined, the triangle AKB is equal to the triangle AKD.

PROPOSITION 44.   PROBLEM.

To a given straight line to apply a parallelogram which shall be equal to a given triangle, and have one of its angles equal to a given angle.

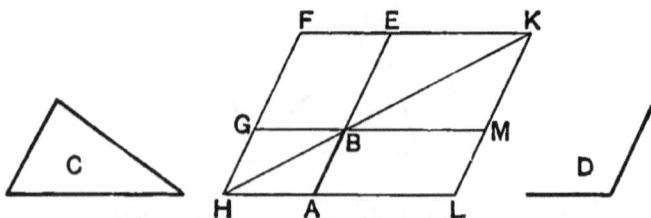

Let **AB** be the given straight line, **C** the given triangle, and **D** the given angle.

It is required to apply to the straight line **AB** *a parallelogram equal to the triangle* **C**, *and having an angle equal to the angle* **D**.

Construction.   On **AB** produced describe a parallelogram **BEFG** equal to the triangle **C**, and having the angle **EBG** equal to the angle **D**.   I. 22 *and* I. 42*.

Through **A** draw **AH** parallel to **BG** or **EF**, to meet **FG** produced in **H**.   I. 31.

Join HB.

Then because **AH** and **EF** are parallel, and **HF** meets them,

∴ the angles **AHF**, **HFE** together = two right angles.   I. 29.

Hence the angles **BHF**, **HFE** are together less than two right angles;

∴ **HB** and **FE** will meet if produced towards **B** and **E**. *Ax.* 12.

Produce HB and FE to meet at K.

Through **K** draw **KL** parallel to **EA** or **FH**;   I. 31. and produce **HA**, **GB** to meet **KL** in the points **L** and **M**.

Then shall **BL** *be the parallelogram required.*

Proof. Now FHLK is a parallelogram, *Constr.*
and LB, BF are the complements of the parallelograms about the diagonal HK :
∴ the complement LB = the complement BF. I. 43.
But the triangle C = the figure BF ; *Constr.*
∴ the figure LB = the triangle C.
Again the angle ABM = the vertically opposite angle GBE ;
also the angle D = the angle GBE ; *Constr.*
∴ the angle ABM = the angle D.
Therefore the parallelogram LB, which is applied to the straight line AB, is equal to the triangle C, and has the angle ABM equal to the angle D. Q.E.F.

* This step of the construction is effected by first describing on AB produced a triangle whose sides are respectively equal to those of the triangle C (I. 22) ; and by then making a parallelogram equal to the triangle so drawn, and having an angle equal to D (I. 42).

QUESTIONS FOR REVISION.

1. Quote Euclid's Twelfth Axiom. What objections have been raised to it, and what substitute for it has been suggested?

2. Which of Euclid's Propositions, dealing with parallel straight lines, depends on Axiom 12? Furnish an alternative proof.

3. *Straight lines which are parallel to the same straight line are parallel to one another* [Prop. 30]. Deduce this from Playfair's Axiom.

4. Define a *parallelogram*, an *altitude* of a triangle, a *median* of a triangle, *parallelograms about the diagonal of a parallelogram*.

5. What is meant by *superposition*? On what Axiom does this method depend? Give instances of figures which are equal *in area*, but which cannot be superposed.

6. In fig. 2 of Prop. 35 shew how one parallelogram may be cut into pieces, which, when fitted together in other positions, make up the other parallelogram.

PROPOSITION 45. PROBLEM.

To describe a parallelogram equal to a given rectilineal figure, and having an angle equal to a given angle.

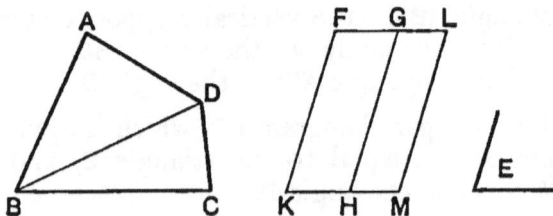

Let ABCD be the given rectilineal figure, and E the given angle.

It is required to describe a parallelogram equal to ABCD, *and having an angle equal to* E.

Suppose the given rectilineal figure to be a quadrilateral.

Construction.   Join BD.
Describe the parallelogram FH equal to the triangle ABD, and having the angle FKH equal to the angle E.   I. 42.
To GH apply the parallelogram GM, equal to the triangle DBC, and having the angle GHM equal to E.   I. 44.

Then shall FKML *be the parallelogram required.*

Proof.   Because each of the angles GHM, FKH = the angle E;
∴ the angle FKH = the angle GHM.

To each of these equals add the angle GHK;
then the angles FKH, GHK together = the angles GHM, GHK.

But since FK, GH are parallel, and KH meets them;
∴ the angles FKH, GHK together = two right angles; I. 29.
∴ also the angles GHM, GHK together = two right angles;
∴ KH, HM are in the same straight line.   I. 14.

Again, because KM, FG are parallel, and HG meets them,
∴ the angle MHG = the alternate angle HGF.   I. 29.

To each of these equals add the angle HGL;
then the angles MHG, HGL together = the angles HGF, HGL.

But because HM, GL are parallel, and HG meets them,
∴ the angles MHG, HGL together = two right angles : I. 29.
∴ also the angles HGF, HGL together = two right angles :
∴ FG, GL are in the same straight line.   I. 14.

And because KF and ML are each parallel to HG,   *Constr.*
therefore KF is parallel to ML;   I. 30.
and KM, FL are parallel;   *Constr.*
∴ FKML is a parallelogram.   *Def.* 36.

Again, because the parallelogram FH = the triangle ABD,
and the parallelogram GM = the triangle DBC; *Constr.*
∴ the whole parallelogram FKML = the whole figure ABCD;
and it has the angle FKM equal to the angle E.

By a series of similar steps, a parallelogram may be
constructed equal to a rectilineal figure of more than four
sides.   Q.E.F.

The following Problem is important, and furnishes a useful application of the principles of the foregoing propositions.

ADDITIONAL PROBLEM.

To describe a triangle equal in area to a given quadrilateral.

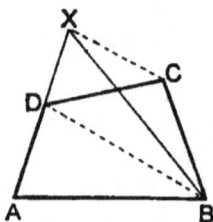

Let **ABCD** be the given quadrilateral.

It is required to describe a triangle equal to **ABCD** *in area.*

Construction. Join **BD**.

Through C draw **CX** parallel to **BD**, meeting AD produced in **X**.

Join **BX**.

Then **XAB** shall be the required triangle.

Proof. Now the triangles **XDB**, **CDB** are on the same base **DB** and between the same parallels **DB**, **XC** ;

∴ the triangle **XDB** = the triangle **CDB** in area.   I. 37.

To each of these equals add the triangle ADB ;

then the triangle **XAB** = the figure **ABCD**.

EXERCISE.

Construct a rectilineal figure equal to a given rectilineal figure, and having fewer sides by one than the given figure.

Hence shew how to construct a triangle equal to a given rectilineal figure.

PROPOSITION 46. PROBLEM.

To describe a square on a given straight line.

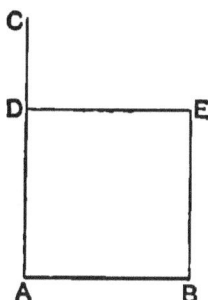

Let AB be the given straight line.
It is required to describe a square on AB.

Constr. From A draw AC at right angles to AB ;   I. 11.
         and make AD equal to AB.   I. 3.
         Through D draw DE parallel to AB ;   I. 31.
and through B draw BE parallel to AD, meeting DE in E.
Then shall ADEB *be a square.*

Proof. For, by construction, ADEB is a parallelogram :
           ∴. AB = DE, and AD = BE.   I. 34.
           But AD = AB ;   *Constr.*
∴. the four straight lines AB, AD, DE, EB are all equal ;
    that is, the figure ADEB is equilateral.

Again, since AB, DE are parallel, and AD meets them,
∴. the angles BAD, ADE together = two right angles ; I. 29.
        but the angle BAD is a right angle ;   *Constr.*
      . ∴. also the angle ADE is a right angle.

And the opposite angles of a parallelogram are equal ; I. 34.
   ∴. each of the angles DEB, EBA is a right angle :
        that is the figure ADEB is rectangular.
    Hence it is a square, and it is described on AB.

                                            Q.E.F.

COROLLARY. *If one angle of a parallelogram is a right
angle, all its angles are right angles.*

Proposition 47.   Theorem.

In a right-angled triangle the square described on the hypotenuse is equal to the sum of the squares described on the other two sides.

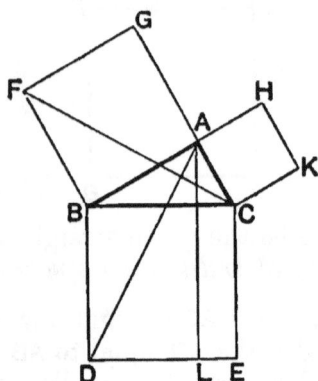

Let ABC be a right-angled triangle, having the angle BAC a right angle.

Then shall the square described on the hypotenuse BC *be equal to the sum of the squares described on* BA, AC.

Construction.   On BC describe the square BDEC;   I. 46.
and on BA, AC describe the squares BAGF, ACKH.
Through A draw AL parallel to BD or CE ;   I. 31.
and join AD, FC.

Proof.   Then because each of the angles BAC, BAG is a right angle,
∴ CA and AG are in the same straight line.   I. 14.

Now the angle CBD = the angle FBA,
for each of them is a right angle.
Add to each the angle ABC :
then the whole angle ABD = the whole angle FBC.

Then in the triangles ABD, FBC,

Because
$\begin{cases} \qquad \text{AB} = \text{FB}, \\ \qquad \text{and BD} = \text{BC}, \\ \text{also the angle ABD} = \text{the angle FBC} ; \end{cases}$ *Proved.*

∴ the triangle ABD = the triangle FBC.   I. 4.

Now the parallelogram BL is double of the triangle ABD, being on the same base BD, and between the same parallels BD, AL.   I. 41.

And the square GB is double of the triangle FBC, being on the same base FB, and between the same parallels FB, GC.   I. 41.

But doubles of equals are equal :   *Ax.* 6.

therefore the parallelogram BL = the square GB.

Similarly, by joining AE, BK it can be shewn that the parallelogram CL = the square CH.

Therefore the whole square BE = the sum of the squares GB, HC :

that is, the square described on the hypotenuse BC is equal to the sum of the squares described on the two sides BA, AC.   Q.E.D.

NOTE.   It is not necessary to the proof of this Proposition that the three squares should be described *external* to the triangle ABC ; and since *each* square may be drawn either *towards* or *away from* the triangle, it may be shewn that there are $2 \times 2 \times 2$, or *eight*, possible constructions.

Obs.   The following properties of a square, though not formally enunciated by Euclid, are employed in subsequent proofs. [See I. 48.]

(i)   *The squares on equal straight lines are equal.*

(ii)   *Equal squares stand upon equal straight lines.*

EXERCISES ON PROPOSITION 47.

1. In the figure of this Proposition, shew that

    (i)   If BG, CH are joined, these straight lines are parallel ;

    (ii) The points F, A, K are in one straight line ;

    (iii) FC and AD are at right angles to one another ;

    (iv) If GH, KE, FD are joined, the triangle GAH is equal
          to the given triangle in all respects ; and the triangles
          FBD, KCE are each equal in area to the triangle ABC.
          [See Ex. 9, p. 79.]

2. On the sides AB, AC of *any* triangle ABC, squares ABFG,
ACKH are described both toward the triangle, or both on the side
remote from it : shew that the straight lines BH and CG are equal.

3. On the sides of any triangle ABC, equilateral triangles BCX,
CAY, ABZ are described, all externally, or all towards the triangle :
shew that AX, BY, CZ are all equal.

4. *The square described on the diagonal of a given square, is
double of the given square.*

5. *ABC is an equilateral triangle, and AX is the perpendicular
drawn from A to BC : shew that the square on AX is three times the
square on BX.*

6. Describe a square equal to the sum of two given squares.

7. From the vertex A of a triangle ABC, AX is drawn perpendicular to the base : shew that the difference of the squares on the
sides AB and AC, is equal to the difference of the squares on BX and
CX, the segments of the base.

8. If from any point O within a triangle ABC, perpendiculars
OX, OY, OZ are drawn to the sides BC, CA, AB respectively : shew
that the sum of the squares on the segments AZ, BX, CY is equal to
the sum of the squares on the segments AY, CX, BZ.

9. ABC is a triangle right-angled at A ; and the sides AB, AC
are intersected by a straight line PQ, and BQ, PC are joined.
Prove that the sum of the squares on BQ, PC is equal to the sum
of the squares on BC, PQ.

10. In a right-angled triangle four times the sum of the squares
on the two medians drawn from the acute angles is equal to five
times the square on the hypotenuse.

NOTES ON PROPOSITION 47.

It is believed that Proposition 47 is due to Pythagoras, a Greek philosopher and mathematician, who lived about two centuries before Euclid.

Many experimental proofs of this theorem have been given by means of actual *dissection* : that is to say, it has been shewn how the squares on the sides containing the right angle may be cut up into pieces which, when fitted together in other positions, exactly make up the square on the hypotenuse. Two of these methods of dissection are given below.

I. In the adjoining diagram ABC is the given right-angled triangle, and the figures AF, HK are the squares on AB, AC, placed side by side.

FD is made equal to EH or AC;

and the two squares AF, HK are cut along the lines ED, DB.

Then it will be found that the triangle EHD may be placed so as to fill up the space CAB ; and the triangle BFD may be made to fill the space CKE.

Hence the two squares AF, HK may be fitted together so as to form the single figure CBDE, which will be found to be a perfect square, namely the square on the hypotenuse BC.

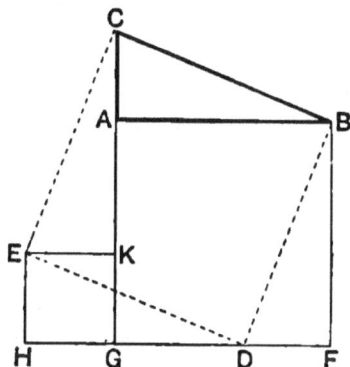

II. In the figure of I. 47, let DB and EC be produced to meet FG and AH in L and N respectively ; and let LM be drawn parallel to BC.

Then it will be found that the several parts of the two squares FA, AK can be fitted together (in the places bearing corresponding numbers) so as exactly to fill up the square DC.

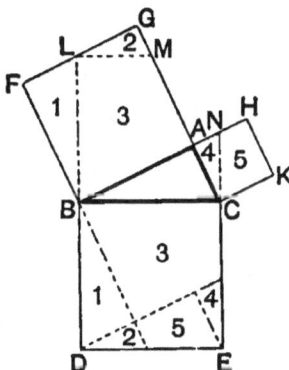

Proposition 48.   Theorem.

If the square described on one side of a triangle be equal to the sum of the squares described on the other two sides, then the angle contained by these two sides shall be a right angle.

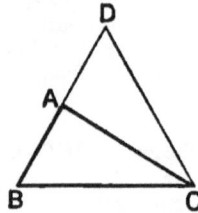

Let ABC be a triangle ; and let the square described on BC be equal to the sum of the squares described on BA, AC.

Then shall the angle BAC be a right angle.

Construction.   From A draw AD at right angles to AC;   I. 11.
and make AD equal to AB.   I. 3.
Join DC.

Proof.   Then, because AD = AB,   *Constr.*
∴ the square on AD = the square on AB.
To each of these add the square on CA ;
then the sum of the squares on CA, AD = the sum of the squares on CA, AB.

But, because the angle DAC is a right angle, *Constr.*
∴ the square on DC = the sum of the squares on CA, AD. I. 47.
And, by hypothesis, the square on BC = the sum of the squares on CA, AB ;
∴ the square on DC = the square on BC :
∴ also the side DC = the side BC.

Then in the triangles DAC, BAC,

Because { DA = BA,   *Constr.*
    and AC is common to both ;
    also the third side DC = the third side BC;   *Proved.*

∴ the angle DAC = the angle BAC.   I. 8.
But DAC is a right angle.   *Constr.*
Therefore also BAC is a right angle.   Q.E D.

THEOREM A. *Two right-angled triangles which have their hypotenuses equal, and one side of one equal to one side of the other, are identically equal.*

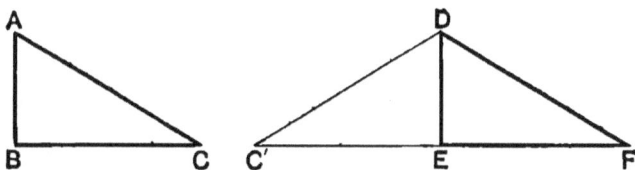

Let ABC, DEF be two △ˢ right-angled at B and E, having AC equal to DF, and AB equal to DE.

Then shall the △ ABC be equal to the △ DEF in all respects.

For apply the △ ABC to the △ DEF, so that AB may coincide with the equal line DE, and C may fall on the side of DE remote from F. Let C' be the point on which C falls.

Then DEC' represents the △ ABC in its new position.
Now each of the ∠ˢ DEF, DEC' is a rt. ∠ ;   *Hyp.*
∴ EF and EC' are in one st. line.   I. 14.

Then in the △ C'DF, because DF = DC' (*i.e.* AC),   *Hyp.*
∴ the ∠ DFC' = the ∠ DC'F.   I. 5.

Hence in the two △ˢ DEF, DEC',
Because { the ∠ DEF = the ∠ DEC', being rt. ∠ˢ ;
and the ∠ DFE = the ∠ DC'E ;   *Proved.*
also the side DE is common to both ;
∴ the △ˢ DEF, DEC' are equal in all respects ;   I. 26.
that is, the △ˢ DEF, ABC are equal in all respects.   Q.E.D.

Alternative Proof. Since the ∠ ABC is a rt. angle ;
∴ the sq. on AC = the sqq. on AB, BC.   I. 47.
Similarly, the sq. on DF = the sqq. on DE, EF ;   I. 47.
But the sq. on AC = the sq. on DF (since AC = DF, *Hyp.*) ;
∴ the sqq. on AB, BC = the sqq. on DE, EF.

And of these, the sq. on AB = the sq. on DE (since AB = DE, *Hyp.*);
∴ the sq. on BC = the sq. on EF ;   *Ax.* 3.
∴ BC = EF.

Hence the three sides of the △ ABC are respectively equal to the three sides of the △ DEF ;
∴ the △ ABC = the △ DEF in all respects.   I. 8.

THEOREM B. *If two triangles have two sides of the one equal to two sides of the other, each to each, and have likewise the angles opposite to one pair of equal sides equal, then the angles opposite to the other pair of equal sides shall be either equal or supplementary, and in the former case the triangles shall be equal in all respects.*

  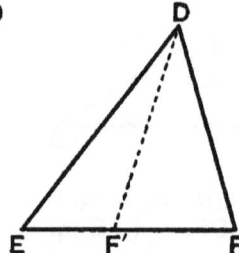

Fig. 1.   Fig. 2.   Fig. 3.

Let ABC, DEF be two triangles, in which
the side AB = the side DE,
the side AC = the side DF,
and the ∠ ABC = the ∠ DEF.

Then shall the ∠ˢ ACB, DFE be either equal or supplementary, and in the former case the triangles shall be equal in all respects.

If the ∠ BAC = the ∠ EDF,   [*Figs. 1 and 2.*]
then the ∠ ACB = the ∠ DFE, and the triangles are equal in all respects.   I. 4.

But if the ∠ BAC be not equal to the ∠ EDF, [*Figs. 1 and 3.*]
let the ∠ EDF be greater than the ∠ BAC.

At D in ED make the ∠ EDF′ equal to the ∠ BAC.
Then the △ˢ BAC, EDF′ are equal in all respects.   I. 26.
∴ AC = DF′;
but AC = DF ;   *Hyp.*
∴ DF = DF′,
∴ the angle DFF′ = the ∠ DF′F.   I. 5.

But the ∠ˢ DF′F, DF′E are supplementary,   I. 13.
∴ the ∠ˢ DFF′ DF′E are supplementary :
that is, the ∠ˢ DFE, ACB are supplementary.

Q.E.D.

COROLLARIES.   Three cases of this theorem deserve special attention.

It has been proved that if the angles ACB, DFE are not *supplementary* they are *equal :*

and angles which are both acute or both obtuse cannot be supplementary ; hence

   (i)   If the angles ACB, DFE opposite to the two equal sides AB, DE are both acute or both obtuse they cannot be supplementary, and are therefore equal ; or if one of them is a right angle, the other must also be a right angle (whether considered as supplementary or equal to it) :

in either case the triangles are equal in all respects.

   (ii)   If the two given angles are right angles or obtuse angles, it follows that the angles ACB, DFE must be both acute, and therefore equal, by (i) :

so that the triangles are equal in all respects.

   (iii)   If in each triangle the side opposite the given angle is not less than the other given side ; that is, if AC and DF are not less than AB and DE respectively, then the angles ACB, DFE cannot be greater than the angles ABC, DEF respectively ;

      therefore the angles ACB, DFE are both acute ;

      hence, as above, they are equal ;

and the triangles ABC, DEF are equal in all respects.

THEOREM C. *The straight line drawn through the middle point of a side of a triangle parallel to the base, bisects the remaining side.*

Let ABC be a $\triangle$, and Z the middle point of the side AB. Through Z, ZY is drawn par¹ to BC.

Then shall Y be the middle point of AC.

Through Z draw ZX par¹ to AC.   I. 31.

   Then in the $\triangle^s$ AZY, ZBX,
because ZY and BC are par¹,
   $\therefore$ the $\angle$ AZY = the $\angle$ ZBX ;   I. 29.
and because ZX and AC are par¹,
   $\therefore$ the $\angle$ ZAY = the $\angle$ BZX ;   I. 29.
   also AZ = ZB :   *Hyp.*
      $\therefore$ AY = ZX.   I. 26.

But ZXCY is a par^m by construction ;
      $\therefore$ ZX = YC.   I. 34.
   Hence AY = YC ;
   that is, AC is bisected at Y.   Q.E.D.

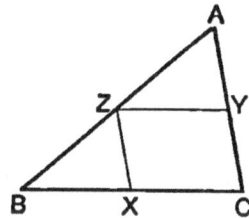

THEOREM D.   *The straight line which joins the middle points of two sides of a triangle, is parallel to the third side.*

Let ABC be a △ , and Z, Y the middle points of the sides AB, AC.

Then shall ZY *be par^l to* BC.

Produce ZY to V, making YV equal to ZY.

Join CV.

Then in the △^{s} AYZ, CYV,

Because $\Bigg\{$

      AY = CY,   *Hyp.*

    and YZ = YV,   *Constr.*

  and the ∠ AYZ = the vert. opp. ∠ CYV ;   I. 15.

∴ AZ = CV,.   I. 4.

and the ∠ ZAY = the ∠ VCY ;

hence CV is par^l to AZ.   I. 27.

But CV is equal to AZ, that is, to BZ   *Hyp.*

∴ CV is equal and par^l to BZ :

∴ ZV is equal and par^l to BC :   I. 33.

that is, ZY is par^l to BC.   Q.E.D.

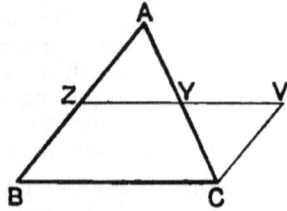

[A second proof of this proposition may be derived from I. 38, 39.]

DEFINITION.   Three or more straight lines are said to be **concurrent** when they meet in one point.

THEOREM E.   *The perpendiculars drawn to the sides of a triangle from their middle points are concurrent.*

Let ABC be a △, and X, Y, Z the middle points of its sides.

Then shall the perp^s drawn to the sides from X, Y, Z *be concurrent.*

From Z and Y draw perp^s to AB, AC ; these perp^s, since they cannot be parallel, will meet at some point O.   *Ax.* 12.

Join OX.

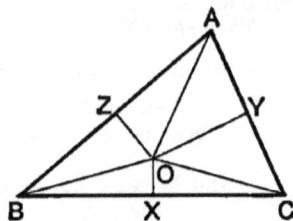

It is required to prove that OX *is perp. to* BC.

Join OA, OB, OC.

In the △^{s} OYA, OYC,

Because $\Bigg\{$

      YA = YC,   ·   *Hyp.*

  and OY is common to both ;

 also the ∠ OYA = the ∠ OYC, being rt. ∠^{s} ;

∴ OA = OC.   I. 4.

Similarly, from the △ˢ OZA, OZB,
it may be proved that OA=OB.
Hence OA, OB, OC are all equal.

Again, in the △ˢ OXB, OXC

Because $\begin{cases} \text{BX=CX,} & \textit{Hyp.} \\ \text{and XO is common to both ;} \\ \text{also OB=OC :} & \textit{Proved.} \end{cases}$

∴ the ∠ OXB=the ∠ OXC ;   I. 8.
but these are adjacent ∠ˢ ;
∴ they are rt. ∠ˢ ;   *Def.* 10.
that is, OX is perp. to BC.

Hence the three perpˢ OX, OY, OZ meet in the point O.

Q. E. D.

THEOREM F.   *The bisectors of the angles of a triangle are concurrent.*

Let ABC be a △.
Then shall the bisectors of the ∠ˢ ABC,
BCA, BAC *be concurrent.*

Bisect the ∠ˢ ABC, BCA, by straight
lines which must meet at some point
O.   *Ax.* 12.

Join AO.

It is required to prove that AO *bisects the*
∠ BAC.

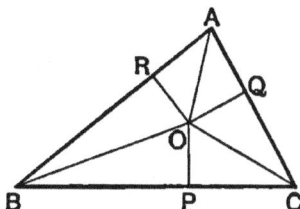

From O draw OP, OQ, OR perp. to the sides of the △.

Then in the △ˢ OBP, OBR,

Because $\begin{cases} \text{the ∠ OBP=the ∠ OBR,} & \textit{Constr.} \\ \text{and the ∠ OPB=the ∠ ORB, being rt. ∠ˢ,} & \textit{Constr.} \\ \text{and OB is common ;} \end{cases}$

∴ OP=OR.   I. 26.

Similarly from the △ˢ OCP, OCQ,
it may be shewn that OP=OQ,
∴ OP, OQ, OR are all equal.

Again in the △ˢ ORA, OQA,

Because $\begin{cases} \text{the ∠ˢ ORA, OQA are rt. ∠ˢ,} & \textit{Constr.} \\ \text{and the hypotenuse OA is common,} \\ \text{also OR=OQ ;} & \textit{Proved.} \end{cases}$

∴ the ∠ RAO=the ∠ QAO.   *Theorem* A.

That is, AO is the bisector of the ∠ BAC.

Hence the bisectors of the three ∠ˢ meet at the point O.

Q. E. D.

THEOREM G.   *The medians of a triangle are concurrent.*

Let ABC be a △.
Then shall its three medians be concurrent.

Let BY and CZ be two of the medians,
and let them intersect at O.
            Join AO,
and produce it to meet BC in X.
It is required to shew that AX *is the remaining median of the △.*

Through C draw CK parallel to BY :
    produce AX to meet CK at K.
            Join BK.

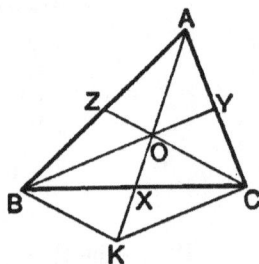

In the △ AKC,
because Y is the middle point of AC, and YO is parallel to CK,
        ∴ O is the middle point of AK.   *Theorem* C.

Again in the △ ABK,
since Z and O are the middle points of AB, AK,
        ∴ ZO is parallel to BK,   *Theorem* D.
        that is, OC is parallel to BK :
        ∴ the figure BKCO is a par^m.

But the diagonals of a par^m bisect one another,   Ex. 5, p. 70.
        ∴ X is the middle point of BC.
        That is, AX is a median of the △.

Hence the three medians meet at the point O.   Q.E.D.

COROLLARY.   *The three medians of a triangle cut one another at a point of trisection, the greater segment in each being towards the angular point.*

For in the above figure it has been proved that
            AO=OK,
        also that OX is half of OK ;
        ∴ OX is half of OA :
        that is, OX is one third of AX.
        Similarly OY is one third of BY,
            and OZ is one third of CZ.   Q. E. D.

By means of this Corollary it may be shewn that in any triangle the shorter median bisects the greater side.

[The point of intersection of the three medians of a triangle is called the **centroid**. It is shewn in mechanics that a thin triangular plate will balance in any position about this point ; therefore the centroid of a triangle is also its centre of gravity.]

THEOREM H. *The perpendiculars drawn from the vertices of a triangle to the opposite sides are concurrent.*

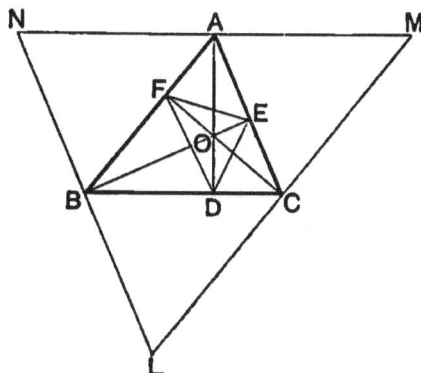

Let ABC be a △, and AD, BE, CF the three perp^s drawn from the vertices to the opposite sides.

Then shall the perp^s AD, BE, CF be concurrent.

Through A, B, and C draw straight lines MN, NL, LM parallel to the opposite sides of the △.

Then the figure BAMC is a par^m.   *Def.* 36.
∴ AB = MC.   I. 34.
Also the figure BACL is a par^m.
∴ AB = LC,
∴ LC = CM :
that is, C is the middle point of LM.

So also A and B are the middle points of MN and NL.

Hence AD, BE, CF are the perp^s to the sides of the △ LMN from their middle points.   Ex. 3, p. 60.

But these perp^s meet in a point :   *Theorem* E.
that is, the perp^s drawn from the vertices of the △ ABC to the opposite sides meet in a point.   Q.E.D.

DEFINITIONS.

(i) The intersection of the perpendiculars drawn from the vertices of a triangle to the opposite sides is called its **orthocentre.**

(ii) The triangle formed by joining the feet of the perpendiculars is called the **pedal triangle.**

EXERCISES ON BOOK I.

ON THE IDENTICAL EQUALITY OF TRIANGLES.

1. If in a triangle the perpendicular from the vertex on the base bisects the base, then the triangle is isosceles.

2. If the bisector of the vertical angle of a triangle is also perpendicular to the base, the triangle is isosceles.

3. If the bisector of the vertical angle of a triangle also bisects the base, the triangle is isosceles.

[Produce the bisector, and complete the construction after the manner of I. 16.]

4. If in a triangle a pair of straight lines drawn from the extremities of the base, making equal angles with the other sides, are equal, the triangle is isosceles.

5. If in a triangle the perpendiculars drawn from the extremities of the base to the opposite sides are equal, the triangle is isosceles.

6. Two triangles ABC, ABD on the same base AB, and on opposite sides of it, are such that AC is equal to AD, and BC is equal to BD : shew that the line joining the points C and D is perpendicular to AB.

7. ABC is a triangle in which the vertical angle BAC is bisected by the straight line AX : from B draw BD perpendicular to AX, and produce it to meet AC, or AC produced, in E ; then shew that BD is equal to DE.

8. In a quadrilateral ABCD, AB is equal to AD, and BC is equal to DC : shew that the diagonal AC bisects each of the angles which it joins.

9. In a quadrilateral ABCD the opposite sides AD, BC are equal, and also the diagonals AC, BD are equal : if AC and BD intersect at K, shew that each of the triangles AKB, DKC is isosceles.

10. If one angle of a triangle be equal to the sum of the other two, the greatest side is double of the distance of its middle point from the opposite angle.

ON PARALLELS AND PARALLELOGRAMS.

11. If a straight line meets two parallel straight lines, and the two interior angles on the same side are bisected; shew that the bisectors meet at right angles. [I. 29, I. 32.]

12. The straight lines drawn from any point in the bisector of an angle parallel to the arms of the angle, and terminated by them, are equal; and the resulting figure is a rhombus.

13. The middle point of any straight line which meets two parallel straight lines, and is terminated by them, is equidistant from the parallels.

14. A straight line drawn between two parallels and terminated by them, is bisected; shew that any other straight line passing through the middle point and terminated by the parallels, is also bisected at that point.

15. If through a point equidistant from two parallel straight lines, two straight lines are drawn cutting the parallels, the portions of the latter thus intercepted are equal.

16. AB and CD are two given straight lines, and X is a given point in AB : find a point Y in AB such that YX may be equal to the perpendicular distance of Y from CD.

17. ABC is an isosceles triangle; required to draw a straight line DE parallel to the base BC, and meeting the equal sides in D and E, so that BD, DE, EC may be all equal.

18. ABC is any triangle; required to draw a straight line DE parallel to the base BC, and meeting the other sides in D and E, so that DE may be equal to the sum of BD and CE.

19. If two straight lines are parallel to two other straight lines, each to each; and if the angles contained by each pair are bisected, shew that the bisecting lines are either parallel or perpendicular to one another.

20. The straight line which joins the middle points of two sides of a triangle is equal to half the third side. [See Theorem D.]

21. Shew that the three straight lines which join the middle points of the sides of a triangle, divide it into four triangles which are identically equal.

22. Any straight line drawn from the vertex of a triangle to the base is bisected by the straight line which joins the middle points of the other two sides of the triangle.

23. AB, AC are two given straight lines, and P is a given point between them; required to draw through P a straight line terminated by AB, AC, and bisected by P.

24. ABCD is a parallelogram, and X, Y are the middle points of the opposite sides AD, BC: shew that BX and DY trisect the diagonal AC.

25. If the middle points of adjacent sides of any quadrilateral be joined, the figure thus formed is a parallelogram.

26. Shew that the straight lines which join the middle points of opposite sides of a quadrilateral, bisect one another.

ON AREAS.

27. Shew that a parallelogram is bisected by any straight line which passes through the middle point of one of its diagonals. [I. 29, 26.]

28. Bisect a parallelogram by a straight line drawn through a given point.

29. Bisect a parallelogram by a straight line drawn perpendicular to one of its sides.

30. Bisect a parallelogram by a straight line drawn parallel to a given straight line.

31. ABCD is a trapezium in which the side AB is parallel to DC. Shew that its area is equal to the area of a parallelogram formed by drawing through X, the middle point of BC, a straight line parallel to AD. [I. 29, 26.]

32. If two straight lines AB, CD intersect at X, and if the straight lines AC and BC, which join their extremities are parallel, shew that the triangle AXD is equal to the triangle BXC.

33. If two straight lines AB, CD intersect at X, so that the triangle AXD is equal to the triangle XCD, then AC and BD are parallel.

34. ABCD is a parallelogram, and X any point in the diagonal AC produced; shew that the triangles XBC, XDC are equal. [See Ex. 13, p. 70.]

35. If the middle points of the sides of a quadrilateral be joined in order, the *parallelogram* so formed [see Ex. 25] is equal to half the given figure.

MISCELLANEOUS EXERCISES.

36.   A is the vertex of an isosceles triangle ABC, and BA is produced to D, so that AD is equal to BA ; if DC is drawn, shew that BCD is a right angle.

37.   The straight line joining the middle point of the hypotenuse of a right-angled triangle to the right angle is equal to half the hypotenuse.

38.   From the extremities of the base of a triangle perpendiculars are drawn to the opposite sides (produced if necessary) ; shew that the straight lines which join the middle point of the base to the feet of the perpendiculars are equal.

39.   In a triangle ABC, AD is drawn perpendicular to BC ; and X, Y, Z are the middle points of the sides BC, CA, AB respectively : shew that each of the angles ZXY, ZDY is equal to the angle BAC.

40.   In a right-angled triangle, if a perpendicular be drawn from the right angle to the hypotenuse, the two triangles thus formed are equiangular to one another.

41.   Given the three middle points of the sides of a triangle, construct the triangle. ´ [See Theorem D.]

42.   If three parallel straight lines make equal intercepts on a fourth straight line, they will also make equal intercepts on any other straight line which meets them.

43.   Shew that the bisectors of two exterior angles of a triangle meet on the bisector of the third angle.   [See Theorem F.]

44.   If in a right-angled triangle one of the acute angles is double of the other, shew that the hypotenuse is double of the shorter side.

45.   In a triangle ABC, if AC is not greater than AB, shew that any straight line drawn through the vertex A, and terminated by the base BC, is less than AB.

46.   ABC is a triangle, and the vertical angle BAC is bisected by a straight line which meets the base BC in X ; shew that BA is greater than BX, and CA greater than CX.   Hence obtain a proof of I. 20.

47.   The perpendicular is the shortest straight line that can be drawn from a given point to a given straight line ; and of others, that which is nearer to the perpendicular is less than the more remote ; and two, and only two equal straight lines can be drawn from the given point to the given straight line, one on each side of the perpendicular.

E.C.   Y

48.　The sum of the distances of any point from the three angular points of a triangle is greater than half its perimeter.

49.　The sum of the distances of any point within a triangle from its angular points is less than the perimeter of the triangle.

50.　In the figure of I. 47, shew that
- (i)　the sum of the squares on AB and AE is equal to the sum of the squares on AC and AD;
- (ii)　the square on EK is equal to the square on AB with four times the square on AC;
- (iii)　the sum of the squares on EK and FD is equal to five times the square on BC.

HARDER MISCELLANEOUS EXERCISES.

51.　The perimeter of a quadrilateral is greater than the sum of its diagonals.

52.　In a triangle any two sides are together greater than twice the *median* which bisects the remaining side.　[See Def., p. 79.]

[Produce the median, and complete the construction after the manner of I. 16.]

53.　Use the properties of the equilateral triangle to trisect a given finite straight line.

54.　Construct a triangle having given the base, one of the angles at the base, and the sum of the remaining sides.

55.　Construct a triangle having given the base, one of the angles at the base, and the difference of the remaining sides.

[Two cases arise, according as the side opposite to the given angle is greater or less than the other.]

56.　Prove that the straight line which joins the middle points of the oblique sides of a trapezium [see note, Def. 34] is
- (i)　parallel to the two parallel sides;
- (ii)　equal to half the sum of the parallel sides.

57.　The sum of the perpendiculars drawn from any point in the base of an isosceles triangle to the equal sides is equal to the perpendicular drawn from either extremity of the base to the opposite side.

58.　Shew that the difference of the perpendiculars drawn to the equal sides of an isosceles triangle from any point in the base produced is *constant*, *i.e.* the same whatever point is taken.

59. In any triangle if a perpendicular be drawn from one extremity of the base to the bisector of the vertical angle ; then (i) it will make with either of the sides containing the vertical angle an angle equal to half the sum of the angles at the base ; (ii) it will make with the base an angle equal to half the difference of the angles at the base.

60. In any triangle the angle contained by the bisector of the vertical angle and the perpendicular from the vertex to the base is equal to half the difference of the base angles.

61. In any triangle the sum of the medians is less than the perimeter. [See Ex. 52.]

62. If the vertical angle of a triangle is contained by unequal sides, then (i) the median drawn from the vertex lies within the angle contained by the bisector of the vertical angle and the longer side ; and (ii) the median is greater than the bisector of the vertical angle.

63. Prove that a trapezium is equal to a parallelogram whose base is half the sum of the parallel sides of the given figure, and whose altitude is equal to the perpendicular distance between them. [See Ex. 56.]

64. ABC is a triangle, and D is any point in AB : it is required to draw through D a straight line DE to meet BC produced in E, so that the triangle DBE may be equal to the triangle ABC.

[Join DC. Through A draw AE parallel to DC ; and join DE.]

65. ABCD is a quadrilateral : it is required to construct a triangle equal in area to ABCD, having its vertex at a given point X in DC, and its base in the same straight line as AB.

66. Bisect a triangle ABC by a straight line drawn through a given point P in one of its sides.

[Bisect AB at Z ; and join CZ, CP. Through Z draw ZQ parallel to CP. Join PQ.]

67. ABC is a triangle right-angled at A ; the sides AB, AC are intersected by a straight line PQ, and BQ, PC are joined. Prove that the sum of the squares on BQ, PC is equal to the sum of the squares on BC, PQ.

68. Divide a straight line into two parts so that
  (i) the sum of their squares shall be equal to a given square ;
  (ii) the square on one part shall be double the square on the other part.

APPENDIX.

A.

1. Reduce to the simplest form as a mixed number:

(a) $\dfrac{2\frac{1}{4} - \frac{2}{3} \text{ of } 1\frac{5}{6}}{\frac{1}{5} \text{ of } 3\frac{1}{3} + \frac{7}{9}} \div \dfrac{2\frac{1}{2} - \frac{1}{3} \div \frac{4}{13}}{1\frac{1}{4} \text{ of } 8\frac{1}{2}} + \dfrac{1}{1 + \frac{5}{13}}$;

(b) $\dfrac{0\cdot17 + 0\cdot65333\ldots}{0\cdot247 \div 2\cdot21}$.   (8)

2. (a) Reduce £2. 7s. 6d. to a decimal of £5. 10s.

(b) Find the number of yards in 0·217 of a mile and a half. (6)

3. A square lawn is bordered by a path 4 ft. 6 in. wide, the path and the lawn together occupying one-tenth of an acre. Find the expense of covering the path with gravel at a cost of $7\frac{1}{2}d.$ a square yard.   (10)

4. The larger of two rooms is 47 ft. long, 30 ft. wide, and 25 ft. high; the smaller is 25 ft. long, 20 ft. wide, and 18 ft. high; compare their cubic contents.

If the four walls of the larger room are painted at a cost of 1s. 3d. a square yard, and the four walls and ceiling of the smaller room at a cost of 1s. $4\frac{1}{2}d.$ a square yard, compare the expenses of painting the rooms.   (8)

5. A man realised £2730 by selling a 3 per cent. stock at $113\frac{3}{4}$, and invested one-third of this sum in a 4 per cent. stock at 130, and the remainder in a $2\frac{3}{4}$ per cent. stock at $110\frac{5}{8}$; find how much of each kind of stock he bought, and the difference it made in his income.   (10)

6. (a) Find to the second place of decimals the number of square yards in 7280 square metres. (b) Find to the nearest penny the value of 900 kilogrammes of a material which costs £25. 14s. 6d. a ton. N. B.—1 metre = 3·28092 ft., 1 kilogramme = 2·2046 lbs.   (10)

B.

7. If at a point in a straight line two other straight lines, on opposite sides of it, make the adjacent angles together equal to two right angles, shew that these two straight lines are in the same straight line.

Equilateral triangles BAD, CAE are described on the sides AB, AC of an equilateral triangle ABC; shew that DA is in the same straight line with AE. (10)

8. Shew that the greater side of a triangle is opposite to the greater angle.

The angles B and C of a triangle ABC are acute angles, and C is greater than B; P is a point in BC. Shew that AP is shorter than AB, and find for what positions of P, AP will also be shorter than AC. (12)

9. Define parallel straight lines and alternate angles.

If a straight line be drawn across two parallel straight lines, shew that it makes the alternate angles equal.

ABC is an equilateral triangle whose angles B and C are bisected by BD and CD respectively; DE is drawn parallel to AB to meet BC in E, and DF is drawn parallel to AC to meet BC in F. Shew that BE, EF, FC are all equal. (12)

10. Shew that the complements, which are about a diameter of any parallelogram, are equal.

Given a rectangle and a straight line, shew how to construct a rectangle, equal to the given rectangle, and having a side equal to the given line. (12)

11. AB is the hypotenuse of a right-angled triangle ABC; in AB take a point D, such that BD equals BC; draw CE at right angles to AB, and meeting it in E; shew that CD bisects the angle ACE. (12)

12. Shew how to construct a square which shall have two adjacent sides passing through two given points, and the intersection of the diagonals at a third given point.

Shew that there are generally two solutions. (16)

C.

13. Define a factor, a term, a power, a root, as used in Algebra.

From a rod a ft. long, $b-c$ ft. are cut off; express in two ways, with brackets, and without brackets, the number of feet that are left.

Find the numerical values of the following expressions when $x=-1$, $y=-2$, $z=\frac{1}{2}$:

(a) $2x-\{9y-8x+2z-(4x+y)\}$.

(b) $(x+y-z)^2+(x+y)^2(x-y+z)+(x-y)^3$.   (12)

14. (a) Simplify $(xy-1)(x^2y^2+xy+1)+(xy+1)(x^2y^2-xy+1)$.

(b) Divide $\frac{1}{8}x^3+\frac{1}{27}y^3$ by $\frac{1}{2}x+\frac{1}{3}y$.   (10)

15. (a) Simplify $\dfrac{x+1}{x^2-x}-\dfrac{x+2}{x^2-1}-\dfrac{1}{x^3+1}$.

(b) Substitute $\dfrac{a}{a-1}$ for x, and $\dfrac{2a}{2a+1}$ for y, in $\dfrac{xy}{x-y}$, and reduce the result to its simplest form.   (12)

16. Write down the following expressions in factors :

(a) $15x^2-16xy-15y^2$.

(b) $(a+b+c)^2-4(b-c)^2$.

(c) $(1+x)^3-(1-x)^3$.   (12)

17. Solve the following equations :

(a) $\dfrac{x}{x-2}-\dfrac{x}{x+2}=\dfrac{1}{x-2}-\dfrac{4}{x+2}$.

(b) $a(x-a)=b(x+b)-2ab$.

(c) $3x-\dfrac{y}{2}=5,\quad \dfrac{x}{3}+\dfrac{y}{4}=3$.   (14)

18. Two boys, A and B, have money consisting of shillings and pennies. A has twice as many pennies as shillings ; B, who has fivepence more than A, has three times as many pennies as shillings ; together they have two and a half times as many pennies as they have shillings. How much has each ?   (14)

GLASGOW : PRINTED AT THE UNIVERSITY PRESS BY ROBERT MACLEHOSE AND CO.

www.ingramcontent.com/pod-product-compliance
Lightning Source LLC
Chambersburg PA
CBHW021404210326
41599CB00011B/1003